Dreams of Other Worlds

Dreams of Other Worlds

The Amazing Story of Unmanned Space Exploration

Chris Impey and Holly Henry

Princeton University Press
Princeton and Oxford

Published by Princeton University Press,
41 William Street, Princeton, New Jersey 08540

In the United Kingdom: Princeton University Press,
6 Oxford Street, Woodstock, Oxfordshire OX20 1TW

press.princeton.edu

Jacket Illustrations: Planet with sunrise on the background of stars. © Molodec. Courtesy of Shutterstock. Artist's rendering of the planet Kepler-22b, located in the "habitable zone" of the Kepler-22 star system. Courtesy of NASA/Ames/JPL-Caltech.

ISBN 978-0-691-14753-6

Library of Congress Control Number: 2013939381

British Library Cataloging-in-Publication Data is available

This book has been composed in Sabon and Helvetica Neue

Printed on acid-free paper ∞

Printed in the United States of America

10 9 8 7 6 5 4 3 2 1

Contents

Dreams of Other Worlds

1 Introduction

SOMEONE WHO "MISSED" the late part of the twentieth century, perhaps by being in a coma or a deep sleep, or by being marooned on a desert island, would have many adjustments to make upon rejoining civilization. The largest would probably be the galloping progress in computers and telecommunications and information technology. But if their attention turned to astronomy, they would also be amazed by what had been learned in the interim. In the last third of the century, Mars turned from a pale red disk as seen through a telescope to a planet with ancient lake beds and subterranean glaciers. The outer Solar System went from being frigid and uninteresting real estate to being a place with as many as a dozen potentially habitable worlds. They would be greeted by a cavalcade of exoplanets, projecting to billions across the Milky Way galaxy. Their familiar view of the sky would now be augmented by images spanning the entire electromagnetic spectrum, revealing brown dwarfs and black holes and other exotic worlds. Finally, they would encounter a cosmic horizon, or limit to their vision, that had been pushed back to within an iota of the big bang, and they would be faced with the prospect that the visible universe might be one among many universes.

This book is a story of those discoveries, made by planetary probes and space missions over the past forty years. The word "world" means "age of man" in the old Germanic languages, and that proximate perspective took centuries to expand into a universe filled with galaxies and stars and their attendant planets. The missions at the heart of this narrative have not only transformed our view of the physical universe, they've also become embedded in culture and inspired the imagination—this book is also a story about that relationship. But people were dreaming of other worlds long before the space program and modern astronomy.

Almost nothing written by Anaxagoras has survived, so we can only imagine his dreams. He was born around 500 BC in Clazomenae in Ionia, a bustling port city on the coast of present-day Turkey. Before he moved to Athens and helped to make it the intellectual center of the ancient world, and long before he was sentenced to death for his heretical ideas, we can visualize him as an intense and austere young man. Anecdotes suggest someone who was far removed from the concerns of everyday life. He believed that the opportunity to understand the universe was the fundamental reason why it was better to be born than to not exist.[1]

Anaxagoras' mind was crowded with ideas. Philosophy is based on abstraction—the power to manipulate concepts and retain aspects of the physical world in your head. He believed that the Sun was a mass of fiery metal, that the Moon was made of rock like the Earth and did not emit its own light, and that the stars were fiery stones. He offered physical explanations for eclipses, for the solstices and the motions of the stars, and for the formation of comets. He thought the Milky Way represented the combined light of countless stars.[2] We imagine him standing on the rocky Ionian shore at night, with starlight glittering on dark water, gazing up into the sky and sensing the vastness of the celestial vault. The dreams of such a powerful and original thinker were probably suffused with the imagery of other worlds.

This is speculation. As with most of the Greek philosophers, and especially the pre-Socratics, very little of their writing has come down to us unaltered. Typically, there are only isolated fragments and commentaries, sometimes by contemporaries and often written centuries later. Each historical interpreter added their own predilections and biases; the result is a view of the original ideas seen through a gauzy veil.[3] Modern scholars pore over the shards and often come up with strikingly different interpretations. Anaxagoras thought that the original state of the cosmos was undifferentiated, but contained all of its eventual constituents. The cosmos was not limited in extent and it was set in motion by the action of "mind." Out of this swirling, rotating mixture the ingredients for material objects like the Earth, Sun, Moon, and planets separated. Although the nature of the animating agent is not clear from his writings, Anaxagoras was the first person to devise a purely mechanical and natural explanation for the cosmos, without any reference to gods or divine intervention. His theory sets no limit to the scale of this process, so there can be worlds within worlds, without end, either large or small.[4] A case can be made that he believed that our world system is not unique, but is one of many formed out of the initial and limitless mass of ingredients.[5]

Radical ideas often come with a price. For daring to suggest that the Sun was larger than the Peloponnese peninsula, Anaxagoras was charged with impiety.[6] He avoided the death penalty by going into exile in Asia Minor, where he spent the remainder of his life. Pluralism—the idea of a multiplicity of worlds, including the possibility that some of them harbor life—had antecedents in work by Anaximander and Anaximenes, and in speculation passed down by the Pythagorean School. But Anaxagoras was the first to embed the idea in a sophisticated and fully fledged cosmology. By the time of the early atomists Leucippus and Democritus, plurality of worlds was a natural and inevitable consequence of their physics. There were not just other worlds in space, but infinite worlds, some like this world and some utterly unlike it.[7] It was a startling conjecture.

The next two thousand years saw the idea of the plurality of worlds ebb and flow, as different philosophical arguments were presented and were molded to accommodate Christian theology.[8] The pluralist position was countered by the arguments of Plato, and particularly Aristotle, who held that the Earth was unique and so there could be no other system of worlds. European cultures were not alone in developing the idea of plurality of worlds. Babylonians held that the moving planets in the night sky were home to their gods. Hindu and Buddhist traditions assume a multiplicity of worlds with inhabiting intelligences. For example, in one myth the god Indra says, "I have spoken only of those worlds within this universe. But consider the myriad of universes that exist side by side, each with its own Indra and Brahma, and each with its evolving and dissolving worlds."[9] In cultures around the world, dreamers' imaginations soared. The Roman poet Cicero and the historian Plutarch wrote about creatures that might live on the Moon, and in the second century CE, Lucian of Samosata wrote an extraordinary fantasy about an interplanetary romance. *A True Story* was intended to satirize the epic tales of Homer and other travelers, and it began with the advisory that his readers should not believe a single word of it. Lucian and his fellow travelers are deposited by a water spout onto the Moon, where they encounter a bizarre race of humans who ride on the backs of three-headed birds. The Sun, Moon, stars, and planets are locales with specific geographies, human inhabitants, and fantastical creatures. This singular work is considered a precursor of modern science fiction.[10]

For more than a millennium it was dangerous in Europe to espouse the idea of fully fledged worlds in space with life on them. Throughout medieval times, the Catholic Church considered it heresy. There was an obvious problem with this position: if God was really omnipotent, why would he create only one world? Thomas Aquinas resolved the issue by saying that although the Creator had the power to create infinite worlds, he had chosen not to do so, and this became official Catholic doctrine in a pronouncement of the Bishop of Paris in 1177. Nicolas of Cusa sorely tested the bounds of this doctrine. In 1440, he produced a book called *Of Learned Ignorance* where he proposed that men, animals, and plants lived on the Sun, Moon, and stars.[11] He further claimed that intelligent and enlightened creatures lived on the Sun while lunatics lived on the Moon. It's said that friendship with the Pope shielded him from repercussions, and he went on to become a cardinal.

Giordano Bruno was less fortunate. The lapsed Dominican monk had deviated from Catholic orthodoxy in a number of ways, but his espousal of the Copernican system, which displaced the Earth from the center of the universe, brought him extra scrutiny. He believed that the stars were infinite in number, and that each hosted planets and living creatures.[12] Bruno was incarcerated for seven years before his trial and was eventually convicted of heresy.[13] A statue in the Campo de' Fiori in Rome marks the place where he was burned at the stake in 1600 as an "impenitent and pertinacious heretic." Religion had cast an ominous shadow over the idea of the plurality of worlds.

The same year Bruno was put to death, a twenty-nine-year-old mathematician named Johannes Kepler, an assistant to Tycho Brahe, was working with data that would cement the Copernican model of the Solar System. As he published his work on planetary motion in 1609, he dusted off a student dissertation he had written sixteen years earlier, where he defended the Copernican idea by imagining how the Earth might look when viewed from the Moon. Kepler elaborated on his youthful paper and added a dream narrative to turn it into a sophisticated scientific fantasy: *Somnium*.[14] Kepler was inspired by Lucian and Plutarch's earlier work, but unlike them, and unlike the mystic Bruno, he was a rational scientist who wanted to realistically envisage space travel and aliens. His narrative is rich with comments on the problems created by acceleration and varying gravity. The geography and geology of the Moon are realistically rendered. He even speculates on the effect of the physical environment on lunar creatures, foreshadowing Darwin and Lyell.[15] Kepler had every reason to take refuge in a dream. He was frail and bow-legged, covered in boils, and was cursed with myopia severe enough that he would never see the celestial phenomena he enunciated so elegantly. *Somnium* was known to Jules Verne and H. G. Wells, and it's a crucial step in the progression toward rational speculation about other worlds.

The Copernican revolution was not a single event; it was a series of realizations over a period of a century that the cozy idea of Earth as a singular place at the center of the universe was wrong. Displacing the Earth into motion around the Sun was the first wrenching step, but another was recognizing that the Earth was one of many worlds in space. The Copernican principle is more than just a cosmological model; it's a statement that the Earth is not in any central or favored position in the universe. A heuristic that extends from the work of Copernicus is the principle of mediocrity, which goes much further by supposing that there's nothing special or unusual about the situation of the Earth, or by extension, the fact that humans exist on this planet. That is of course a central tenet of modern astrobiology, but four hundred years ago it was a radical idea.

The Scientific Revolution recast the debate over the plurality of worlds. Within months of Kepler's dream piece, Galileo pointed his telescope at the Moon and affirmed it as a geological world, with topography similar in scale to the Earth. He also showed that Jupiter had orbiting moons and that the Milky Way resolved into points of light that seemed to be more distant versions of the bright stars.[16] The word *world* was no longer confused with *kosmos*; it meant a potentially life-bearing planet orbiting the Sun or, hypothetically, a distant star.[17] Speculation about life on the Moon became routine, almost mundane. However, theology and philosophy still colored the debate in several ways. One theological concept was the principle of plenitude—everything within God's power must have been realized, so inhabited worlds should be abundant. Another was the strong influence of teleology—purpose and direction in nature that implies a Creator, who would surely not have gone to the trouble of creating uninhabited worlds.[18]

For a long time, scientific arguments could do no more than support the general plausibility of the plurality of worlds. Telescopes could easily track the motion of stars and planets, but gaining a physical understanding was much more challenging. The blurring effect of Earth's atmosphere prevented astronomers from resolving anything smaller than continent-sized surface features on any Solar System body other than the Moon. Even the nearest stars are a hundred thousand times farther from us than the size of the Solar System. In addition, planets do not emit their own light, so astronomers must gather the hundred million times dimmer light that they reflect from their parent stars. Three centuries of improvements in telescope design after Galileo yielded only two new planets, a dozen or so moons, and no success in detecting worlds beyond the Solar System.

And so the dreamers held sway. Many of them were grounded in science so they advanced the Copernican idea that our situation in the universe was not special.[19] One striking work from the beginning of the Age of Enlightenment was *Conversations about the Plurality of Worlds* by Bernard de Fontenelle, published in 1686.[20] He wrote about intelligent beings inhabiting worlds beyond the Earth, and incorporated the biological argument that their characteristics would be shaped by their environment. Fontenelle also followed Galileo's lead by writing in his native language, French, rather than the scholarly language of Latin, and he was forward-looking in having a female protagonist and explicitly addressing female readers.[21] A much later high-water mark was *On the Plurality of Habitable Worlds* by Camille Flammarion, which reached a wide audience in 1862.[22] By the early twentieth century, scientific speculations and fictional accounts of worlds beyond the Earth proliferated, but technology and research weren't able to address such conjectures.[23] There's an unbroken thread between earliest Greek thinkers and more recent explorations of science fiction writers. Anaxagoras was a visionary, but it would probably have taken his breath away to know that one day we would actually visit other worlds.

Isaac Newton's *Mathematical Principles of Natural Philosophy*, a three-volume masterwork published in 1687, is a landmark in the history of science. *Principia*, as it is known, laid down the foundations of classical mechanics and gravitation.[24] Tucked away in one of the volumes is the drawing of a cannonball being launched horizontally from a tall mountaintop. This "thought experiment" sustained the dreams of space travel for nearly three centuries. October 4, 1957 was a pivotal moment in the history of the human race; on that day a metal sphere, no bigger than a beach ball and no heavier than an adult, was launched into orbit. The world was transfixed, and amateur radio operators monitored Sputnik's steady "beep" for three weeks until its battery expired.[25] Within two years the Soviets had crashed a probe into the Moon—the first manmade object to reach another world—and the Space Age was in full flight. Humans have never been any farther than the Moon but we've sent our robotic sentinels through most of the Solar System and slightly beyond.

For the universe beyond our backyard in the Solar System, we have no direct evidence and we cannot gather and analyze physical samples. The data are limited to electromagnetic radiation. Newton improved on Galileo's simple spyglass with a design for a reflecting telescope. All research telescopes are now reflectors. In understanding distant worlds, the complement to direct exploration with spacecraft is remote sensing with telescopes. A succession of larger and larger telescopes over the past century have now expanded our horizons, and extended the Copernican revolution.[26] We know that we orbit a middle-aged, middle-weight star, one of 400 billion in the Milky Way, which is one of 100 billion galaxies in the observable universe. The pivotal moment in the remote sensing of distant worlds happened on October 6, 1995, when Michel Mayor and Didier Queloz announced that they had discovered the first planet beyond the Solar System.[27] We're now "harvesting" Earths from deep space, and our dreams have moved on to the nature of life that might be found there.

This book explores how our concepts of distant worlds have been shaped and informed by space science and astronomy in the past forty years. Scientific understanding of the universe has been intertwined with culture since the time of Anaxagoras, and the popular imagination continues to be fueled by insights from space probes and large telescopes. What follows is not a survey of the many facilities that have furthered our understanding of the cosmos. Rather, it's an exploration of twelve iconic space missions that have opened new windows onto distant worlds. Most are in NASA's portfolio, but all have non-U.S. investigators, as space science and astronomy have become increasingly international.[28] In general, the arc of the book is chronological and moves from the proximate toward the remote. From comets to cosmology, from the Mars rovers to the multiverse, these missions have given us a sense of our cosmic environment and have redefined what it means to be the temporary tenants of a small planet.

The journey starts with Mars. Six years to the day after humans left their first footprints on the Moon—still the only world humans have ever visited in half a century of the Space Age—the first Viking lander touched down on Mars, with its twin reaching the opposite side of Mars six weeks later. The Vikings dashed hopes that Mars might be habitable, but they opened up the modern age of exploration of the red planet. Nearly three decades later, another pair of intrepid machines bounced to a safe landing on their cushioning airbags. The Mars Exploration Rovers were embraced by the public as they trundled across the rocky red soil, inspecting interesting rocks, sending back pictures in 3D, and gathering evidence for a warmer and wetter Mars in the distant past. Mars may have hosted life in the past, and life might still be there in underground aquifers, and it is this oscillation in the popular imagination between hostile and hospitable that makes it an uneasy doppelganger of the Earth.

Next up are two spacecraft that made a grand tour of the outer Solar System during the 1970s. Where before we had had nothing more than rough sketches, the Voyagers painted detailed portraits of the gas giant planets and their moons. These epic missions each ventured billions of miles from home, and they taxed the ingenuity of the scientists and engineers involved, many of whom aged and retired in the years between the first concept and its completion. The successor to the Voyagers was Cassini, which will soon enter its second decade of exploring the Saturn system. Cassini bristles with complex instruments and it dwarfs its predecessors. Together, these three missions have recast our understanding of the frigid realm beyond the asteroid belt. Giant planets may be cold and miniature versions of the Sun, but their moons are anything but dull and lifeless. Some have active geology and liquid water under their crusts. Others spew out sulfur or tiny ice particles. Many of them have distinct "personalities," like the more familiar worlds in the inner Solar System.

The planets and moons of the Solar System are intriguing enough to have earned the names of gods and mythological figures. Yet they are just side shows in a process that concentrates most of the mass into a central sphere of glowing gas. The star is the central character in this drama, and the plot line is alchemy: the creation of the heavy elements that make up the planets and moons. Stardust was the mission that caught not just one but two comets by the tail and in doing so told us how the Solar System was likely to have formed. The story of Stardust is our story, since most of our atoms were forged in the central cauldrons of long-dead stars. The Solar and Heliospheric Observer, by contrast, focused on the Sun itself and taught us what it means to live with a star. Belying its steady light, the Sun leads an active life that manifests in invisible forms of radiation. Distant worlds will also have to deal with the vagaries of their nurturing stars. After that, we drop back to take in a view of the solar neighborhood, from the unsung but impressive Hipparcos satellite. Hipparcos has refined the work of William Herschel over two hundred years ago by placing us accurately within the city of stars we call the Milky Way. If the Copernican principle holds, the "grit" from stellar fusion that gathered to form the Earth is not unique to our region of space, and similar worlds have formed across the galaxy.

The two missions that follow illustrate the revolution in astronomy when astronomers' blinders were removed after centuries of learning about the universe through visible light. Spitzer and Chandra are two of NASA's Great Observatories, straddling the electromagnetic spectrum from waves hundreds of times longer to hundreds of times shorter than the eye can see. Each telescope looks at regions of space that are hidden from view. Spitzer penetrates the murk of gas and dust that permeates interstellar space and reveals new worlds being formed. Young stars and planets are shrouded in placental dust that is opaque to light but nearly transparent to long infrared waves. This is a huge advantage when looking for exoplanets because their contrast relative to the host star is hundreds of times better at infrared than at optical wavelengths. Chandra, by contrast, has revealed the violence of dark objects like neutron stars and black holes, where such tiny worlds distort space-time and accelerate particles beyond any capability of our best accelerators. We would be ignorant of all these phenomena without space-based telescopes.

Closing the book are two missions that venture to the edges of space and time. The Hubble Space Telescope is the only space facility that has embedded itself deeply into the consciousness of the general public, to the level where the prospect of not servicing the telescope generated a backlash and an eventual reversal of NASA's original decision. Hubble has contributed to every area of astrophysics, but in particular it has quantified the limits of our vision, a region spanning 46 billion light-years in any direction, which contains roughly 100 billion galaxies. The inferred hundred thousand billion billion stars, with their attendant (and similar number of) habitable worlds, form the prodigious real estate of the observable universe, a census inconceivable to Anaxagoras and his colleagues. The Wilkinson Microwave Anisotropy Probe was a specialist mission to map the microwave sky and pin down conditions in the infant universe. By gathering exquisitely precise data, this satellite has confirmed the big bang model in great detail. It has also shown that there are likely to be innumerable distant worlds out there whose light hasn't yet had time to reach us since the big bang.

The journey ends with the near future, and efforts to measure realms of the universe that are currently at the edge of our vision. Close to home the goal is to see whether Mars has hosted or could host life—finding Life 2.0 would reset our views of biology beyond the home planet. In the proximate universe, we have the hope of detecting Earth clones and seeing if these worlds have had their atmospheres altered by a metabolism. At the frontier of cosmology, the hope is to test the multiverse concept, where the planets around 10^{23} stars are just part of the story, and a suite of alternate universes may exist, with properties perhaps egregiously different or perhaps uncannily similar to our own. In this extreme version of plenitude, everything that can happen has happened, and the set of events that led to our existence are neither special nor unique.

The authors are grateful to the two Steves—Dick and Garber—from NASA's History Program Office for their careful attention to this project, and to NASA for financial support during the writing of the manuscript. We acknowledge Ingrid Gnerlich at Princeton University Press for her epic patience during the long and winding road that led to the completion of the project, and to the staff at the Press for their assistance during production.

CI is also grateful to the Aspen Center for Physics, which is supported by the National Science Foundation, for providing a congenial setting for substantial work on the manuscript in 2010 and 2011, and to his astronomy colleagues at the University of Arizona for answering questions too numerous to count when he strayed from his expertise. CI also acknowledges the hospitality of his colleagues in the Department of Astrophysical Sciences at Princeton University, where he finished work on the manuscript during an appointment as the Stanley Kelley Visiting Professor for Distinguished Teaching.

HH would like to especially thank NASA for supporting the project and the research. She also wishes to thank the administrators, faculty, and staff at the College of Arts and Letters, the Department of English, the Office of Academic Research and Sponsored Programs, and the Pfau Library at the California State University, San Bernardino, for their assistance throughout the project. HH is extremely grateful to the many colleagues, friends, and family members who discussed and recommended topics and sources. It has been a great pleasure to research and explore the breadth of ideas that inform the study and that affirm our deep connection to the universe around us.

2 Viking

DISCOVERING THE RED PLANET

Sometimes the dream is a nightmare. Mars has always had an ominous mien in myth and culture. Ancient civilizations regarded the planet as a malevolent agent of war and apocalypse. Similar myths emerged around the world.[1] In late Babylonian texts, Mars is identified with Nergal, the fiery god of destruction and war. To the Greeks, Mars was Ares, one of Twelve Olympians and the son of Zeus and Hera. His attendants on the battlefield were Deimos and Phobos, terror and fear, and his sister and companion was Eris, the goddess of discord.[2] Ares was an important but an unlikeable character. In Roman hands he morphed into a virile and noble god, one who facilitated agriculture as well as war. The third month of our year honors him and the time when winter abated enough that Roman legions could begin their military campaigns. In legend, Mars abandoned his children Romulus and Remus and the twins went on to found the city of Rome.[3] The mystique of Mars may have been enhanced by its retrograde motion: the fact that every few years it twice reverses its direction of motion among the stars.[4] All exterior planets show this behavior, but the reversal is more dramatic for Mars than for Jupiter and Saturn. It's curious that such a modest speck of reddish light could exert such power (plate 1).

Fast forward nearly two thousand years and Mars still exerts a grip on the imagination. It's the night before Halloween, on the eve of World War II. Families across America are settling around the radio to hear "The Mercury Theatre on the Air," a weekly program directed by the young Orson Welles and featuring him and a talented ensemble cast. Listeners are enjoying salsa-inflected orchestral music from a hotel in New York City when the announcer breaks in: "Ladies and Gentlemen, we interrupt our program of dance music to bring you a special bulletin from the Intercontinental Radio News."[5] There's a news report about unusual activity observed on the surface of Mars, then back to the music. A few minutes later the announcer breaks in with additional information about Mars. More music. The next interruption has the announcer talking in breathless tones about a meteor that just landed in New Jersey. A little later, on the scene, there's horror in his voice as he describes creatures emerging from the meteor, which is in fact a spaceship. The Martians begin using a heat ray to incinerate bystanders, and as the announcer describes the engulfing flames, his voice is cut off in mid-sentence. Welles deliberately scripts several long seconds of silence, or "dead air," to increase the tension and the verisimilitude.[6] In New Jersey and elsewhere around the country, people panic and many load their belongings into cars to escape the menace.[7]

To the modern ear, Welles's broadcast has the tone of cheesy, B-grade science fiction. But this was a younger, more innocent world, worried about war and ignorant about the improbability of aliens actually visiting Earth. It was nearly twenty years before America would enter the Space Age. In fact, the story of invasion from Mars transcends particulars of time and culture. When H. G. Wells's novel *The War of the Worlds* was published in 1898, it was an instant classic. His words retain their evocative power: "Yet across the gulf of space, intellects vast and cool and unsympathetic, regarded our planet with envious eyes, and slowly and surely drew their plans against us." More than a century later, when Stephen Spielberg adapted the book for a 2005 movie, the basic plot was unchanged.[8] Fear of alien invasion taps into something deep in the human psyche, as primal as dreams themselves.

Evolving Views of Mars

Even to the naked eye, Mars clearly varies in brightness over months and years. Mars is roughly 50 percent farther away from the Sun than the Earth, and its distance from us depends on which side of the Sun each planet is on and the details of their elliptical orbits. At its closest,[9] Mars is only 55 million kilometers away and, at its farthest, it's 400 million kilometers away. This variation corresponds to a factor of 50 in apparent brightness and a factor of 7 in angular size. Only the brightness variation is visible to the naked eye; a telescope is needed to resolve Mars into a pale red disk. Even when it looms closest in the sky, Mars is just 25 arc seconds across, or seventy times smaller than the full Moon.

Following the invention of the telescope, the view of Mars evolved relatively slowly. Galileo began observing Mars in September 1610.[10] He noticed that it changed in angular size and he speculated that the planet had phases. The Dutch astronomer Christian Huygens was first to draw a sketch with surface features, in particular the dark area or "mare" called *Syrtis Major*. Huygens thought Mars might be inhabited, perhaps by intelligent creatures. In the middle of the seventeenth century, Giovanni Cassini and Huygens first spotted the pale polar caps of Mars,[11] and in the early eighteenth century Cassini's nephew Giacomo Maraldi saw variations in the polar caps that he speculated were due to water freezing and melting during the Martian seasons, although he could not rule out varying clouds.[12] William Herschel used his state-of-the-art telescopes for a period of more than eight years beginning in 1777 to bolster the interpretation that the poles were made of frozen water. He had measured the tilt of Mars's spin axis relative to the plane of its orbit so knew it had similar seasons to the Earth. He had also read Huygens's posthumous book *Cosmotheoros* in which the Dutchman speculated about life in the Solar System. In an address to the Royal Society in London, Herschel asserted boldly: "These alterations we can hardly ascribe to any other cause than the variable disposition of clouds and vapors floating in the atmosphere of the planet. . . . Mars has a considerable but modest atmosphere, so that its inhabitants probably enjoy a situation in many respects similar to our own."[13] With respected scientists setting up the expectation of life on Mars so long ago, it's not surprising that the idea had taken deep root by the modern age.

Telescope design continued to improve through the nineteenth century, allowing telescopes to make sharper images and resolve smaller features on Mars. In 1863, the Jesuit astronomer Angelo Secchi saw the *maria* appear to change in color; he fancifully drew them as green, yellow, blue, and brown at different times. He also saw two dark, linear features that he referred to as *canali*, which is Italian for grooves or channels.[14] It was a fateful choice of words, because the literal English translation as canals suggests construction by a technological civilization.

Mars Fever

Our vision of distant worlds has improved immensely since Galileo first pointed his slender spyglass at the night sky. Observational astronomy has moved from naked-eye observing to the use of large-format CCDs. These devices register an image by converting incoming light first into electrons and then into an electrical current, and astronomers typically gather light for several minutes up to an hour before reading out the device and inspecting the image. The CCDs that astronomers use are just larger format versions of the ubiquitous detectors found in digital cameras and cell phones. However, before photography matured, the only detector in astronomy was the unaided eye, and the only way to record an image was to sketch it on paper. Professional and amateur astronomers are familiar with "seeing," the rapid fluctuation of images caused by convective motions in the atmosphere; it's the phenomenon that causes stars to "twinkle." Viewed through a telescope, star images flicker and dance. But there are moments of stillness when the images become crisp.[15] Observers ever since the time of Galileo have learned to swiftly record the view when the seeing is at its best. In those moments when the light is not quite as scrambled by the atmosphere, features become apparent that are otherwise invisible and images seem to snap into focus.

In 1877, Mars was at its closest approach to the Earth, and Giovanni Schiaparelli was prepared to make the best observations of Mars yet. Already a talented observer, he used his skills as a draughtsman to make rapid sketches of the planet during the moments of sharp viewing, and he built up the stamina needed to concentrate intensely in short bursts through a long winter's night. He made detailed maps, naming features as "seas," not because he thought they actually contained water, but by tradition, as had been done with lunar features since the time of Galileo. He saw linear features stretching for hundreds of miles across the surface that were evocative of artificial constructions, although he resisted drawing this conclusion (figure 2.1).[16] Meanwhile, a separate debate raged over whether the atmosphere of Mars contained a significant amount of water vapor. Some observers claimed that it did, but it's very difficult to separate the signature of water around a remote planet from the very much stronger signature of water imprinted on the light by the Earth's atmosphere, and these observations turned out to be flawed.[17] As an Italian, Schiaparelli used the term *canali*, which was once again given an erroneous and literal translation in English-speaking media.

Mars fever began to take hold. The Suez Canal had opened in 1869, so the public was primed to appreciate the engineering achievement implied by canals on Mars. Not every observer could confirm the linear markings, but many of them deferred to Schiaparelli's skill and assumed that their own shortcomings were the obstacle. Amateur astronomer and author William Sheehan has noted the power of this type of thinking, where expectation and projection can shape the sensory experience: "Schiaparelli had taught observers how to see the planet, and eventually it was impossible to see it any other way. Expectation created illusion."[18]

The scene then shifted to northern Arizona. It was 1894, and Percival Lowell was driving his workers hard. He was racing to build a telescope before a particularly close approach of Mars. The patrician Bostonian had left his gilded life to fuel a personal obsession in the thin air of the northern Arizona desert. The previous Christmas, Lowell had been given a copy of *The Planet Mars* by Camille Flammarion as a present—Flammarion was a noted French astronomer and popularizer of science, considered by many the early predecessor of Carl Sagan. Flammarion accepted the interpretation that Martian canals represented intelligent life and in his book wrote: "The actual conditions on Mars are such that it would be wrong to deny that it could be inhabited by human species whose intelligence and methods of action could be far superior to our own. Neither can we deny that they could have straightened the original rivers and built up a system of canals with the idea of producing a planet-wide circulation system."[19] Lowell had a prior

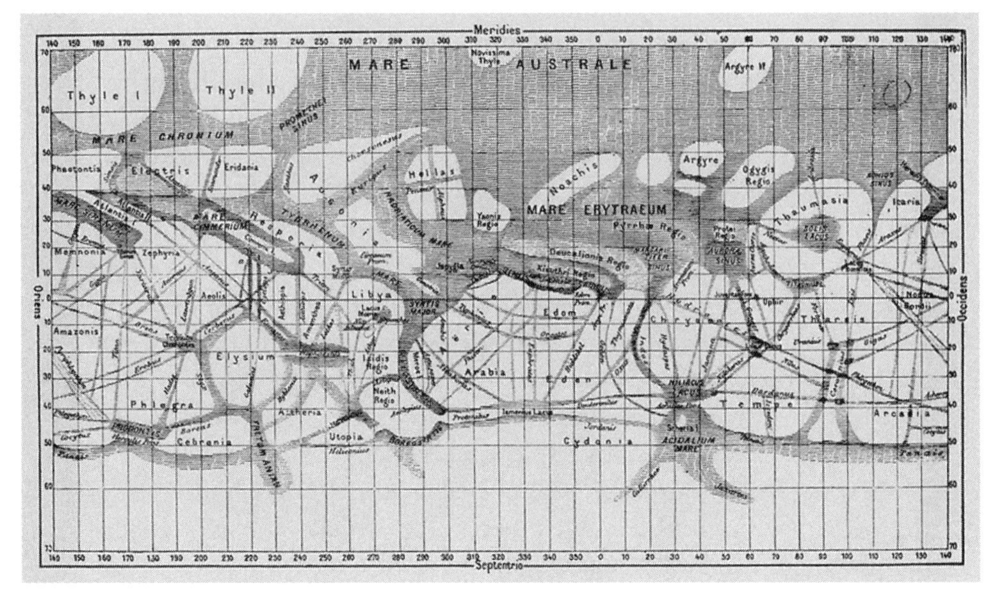

Figure 2.1. Giovanni Schiaparelli's map of Mars, compiled over the period 1877–1886, showed many linear features that Schiaparelli did not interpret as artificial or as signs of intelligent life. However, Percival Lowell strongly attributed the same features to a dying Martian civilization transporting water from the poles to the equator (*The Planet Mars*, Camille Fammarion [1892], Paris: Gauthier-Villars).

interest in astronomy and he correctly judged that the best place to see sharp images was in the high and dry desert air, far from any city lights. The Lowell family motto was "seize your opportunity" and Percival took it to heart, dropping his plans of leisurely travel in Asia to venture into the rugged terrain south of the Grand Canyon.

For fifteen years, Lowell studied Mars diligently and produced a series of drawings of intricate surface markings as he perceived them. To Lowell, the canals were real and they were manifestly artificial. Around his observations he wove a story of a dying race, more intelligent than humans, who had built a network of canals to carry water from the poles to the arid equatorial regions.[20] Professional astronomers were skeptical of the observations and their interpretation, and were generally dismissive of the back story, but Lowell bypassed them with popular books and extensive lecturing. Lowell published his first book on the subject in 1896, titled simply *Mars*. Two years later, H. G. Wells incorporated major elements of Lowell's view of Mars into *The War of the Worlds*, which was very popular and struck a nerve with the public. *The War of the Worlds* was first published in magazine serial form, in the tradition of the novels of Charles Dickens. As a book, it has never been out of print and has so far spawned five movies, a TV series, and numerous imitators. At this point, cultural and scientific views of Mars were closely twined.

Lowell's 1906 book *Mars and Its Canals* met with a strong rebuttal from Alfred Russel Wallace, co-discoverer of the theory of natural selection, who argued that Mars was far too cold to host liquid water. He considered that the polar caps were made of frozen carbon dioxide, not water ice, and he concluded that Mars was uninhabited and uninhabitable. Wallace's critique made no difference in the cultural arena. Ten years later, Edgar Rice Burroughs published *A Princess of Mars*, set on a version of the red planet alive with exotic animals, fierce warriors, and princesses in near-human form. He wrote another ten Mars stories over the following thirty years, inspiring Arthur C. Clarke and Ray Bradbury and launching a grand tradition of Mars science fiction.[21]

Mars fever was resistant to the medicine of improved astronomical observations.[22] Lowell stubbornly defended his position until the end of his life, saying in 1916: "Since the theory of intelligent life on the planet was first enunciated twenty-one years ago, every new fact discovered has been found to be accordant with it. Not a single thing has been detected which it does not explain. This is really a remarkable record for a theory. It has, of course, met the fate of any new idea, which has both the fortune and the misfortune to be ahead of the times and has risen above it. New facts have but buttressed the old, while every year adds to the number of those who have seen the evidence for themselves."[23] By 1938, telescopic remote sensing had demonstrated beyond any reasonable doubt that Mars was a dry, barren, lifeless desert, but that didn't dim the twinkle in Orson Welles's eye as he reeled the public in with his artful hoax.

The fever cooled dramatically in 1965 with Mariner 4. Spurred into existence by a series of firsts for the Soviets in space, NASA was a young government agency with ambitious plans. By the mid-1960s the hardware development for the Apollo program was in full swing, but NASA also wanted to gain the initiative in interplanetary probes.[24] The Mariner series of space probes was designed to investigate the inner Solar System. Space exploration was definitely not for the faint of heart; in the 1960s roughly half of NASA's probes failed. Mariners 1 and 2 were intended for Venus. Mariner 1 veered off-course and had to be destroyed just after launch, while Mariner 2 made it to Venus and transmitted useful data as it flew by. Venus was known to have thick, opaque clouds so there was no camera on board. Mariners 3 and 4 were intended for Mars. Mariner 3 mysteriously lost power eight hours after launch, so all eyes turned to Mariner 4.[25] After seven months and 220 million kilometers of travel and one mid-course correction, it swooped within 10,000 kilometers of the planet's surface.

The spacecraft sent back twenty-one black and white images, the first pictures ever taken of a world beyond the Moon by a space probe. The images were small and grainy, with eight times worse resolution and sixty times fewer pixels than a typical cell phone camera. They showed a barren and cratered surface. Other instruments indicated a sparse atmosphere, daytime high temperatures of –100°C, and no magnetic field that would be needed to protect the planet from harmful cosmic rays.[26] Mars, so deeply rooted in the popular consciousness as a living world, seemed to be Moon-like and lifeless.

The Vikings Reach Mars

On July 20, 1976, a small spacecraft emerged from a cloudless, apricot-colored Martian sky and fell toward the western *Chryse Planitia*, the "Golden Plain." Its heat shield glowed as it buffeted through the tenuous atmosphere.[27] About four miles up, the parachutes deployed, the heat shield was jettisoned, and three landing legs unfolded like a claw. At one mile up, the retrorockets fired, and less than a minute later the Viking 1 lander decelerated to six miles per hour, reaching the surface with a slight jolt.[28] It was a landmark of technological prowess, the first time humans had ever soft-landed an emissary on another planet.

The twin Viking missions were the most complex planetary probes ever designed. Their total price tag was around $1 billion, equivalent to $4 billion today after adjusting for inflation. That can be compared to the $80 million cost of Mariner 4. Mission planners were well aware of the challenges; the Soviets had previously failed four times to soft land on Mars.[29] Each Viking consisted of an orbiter designed to image the planet and a lander equipped to carry out detailed experiments on the surface.[30] For the most part, the hardware worked flawlessly, but there were tense moments for the engineers and scientists on the team. After ten months and 100 million miles of traveling, the Vikings reached Mars two weeks apart. The first landing had been planned for July 4, 1976, the nation's bicentennial, and the landing sites were selected after years of deliberation. But as the twin orbiters started mapping the planet with ten times sharper images than had ever been taken before, mission planners were shocked to see that the planned Viking 1 landing site was not the benign plain they'd expected, but the rock-strewn bottom of what appeared to be a riverbed.

The landing site was abandoned. Gentry Lee, the director of Science Analysis and Mission Planning for Viking, vividly recalled the turmoil the new images caused: "For almost three weeks the Viking Flight Team operated at an unbelievable pace and intensity. Many of the key members of the team, including not just the engineers, but also [Jim] Martin and [Tom] Young and many of the world's foremost planetary scientists, worked fourteen or more hours a day for the entire period. Landing Site Staff meetings, to synthesize the results and look at all the logical options, were held every day. Carl Sagan, Mike Carr, Hal Masursky, and other famous Viking scientists argued eloquently about the safety of each of the candidate landing sites. Finally, the exhausted operations team managed to reach a consensus."[31] A new site at *Chryse Planitia* was selected. Once the lander separated from the orbiter, it would not be possible to redirect the lander with any additional commands. The die was cast. Few team members slept much that night.

Media coverage and public interest were intense. Viking 1 and 2 marked the first close-up glimpse of the red rocky soil of Mars. "Despite the early hour, the von Kármán Auditorium [at NASA's

Jet Propulsion Laboratory] was packed. In addition to 400 journalists from around the world, there were 1,800 invited guests watching a closed-circuit television view showing the control room, with Albert Hibbs, one of the mission planners, providing the commentary."[32] For nineteen agonizing minutes, everyone waited—that was how long it took for telemetry to reach the Earth saying the lander was safe. Its first picture was of its own foot, to see how far it had sunk into the Martian soil. When, on its second day on Mars, Viking 1 sent back the first color panoramic views of the Martian terrain, scientists and public audiences alike recognized a kind of reddish, iron-rich soil familiar to them from the deserts of the American southwest (plate 2).[33] In fact, among the first panoramic photos released to the press was a view of the Martian landscape under blue skies, though JPL scientists quickly realized the sky should be salmon colored. As Paolo Ulivi and David Harland note, "Initially, the image-processing laboratory combined the red, green and blue frames to produce the dark blue-black sky that the thin atmosphere had been expected to yield, but after [the images] had been recalibrated the sky was found to be pinkish-orange."[34]

The missions galvanized global fascination with the stark Martian landscape, and they continue to provide a compelling story of discovery and of the sheer difficulty of trying to do science so far from a conventional laboratory. The feelings were best described by NASA's Gentry Lee: "The Viking team didn't know the Martian atmosphere very well, we had almost no idea about the terrain or the rocks, and yet we had the temerity to try to soft land on the surface. We were both terrified and exhilarated. All of us exploded with joy and pride when we saw that we had indeed landed safely."[35]

What We Learned

Viking 1 was launched from Cape Canaveral on August 20, 1975. Its twin was launched on September 9, 1975, and Viking 2 reached Mars on August 7, 1976, a few weeks after the first triumphant landing. The second lander reached a site several thousand miles away at *Utopia Planitia* on September 3, 1976, after suffering its own small mishaps. The downward looking radar was probably confused by a rock or reflective surface, so the thrusters fired too long, cracking the soil and throwing up dust. It stopped with one leg resting on a rock, tilting the lander by eight degrees. Otherwise, it was unharmed. The hardware was designed to last for ninety days but proved to be very durable.[36] Viking 1's orbiter lasted nearly two years and Viking 2's orbiter lasted just over four years. Meanwhile, on the surface, the Viking 2 lander ceased operating when its battery failed after three and a half years, while the indomitable Viking 1 lander was going strong after more than six years, when simple human error during a software update made the antenna retract and communication with the Earth was lost.[37]

Public attention focused on the landers, with their life detection experiments and "you are there" images, but the orbiters were also very important in shaping a modern view of Mars. The orbiters carried the landers to Mars, scouted for landing sites, and relayed lander data back to Earth. Each equipped with optical and infrared imagers, they mapped 97 percent of the surface and sent back over 46,000 images. They could see features 150 to 300 meters across anywhere on the planet, and in selected areas they could resolve features the size of a small house.[38] Whereas Mariner had only seen old, cratered terrain, the Vikings saw a rich and varied topography and geology. There were immense volcanoes, corrugated lava plains, deep canyons, and wind-carved features. The planet was divided into northern low plains and southern highlands that were pock-marked with craters. Mars had extensive elevated regions of volcanism, although no areas of fresh lava. There was weather: dust storms, pressure variations, and gas circulation between the poles. As NASA's Thomas Mutch said of the Viking orbiter images, "They show Mars as an extremely diverse planet. . . . It is difficult to avoid the conclusion that, though Viking contributed immeasurably to breaking the code of the Mars enigma, we do not yet confidently understand its dramatic and turbulent past."[39]

Most excitingly, Viking provided indirect but compelling evidence for water. Not currently—the air is so cold and thin that a cup of water placed on the surface would evaporate away in seconds. But the orbiters sent back images of rock formations all over Mars that could only have been produced by the action of

Figure 2.2. Four decades after Mariner 9 saw the first signs that Mars had been wet in the ancient past, the evidence for geological formations that can only result from the action of water has grown much stronger. In these images from the Mars Global Surveyor mission, features are suggestive of episodic eruptions of sub-surface water, altering the landscape before the water evaporates and boils into space (NASA/Jet Propulsion Laboratory).

large amounts of liquid water in the past. Huge river valleys were seen, and places where it looked like rivers had once fanned out into a spider's web of channels that ended in ancient shallow seas. The flanks of volcanoes had grooves that on Earth indicate water erosion. Many craters had shapes that were consistent with the impactor landing in mud. Other regions of chaotic terrain looked as if they had collapsed when underground volcanism melted ice, which then flowed away as water. The water flows implied by these features were equal to the greatest rivers on Earth (figure 2.2).[40]

Searching for Life

The grip that Mars holds on the popular imagination is grounded in the question of whether Mars ever was, or is, alive. The Viking landers were explicitly designed to search for evidence of life in the Martian regolith or perhaps in the planet's geological past.[41]

Figure 2.3. The Viking 2 landing site, named by NASA the Gerald Soffen Memorial Station, as seen in a mosaic of images from the lander. The sequence shows the robotic arm extending out to scoop a small amount of the Martian regolith, then retracting to deposit the sample in one of the biological experiments onboard (NASA/Viking Lander Image Archive).

The short section of the mission science summary on life detection is worth quoting in full because of the surprising ambiguity of the wording: "Three experiments were conducted to test directly for life on Mars. The tests revealed a surprisingly chemically active surface—very likely oxidizing. All experiments yielded results, but these are subject to wide interpretation. No conclusions were reached concerning the existence of life on Mars."[42]

The Viking landers packed a substantial scientific punch in their 200-pound payloads. Power came from a plutonium-238 radioisotope thermal generator, eking out 30 Watts of continuous power. Using late 1970s technology, Viking's data capabilities were even more feeble; the data recorder on each lander could only store 8 Mbytes at a time, thousands of times less than the average memory stick. The landers had cameras that could take 360-degree panoramic images. They had seismometers and instruments to test magnetic fields. They had meteorology booms that measured temperature, pressure, and wind speed and direction. Most importantly, they had robotic arms that could scoop up soil samples and deposit them into temperature-controlled and sealed containers on each spacecraft (figure 2.3).

The biological package contained four instruments. A gas chromatograph and mass spectrometer heated soil samples and measured the molecular weight of each component of the vapor released, down to a concentration of a few parts per billion. The instrument found no significant levels of organic, or carbon-based, molecules. Mars soil had even less carbon than the lifeless soils tested on the Moon by the Apollo missions. This seemed to be prima facie evidence against life. The gas exchange instrument added nutrients, and then water, to a soil sample, and then looked for changes in the concentration of gases such as oxygen and methane in the sealed chamber. The hypothesis was that a living organism would process one of the gases. The result was negative. The pyrolytic release experiment created an "atmosphere" in the chamber using radioactive carbon, in the hope that a photosynthetic organism would incorporate the carbon the way plants do on Earth. After several days of incubation under an artificial Sun (in this case, a xenon arc lamp), the sample was baked at a high temperature to see if any of the radioactive carbon had been converted into biomass. The results were also negative.[43]

The only wild card was the labeled release experiment. A sample of soil had nutrients dissolved in water added to it, and the nutrients were "tagged" with radioactive carbon, which was once again used as a tracer. To the surprise of the instrument team, radioactive carbon dioxide was detected in the air above the samples, suggesting that microbes had metabolized one or more of the ingredients. The same result was seen in both Viking landers. However, when the experiment was repeated a week later, the air was free of radioactive carbon. The data were declared inconclusive.[44]

Overall, the results were disheartening for those who hoped that Mars might be a living world. Terrestrial life is based on complex molecules with a carbon backbone—organic ingredients like carbohydrates, proteins, nucleic acids, and lipids. While organic does not mean biological, all life on Earth is carbon-based and so is made of organic ingredients. The Viking experiments detected virtually no organic compounds in the Martian regolith. This was somewhat surprising, since they are fairly common on small Solar System bodies like comets, asteroids, and meteorites. With no organic material, the biological experiments would have been doomed to failure, since their aim was to detect a metabolism that could incorporate carbon from the atmosphere, as microbes on Earth do. The lander could only gather samples from the top few centimeters of the regolith, and that layer is blasted with ultraviolet radiation and cosmic rays from space (Mars has no protective layer of ozone). The surface is strongly oxidizing, as the "rusty" red color of iron oxide indicates.[45] So the conventional interpretation is that activity seen in the experiments was caused by chemical reactions involving oxidizing molecules in the soil, with no biological explanation required.

Could Viking Have Missed Life?

Gil Levin has never wavered. The principal investigator of Viking's labeled release experiment is now approaching ninety, but he's very active and stays current with research on Mars. Levin had an unusual career path, starting as a sanitation engineer before joining NASA. In addition to authoring 120 scientific papers, he owns fifty patents for items ranging from artificial sweeteners to therapeutic drugs.[46] Levin insists that at two locations on Mars, and in seven out of nine tests, his experiment detected biological activity. No purely chemical reactions have been identified that would fully reproduce the labeled release results, which explains the cautious and equivocal wording of the science mission summary.

Levin is in the minority, but he's not alone in believing the Viking results bear revisiting. Rafael Navarro-Gonzalez and his team went to the highest and driest parts of the world, places like the Atacama Desert and Antarctic dry valleys, and duplicated the tests that Viking did more than thirty-five years ago.[47] In northern Chile, they found arid soils with levels of organic material that would have been undetectable by the Viking instruments, yet there were bacteria in the soil. In other words, Viking was not sensitive enough to detect either organics or life in terrestrial locations that are the closest analogs to Mars.[48] The team speculated that the organic material on Mars might have been too stable to turn into a gas, even at the blistering temperature of 500°C (932°F) reached inside the oven of the Viking experiment. They also noted that iron in the soil might oxidize the organic material and prevent its detection by the mass spectrometer. This would account for the carbon dioxide that was detected in the experiments.

Mars is an alien chemical and physical environment, so it makes sense to think "outside the box" when considering how biology might operate there. If we only look for life as we know it, that's all we'll be able to detect and we may miss life with a different biochemical basis. Dirk Schulze-Makuch and Joup Houtkooper have speculated that Mars might be home to microbes that use water and hydrogen peroxide as the basis for their metabolism.[49] Hydrogen peroxide is a toxic household disinfectant and seems implausible as a basis for life, but the researchers note that it's more life-friendly under the extreme conditions typical of Mars. It attracts water and when mixed with water freezes at –57°C, yet it doesn't form cell-destroying ice crystals at even lower temperatures. Hydrogen peroxide is tolerated by many terrestrial microbes and is the basis of the metabolism for *Acetobacter peroxidans*. It's used as a defensive spray by the bombardier beetle and even performs useful functions in the cells of some mammals. As Schulze-Makuch pointed out in 2007, "We can be absolutely wrong, and there might not be organisms like that at all. But it's a consistent explanation that would explain the Viking results. . . . If the hypothesis is true, it would mean we killed the Martian microbes during our first extraterrestrial contact, by drowning—due to ignorance."[50]

Another speculation was spurred by the discovery of perchlorate in a Martian polar region by the Phoenix lander in 2008.[51] Phoenix carried out a wet chemical analysis of Martian soil, finding it to be alkaline and low in the type of salts found on Earth, but it had enough perchlorate to act as antifreeze and allow the soil to hold liquids for short periods during the summer. Perchlorate is strongly oxidizing and so generally considered to be toxic for life, but some microbes metabolize it.[52] In 2010, the Viking results were reinterpreted in the light of the Phoenix discovery, giving support to Gil Levin's lonely position.[53] When perchlorate was added to Atacama Desert soils and analyzed in the manner of the Viking samples, it released chlorine compounds. When these were seen in the Viking experiments back in the 1970s, they were presumed to be cleaning fluid contaminants from Earth. But if perchlorate is present in the mid-latitude Martian soil, it would explain the data. Since perchlorate becomes a strong oxidant when heated, it would have destroyed any organics and so explain why Viking didn't detect any. It's remarkable that the Viking legacy continues to provide so many surprises and unanswered questions.

A leitmotiv in the reinterpretation of the Viking biological experiments is the amazing range of life on Earth—there are microbes that can tolerate or thrive in conditions that would be fatal to plants and animals. Collectively, forms of life found in physically extreme conditions are called extremophiles. The envelope of habitability is much larger than we thought in the 1970s. There are microbes than can live below the freezing point of water and with toxic and alkaline conditions as seen on Mars. Even high doses of ultraviolet radiation are not an impediment to life, since microbes can survive conditions in the upper stratosphere that match the radiation environment of Mars.[54] In science, proof is the gold standard, but it's a very high bar to clear, usually requiring copious amounts of evidence. While Viking didn't find life on Mars, it was unable to prove the converse hypothesis, that biology is absent. It's premature to declare Mars dead. Surprisingly, this harsh planet inspired a new way of thinking about life on Earth.

Mars and the Birth of Gaia

A substantial legacy of Viking is that scientists have gained a better sense of the finely tuned ballet of biogeochemical cycles that sustain Earth's vibrant biosphere. The Viking missions unexpectedly became integral to the late-twentieth-century view of the Earth's biosphere as a self-sustaining system produced by biota interacting with the planet's geochemistry. Brian Skinner and Barbara Murck assert that future historians will consider discovery of the complex interactions between Earth's biota and geologic, hydrologic, and atmospheric cycles among the most significant scientific contributions of the twentieth century.[55] This sea change in thinking about our planet began to emerge in the 1960s and was sparked by British scientist James Lovelock's Gaia hypothesis. That idea later evolved into Gaia Theory and the academic discipline referred to as Earth System Science.

In the early 1960s, Lovelock was working for NASA's Jet Propulsion Laboratory with other planetary scientists to develop the experimental means for determining whether microbial life might exist on Mars. NASA's plan for a mission to robotically explore Mars in search of life was initially titled Voyager, not to be confused with the interplanetary mission launched in 1977. That plan was subsequently scrubbed and reconfigured into what became the Viking mission. In 1965, while helping to develop life detection experiments for the Mars landers, Lovelock came to the sudden realization that Earth's atmosphere must be a natural extension, and a by-product, of Earth's biota. This became the basis for the Gaia hypothesis and a paper Lovelock published in the prestigious journal *Nature* that year.[56]

In his first book, *Gaia: A New Look at Life on Earth* (1979), Lovelock, then sixty years old, explained that as he mulled over ways one could detect organisms in the Martian soil, he turned to our own planet and began to imaginatively "look at the Earth's atmosphere from the top down, from space." In the opening sentence, Lovelock observed: "As I write, two Viking spacecraft are circling our fellow planet Mars, awaiting landfall instructions from the Earth. Their mission is to search for life, or evidence of life, now or long ago. This book also is about a search for life."[57] The organisms on Lovelock's mind, however, were those on Earth.

Lovelock thought it might be possible to answer the question of whether Mars harbored microbes by simply examining the composition of the atmosphere. If a planet supported life, Lovelock posited, its atmosphere would be shaped in part by biota "bound to use the fluid media—oceans, atmosphere, or both—as conveyer belts for raw materials and waste products. . . . The atmosphere of a life-bearing planet would thus become recognizably different from a dead planet."[58] A mix of reactive atmospheric gases like oxygen and methane, Lovelock surmised, would be a biosignature of life as on Earth. Moreover, in 1965, researchers at the Pic du Midi Observatory in France reported that the atmospheres of Venus and Mars were largely made of carbon dioxide.[59] Lovelock knew that Earth's atmosphere, by comparison, contained reactive gases that dissipate if not continuously replenished by Earth's biota. The Pic du Midi observations seemed a sure confirmation to Lovelock, who was then collaborating and publishing with NASA colleague Dian Hitchcock on analyses of infrared surveys of the Martian atmosphere.[60] Methane in our atmosphere "has been fairly constant as ice-core analyses prove, for the past million years, as has oxygen," notes Lovelock, who highlights the fact that "for such constancy to happen by chance is infinitely improbable" and therefore must be sustained by life.[61] While Lovelock set the wheels in motion for a systems approach to understanding the biosphere, it was his collaboration with microbiologist Lynn Margulis that formalized the Gaia hypothesis.

Margulis was just then developing her theory of symbiogenesis, which posits that cell structures, and ultimately organisms, evolved from symbiotic relationships between progenitor cells or organisms. She is celebrated for her research in early cell evolution and was first to identify bacteria as the antecedent of chloroplasts and mitochondria in eukaryotic cells. Margulis also was interested in how bacteria and other microorganisms might impact their environment. As it happened, Lovelock shared an office at JPL with planetary scientist Carl Sagan (figure 2.4). Lovelock biographers John and Mary Gribbin comment: "Margulis had independently become intrigued by the oddity of the Earth's oxygen-rich atmosphere and asked her former husband, Carl Sagan, whom she ought to discuss the puzzle with. Sagan knew just the man, and put her in touch with Lovelock."[62] Upon Sagan's recommendation, Lovelock and Margulis began exploring the question of how the highly reactive gas oxygen in our atmosphere has been sustained at a consistent level over billions of years.

Gaia Theory and Earth System Science

Together Lovelock and Margulis theorized that Earth's habitability was not simply a function of its orbital position relative to the Sun, but was in large part due to biota metabolizing and cycling atmospheric gases and rock minerals. As Margulis explained, "The metabolism, growth, and multiple interactions of the biota modulate

Figure 2.4. Carl Sagan poses with a model of the Viking lander in Death Valley, California. Sagan was a pivotal figure in the mission planning, leading the argument to include a camera in the design. His research pointed to the intimate connection between the composition of a planetary atmosphere and its habitability (NASA/Jet Propulsion Laboratory).

the temperature, acidity, alkalinity, and, with respect to chemically reactive gases, atmospheric composition at the Earth's surface."[63] While Lovelock considered the impact of all forms of life in using and maintaining Earth's biosphere, Margulis's contribution to Gaia in the early 1970s highlighted the shaping effect of microorganisms relative to atmospheric chemistry, rock weathering, and carbon deposition via phytoplankton onto the seafloor. Lovelock wrote: "Lynn brought her deep understanding of microbiology to what until then had been mainly a system science theory that saw a self-regulating Earth through the eyes of a physical chemist. By stressing the importance of the Earth's bacterial ecosystem and its being the fundamental infrastructure of the planet, Lynn put flesh on the skeleton of Gaia."[64]

However, in the early 1970s, when Lovelock and Margulis floated their ideas regarding Gaia, Earth sciences were largely studied in isolation. "Oxygen, for example, was thought to come solely from the breakdown of water vapour and the escape of hydrogen into space, leaving an excess of oxygen behind," explains

Lovelock. "Life merely borrowed gases from the atmosphere and returned them unchanged. Our contrasting view required an atmosphere which was a dynamic extension of the biosphere itself."[65] It's not surprising that the Gaia hypothesis was initially controversial and slow to be accepted. By 1974, Lovelock and Margulis were publishing scientific papers on Gaia. However, they also published a paper for nonspecialists titled "The Atmosphere, Gaia's Circulatory System" in Stewart Brand's *CoEvolution Quarterly*.[66] New Age connotations and the fact that the theory was championed by many nonscientists didn't help their cause. Another issue was the sheer complexity of the biosphere and its interlocking parts—scientists had trouble making robust and predictive models of its behavior.[67] But contrary to the scientific community's cool reception of the Gaia hypothesis, NASA was very interested in Lovelock's ideas regarding planetary atmospheres.

Gaia Theory proposes that bacteria and other organisms not only consume and replenish atmospheric components like oxygen and methane but also maintain our biosphere's temperature and habitability through rock weathering and the cycling of carbon, oxygen, sulfur, and other chemicals. The atmosphere is understood as a coupled system with the lithosphere, hydrosphere, and the planet's biota, and these geospheres interact in cycling water, minerals, chemicals, and atmospheric gases. The theory, however, evolved over time and represented a spectrum of ideas. At one end of the spectrum was the uncontroversial premise that living organisms have altered Earth's atmosphere and surface geology. Following from that is the idea that the biosphere and the Earth co-evolve in a way that is self-regulatory due to negative feedback loops. The most extreme form of the theory, containing the idea that the entire biosphere is somehow a living organism, with the teleological implication that the continuation of life is a purpose of co-evolution, has been discredited (and Lovelock was always clear about not subscribing to these views). Nor were Lovelock and Margulis the first to consider Earth's biology and its geospheres as intimately coupled. Early proponents of such thinking were the eighteenth-century Scottish geologist James Hutton, the father of modern geology, and the Russian mineralogist Vladimir Vernadsky. Hutton was among the first to theorize how rock cycling impacted our entire planet, referring to the Earth as a "superorganism," while Vernadsky theorized in the early twentieth century that geologic processes contributed to sustaining the biosphere.[68]

Lovelock and Margulis's integrated view supposed that microorganisms not only maintained Earth's habitable atmosphere and climate, but also could alter the environment on a planetary scale. For this and reasons noted above, Gaia hypothesis seemed preposterous to many people. However, as Lovelock explains in *The Vanishing Face of Gaia* (2009), "It is too often wrongly assumed that life has simply adapted to the material environment, whatever it was at the time; in reality life is much more enterprising. When confronted with an unfavorable environment it can adapt, but if that is not sufficient to achieve stability it can also change the environment."[69] Stromatolites in Australia's Shark Bay indicate, for example, that by 2.4 billion years ago the progenitors of cyanobacteria developed oxygenic photosynthesis and began to reconfigure the Earth's early atmosphere from largely nitrogen and carbon dioxide to the nitrogen- and oxygen-rich composition we breathe today. But any correlation between the emergence of oxygen metabolizing life and global atmospheric composition is complex. Measurable levels of oxygen appeared 300 million to a billion years after the first evidence of oxygenic photosynthesis, which indicates other geologic or hydrologic processes were simultaneously stripping oxygen from the atmosphere.[70] The complexity of how microorganisms and geochemical processes combine to sustain or alter our atmosphere remains at the core of Gaia Theory and Earth System Science.

NASA and Earth Observation

"When you cut your teeth on other worlds," Carl Sagan once noted, "you gain a perspective about the fragility of planetary environments and about what other, quite different, environments are possible."[71] Sagan points to James Hansen's research of Venus that led to early climate models to predict how greenhouse gases had been trapped in its atmosphere and might similarly impact Earth's climate, as well as Mario Molina and Sherwood Rowland's investigation of chlorine and fluorine molecules in Venus's atmosphere that led to an understanding of the threat of CFCs to Earth's ozone layer. These investigations of Venus revealed that Earth's dynamic biosphere could lose its ability to support life.

Lovelock has expressed his concern not so much that our carbon footprint will destroy the Earth, but that warming of Earth's atmosphere could mean the end of our civilization as we know it. Suggesting we make plans for "sustainable retreat," Lovelock at the age of eighty-nine comments, "It's really a question of . . . where we will get our food, water. How we will generate energy."[72] This is not a new position for Lovelock. Even from the beginning stages of Gaia Theory, Steven Dick and James Strick comment, "He realized that the gases that living organisms most actively affect, especially carbon dioxide, methane, oxygen and water vapor, are just those gases that most dramatically shape the climate of a planet."[73] "I can only guess the details of the warm spell due," wrote Lovelock in 1988. "Will Boston, London, Venice, and the Netherlands vanish beneath the sea? Will the Sahara extend to cross the equator?"[74] He's not being facetious.

In *The Ice Chronicles*, scientist Paul Mayewski and science writer Frank White discuss the findings from the Greenland Ice Sheet Project 2's (GISP2) ice cores drilled in central Greenland in 1990. The cores capture a 100,000-year record of the chemical composition of Earth's atmosphere and a picture of the rate of thinning of Greenland's glaciers. Of the book, Lynn Margulis noted, "The story of how international science obtained this fund of Pleistocene data from the central Greenland ice sheet reads like a novel."[75] However, this text isn't some action-adventure tale of humans against nature, but a serious assessment of changes in atmospheric composition over long time scales. Captured deep in the Greenland ice are tiny air bubbles that serve as time capsules from thousands of years ago. Scientists can analyze the air in the bubbles to determine the composition of Earth's atmosphere in ages past. Mayewski and White assert unequivocally that increases in methane and carbon dioxide produced through industry, agriculture, and mass transport are dramatically warming our planet's atmosphere. They write: "As with CO_2, methane is increasing at a rate that has never before been seen, and, as with CO_2, the increase correlates strongly with rising temperatures . . . . The result is that we are having a profound impact on the climate system." Mayewski and White note that the ice core data indicate that ecosystems on a global scale can be "easily and quickly disturbed."[76]

Climatologists are concerned that global warming could cause the collapse of the world's ice sheets, which in turn would produce an unprecedented rise in sea level. In 2011, British glaciologist Alun Hubbard reported that an expanse of glacial ice estimated at twice the size of Manhattan was on the brink of breaking free from Greenland. The previous year a stretch of ice four times the size of Manhattan, spanning 112 square miles, broke off the Greenland Ice Sheet and was adrift at sea in large chunks. "The freshwater stored in this ice island could keep the Delaware or Hudson rivers flowing for more than two years," estimated Andreas Muenchow at the University of Delaware. "It could also keep all U.S. public tap water flowing for 120 days."[77] Though Muenchow emphasizes that ocean scientists, glaciologists, and climatologists cannot yet predict the effects of melting glacial ice, he is nevertheless concerned, and he's not alone. The 2001 Amsterdam Declaration on Global Change reads:

> Global change cannot be understood in terms of a simple cause-effect paradigm. Human-driven changes cause multiple effects that cascade through the Earth System in complex ways. These effects interact with each other and with local- and regional-scale changes in multidimensional patterns that are difficult to understand and even more difficult to predict. . . . In terms of some key environmental parameters, the Earth System has moved well outside the range of the natural variability exhibited over the last half million years at least. The nature of changes now occurring simultaneously in the Earth System, their magnitudes and rates of change are unprecedented. The Earth is currently operating in a no-analogue state.[78]

It's no coincidence that Gaia Theory emerged in the late 1960s and early 1970s at the height of the Space Age, precisely as space travel and satellite observation began offering consistent and detailed global views of Earth (figure 2.5). NASA's Landsat 1, launched in 1972, was the first satellite dedicated solely to Earth observation, particularly of landmasses.[79] Nor is it surprising that

Figure 2.5. Earthrise in the orientation from which the *Apollo 8* crew saw it from lunar orbit. In zero gravity, the normal senses of up, down, and the horizon have no particular meaning. Just two dozen astronauts have ever experienced the extraordinary sight of their home planet rising over the horizon of another world (NASA/Apollo 8).

the original cover of Lovelock's book *Gaia: A New Look at Life on Earth* (1979) was illustrated with *Apollo 17*'s *Whole Earth* photograph, considered among photo historians as the most reproduced image in history. While "Gaia theory forces a planetary perspective," as Lovelock points out, that invaluable perspective was only possible with a serious commitment to space exploration. "It took the view of the Earth from space," writes Lovelock, "either directly through the eyes of an astronaut, or vicariously through visual media, to let us sense a planet on which living things, the air, the oceans, and rocks all combine in one as Gaia."[80]

By 1983, NASA had established the Earth System Sciences Committee (ESSC) to publish recommendations for a concerted science program in global Earth studies and satellite observing programs. NASA's planetary science and recent Earth observing missions have offered scientists their best understanding yet of the complex and interconnected Earth systems that sustain our biosphere. Several missions are specifically focused on understanding and mitigating, if possible, the effects of industrialized human activity.

NASA's GOES satellites offer real-time data regarding Earth's weather systems. The Aqua mission measures all aspects of Earth's water cycle in the solid, liquid, and gas states as well as global water temperatures, vegetation cover, and phytoplankton, which construct their shells from carbon dioxide and deposit that carbon onto the seafloor when they expire. The joint NASA and Argentinian Aquarius mission catalogs the salinity of Earth's oceans, while NASA's Aura mission monitors changes to Earth's ozone layer. In collaboration with the German Aerospace Center, NASA supports the GRACE mission, which has observed a large drawdown of fresh water globally, from Northern Africa to the large agricultural San Joaquin valley in California.[81] NASA and the French Space Agency together support the CALIPSO mission to track airborne particles or aerosols to determine their impact on global climate. These are only a handful of the varied and complex missions NASA supports in observing Earth's biological, hydrological, and geochemical systems. What we have learned from missions like these not only helps us to better understand climate change and its impact on life on Earth but also has prepared us to search for life on moons in the outer solar system and on exoplanets orbiting nearby stars. When telescopic detectors become sensitive enough, astrobiologists will scour the spectra of exoplanets light-years away searching for the biomarkers of oxygen or ozone, as these gases have no other significant source on Earth than its living organisms.

In 2006, on the thirtieth anniversary of the Viking Mission, Joel Levine, principal investigator of the proposed Mars Ares mission to fly a robotic plane over Mars to search for evidence of life, noted that scholars are still researching Viking's nearly 60,000 images, which remain "one of the most exciting, one of the most productive data sets we have ever obtained."[82] Venus and Mars, the closest worlds to Earth, retain their allure and mystery. Venus is a near twin of the Earth, our true sister planet, yet a runaway greenhouse effect has heated it to the point where life isn't possible. It's a salutary reminder of what might happen if we don't get our house in order in terms of carbon emissions. We wince and turn away. Mars, meanwhile, is the lightweight cousin of the Earth. In 2009, scientists at the India Space Research Organization (ISRO) reported having discovered thriving in Earth's stratosphere three previously unidentified species of bacteria, despite consistent exposure to ultraviolet radiation that usually kills such organisms. Given that conditions in the Earth's stratosphere closely resemble the atmosphere on Mars, the finding profoundly recalls Viking's controversial experiments conducted on Martian soil decades ago. If there is biology on Mars, it leads a tenuous existence in hospitable pockets below the surface. Until we definitively discover life, either on Mars or elsewhere, we invest these worlds with our full imagination, and project onto them our hopes, fears, and longings, anticipating the time when our isolation ends and we find companionship in the cosmos—if only in a colony of microbes.

3 MER

THE LITTLE ROVERS THAT COULD

The essay is short and very simple; the words are almost heartbreaking: "I used to live in an orphanage. It was dark and cold and lonely. At night, I looked up at the sparkly sky and felt better. I dreamed I could fly there. In America, I can make all my dreams come true. . . . Thank you for the Spirit and the Opportunity."[1] Sofi Collis was abandoned at birth into a Siberian orphanage and brought by her adoptive parents to live in Scottsdale, Arizona. In 2003, at age nine, she was writing in response to a call from NASA for names for its upcoming Mars rovers. A team of judges selected by the Lego Company and the nonprofit Planetary Society painstakingly whittled 10,000 entries down to thirty-three, and NASA made the final selection. Sofi unveiled the names at a pre-launch press conference hosted by Sean O'Keefe, NASA's administrator, who noted that her story twinned the two original spacefaring countries. It's safe to say her dreams are as boundless as space itself.

Six years later, in language that was formal, cumbersome, and numbingly prosaic relative to the lofty sentiment it conveyed, the U.S. House of Representatives gave a formal nod to these remarkable robotic emissaries with the following resolution, number 67 from the first session of the 111th session of Congress, adopted unanimously:

> Whereas the Mars Exploration Rovers Spirit and Opportunity successfully landed on Mars on January 3, 2004, and January 24, 2004, respectively, on missions to search for evidence indicating that Mars

once held conditions hospitable to life; whereas NASA's Jet Propulsion Laboratory (JPL), managed by the California Institute of Technology (Caltech), designed and built the Rovers, Spirit and Opportunity; whereas Cornell University led the development of advanced scientific instruments carried by the 2 Rovers, and continues to play a leading role in the operation of the 2 Rovers and the processing and analysis of the images and other data sent back to Earth; whereas the Rovers relayed over a quarter million images taken from the surface of Mars; whereas studies conducted by the Rovers have indicated that early Mars was characterized by impacts, explosive volcanoes, and subsurface water; whereas each Rover has discovered geological evidence of ancient Martian environments where habitable conditions may have existed; whereas the Rovers have explored over 21 kilometers of Martian terrain, climbed Martian hills, descended deep into large craters, survived dust storms, and endured three cold, dark Martian winters; and whereas Spirit and Opportunity will have passed 5 years of successful operation on the surface of Mars on January 3, 2009, and January 24, 2009, respectively, far exceeding the original 90-Martian day mission requirement by a factor of 20, and are continuing their missions of surface exploration and scientific discovery: Now therefore be it resolved, that the House of Representatives commends the engineers, scientists, and technicians of the Jet Propulsion Laboratory and Cornell University for their successful execution and continued operation of the Mars Exploration Rovers, Spirit and Opportunity; and recognizes the success and significant scientific contributions of NASA's Mars Exploration Rovers.[2]

Geographies of Other Worlds

By the late twentieth century, Mars emerged as the familiar landscape we recognize and cherish. Unlike Mercury, Venus, or even our own Moon, Mars has come into full relief. It is as if a Mercator's projection of Mars has lifted from its paper and rounded into a globe with tangible polar ice caps, soaring volcanoes, immense rift valleys, recognizable craters, and plains of crescent barchan dunes. Our robotic partners have added these features of Martian terrain to the iconic landscapes that preoccupy, and persist in, the human imagination. Images sent back by Mariner, the Viking Orbiters, Mars Global Surveyor, and Mars Reconnaissance Orbiter allow us to turn the globe of Mars in our hands, so to speak, and trace the planet's volcanoes, its broken and unzipped canyons, and its ancient and desiccated river beds.[3] With the Mars Exploration Rovers Spirit and Opportunity, we've explored in fine detail Victoria, Gusev, and Endeavor craters and captured our closest view yet of the red planet. Asked to name a landmark on the Moon, most people would probably answer the Sea of Tranquility, the landing site of *Apollo 11*. A few might mention the lunar Apennines, landing site of *Apollo 15*. But ask most fifth graders about Olympus Mons, Tharsis Bulge, or *Valles Marineris* and they not only readily reply, but also immediately envision these remarkable topographic features.

With a fleet of NASA planetary missions perusing the Solar System, by the late twentieth century "near space" emerged as a set of familiar landscapes. Serge Brunier comments that "the deserts of Mars and the rings of Saturn are as familiar to us today as the most awe-inspiring landscapes of our own planet."[4] Cultural geographer Denis Cosgrove similarly has observed that the "differentiated surfaces" of the Moon, Mercury, and Mars "are increasingly present" to us and have become a valued aesthetic to scientists and general audiences alike. As the planetary landscapes of our Solar System settle into the imagination, Cosgrove explains that this terrain has evolved into places of "detailed human understanding and care."[5] He equates our familiarity with lunar or Martian panoramas or the surfaces of Saturnian and Jovian moons with the crude maps and charted rocks that seafarers used to cross the open ocean. For the earliest mariners, horizon markers, if only the forbidding and jagged edge of a dangerous coast, would have been a welcome sight, waypoints indicating location and direction. Even a handful of recognizable land or sea features made the seemingly blank and formidable marine expanse less threatening and more navigable. Passes and shorelines, mapped and remembered, altered our relation to Earth's globe. The currently unfolding planetary geographies are again reconfiguring our sense of place, this time in relation to the Solar System itself.

Of all the planetary geographies beyond those of Earth, Mars's iconic features are perhaps the most familiar and deeply embedded. Viking 1's initial glimpses of a rock-strewn plain rising to a salmon-colored sky surprised and delighted us, as did the frost-covered rubble of *Utopia Planitia* photographed by the Viking 2 Lander. With its Arizona-like desert terrain, and dust devils dancing lazily over seemingly endless rock fields, Mars has become a place we know, remember, and dream of exploring. Even now, teams of scientists practice working in Mars-analog environments such as the Mars Society's Desert Research Station in Utah, or the barren and desiccated salt pans of Chile's Atacama Desert. In a valley near the Atlas Mountains in Morocco, scientists are studying the landscape in preparation for ESA's ExoMars mission.[6] Research at these sites is focused on developing new technologies necessary for exploring Mars with a geologist's hands and eyes.[7] Until a mission manages to place astronauts on the Martian surface, we linger over coffee table volumes from which spill its alien terrain of jagged and scalloped craters, hanging canyons, and high clouds clinging to volcanic summits.

Jim Bell's *Postcards from Mars* vividly illustrates why we are so captivated by the red planet. Multi-page panoramas tumble open to a sunset glimpsed by Spirit from the rim of Gusev Crater, and of Opportunity's sweeping view of Endurance Crater, with its angular dunes a meter in height. A day on Mars, or Martian *sol*, is not so different from that on Earth, about 24 hours and 37 minutes. Like Earth, Mars has seasons due to its similar axial tilt, polar ice caps, and a year of 687 days. Bell, however, clarifies that while its landscape in appearance resembles Earth's desert terrain, atmospheric conditions on Mars are nothing like the world we inhabit. Having served as the landscape photographer and primary camera operator for the MER rovers, Bell was responsible for processing and interpreting images from Spirit and Opportunity's panoramic cameras or pancams. Of the Mars he has so patiently and tirelessly photographed and rendered, Bell writes: "There's an 'I've seen that place before' feel of looking out the window across a long drive in the desert somewhere. Rocks, hill, sky—it's all very Earthlike and comforting, in a way. But it's an illusion. It's 30 to 50 degrees below zero (°C or °F, it doesn't matter) on average out there; the air is almost entirely carbon dioxide, with only a trace of oxygen; and it hasn't rained in something like 2 to 3 billion years, if ever."[8]

Panoramas taken by the rovers' pancams, specially designed to capture Mars in spectacular detail, called for hundreds of photos to be carefully stitched together for a single panoramic view.[9] "I wanted them to be postcards—views showcasing the beauty of the natural environment that we now found ourselves in," writes Bell. "We even set the first few mosaics up to be rectangular in shape, just like postcards." Photocomposition requires proper interpretation with regard to color under Martian sunlit conditions and is driven by both aesthetic and scientific objectives.[10] For instance, in the first image Spirit took of its landing site, Bell noted scuff marks in the Martian soil made by cables that retracted the lander's airbags: "The scratches looked very strange, though, like someone had taken a carpenter's plane and dragged it across the ground, so that pieces of soil were lifted and curled up like wood shavings. A place that at first glance seems commonplace turns out to be quite alien after all." Large, sweeping vistas taken by Spirit and Opportunity have afforded greater understanding of water deposition, erosion, as well as aeolian processes, and further demonstrate that Mars is a planet of ever-evolving geomorphology.[11] When, for instance, the rover team wondered about the frequency of dust devils, Bell ordered the rovers to lie in wait and record what might pass the camera lens: "Within days we started catching dust devils moving across the plains in the images. Over the course of months of occasional monitoring we saw hundreds of them. . . . As the plains got warmer with the advance of the seasons, it became clear that these mini-storms are a major way that dust gets moved around on Mars."[12] Fortuitously, the dust devils happened to clean the rovers' solar arrays, inadvertently extending their lives for years.

Driving the rovers over broad stretches of terrain reminds us of the grave challenges of Martian exploration, for robotic or human explorers. During the fall of 2004, when the Sun had shifted farther North and its light was less direct, Spirit was trekking as far as possible each day, depending on battery reserves. Bell recalls, "When we drove or parked, we had to try to bask, lizard-like, with the panels tilted as much into the Sun as possible to maximize our power supply." He recounts how the rover team targeted preselected "lily pads" on slopes that would catch the most sunlight. Late in 2005, while parked on Husband Hill, Spirit took some time away from its geological scrutiny to look upward and make a few astronomical observations. There, along with the rover, Bell recorded "curving star trails, potato-shaped moons moving in the night, shooting stars, the setting Sun, and the rising Earth [as] familiar, yet alien and evocative."[13]

It was detailed images of the Martian landscape sent back by the Viking orbiters and landers that enticed Steve Squyres to become a planetary scientist. Principal investigator for the MER mission, Squyres recounts how in 1977, as an undergraduate at Cornell University, he became entranced with Mars's stark terrain. He signed up for a graduate seminar to be taught by one of the Viking project scientists and ended up gaining access to the "Mars Room" in Clark Hall that housed the pictures streaming in from Viking. While some of the images had been collected into notebooks, the majority, Squyres noted, "were on long rolls of photographic paper, stacked on the floor or still in their shipping cartons." Looking to uncover in a matter of minutes some hook for a course assignment, Squyres quickly became captivated by the alien landscape unfolded before him in photos that he realized even scientists had not yet digested. He found himself poring over image after image—for the next four hours. "The planet that I saw in those pictures is a beautiful, terrible, desolate place," wrote Squyres, who recalls stepping "out of that room knowing exactly what I wanted to do with the rest of my life."[14] Squyres's journey to exploring Mars resulted in NASA's twin rovers Spirit and Opportunity that in turn have captivated the world's next generation of planetary scientists and explorers.

Decoding the Red Planet

As we saw in the last chapter, Mars seems dead to the orbiters that daily send back images of the surface. The atmosphere is tenuous, ultraviolet radiation and cosmic rays scorch the soil, and it rarely gets above freezing even on the balmiest summer day.[15] It's unlikely any form of life could exist on the surface now, but Mars has not always been so inhospitable. NASA's strategy in searching for life in the Solar System is to "follow the water," and even if there's no surface water now, there was in the past. Each of the Mars Exploration Rovers, Spirit and Opportunity, was designed for just a ninety-day mission. In the end, they have vastly exceeded expectations with their indomitable traverses of the forbidding Martian terrain. Think of them as twin robotic field geologists whose primary goal is to search for the signposts of water.[16] The record of past water can be found in the rocks, minerals, and landforms on Mars, particularly those that could only have formed in the presence of water.

Spirit and Opportunity were not designed to detect current or past life in the Martian soil,[17] but they can do the detective work needed to say whether there have been stable bodies of water that could have supported life in the past. Surface rocks reveal evidence of previous water in the way that they formed, by processes like precipitation, evaporation, and sedimentation. They can also hold clues for the possibility that water currently exists under the surface. The rovers have helped scientists diagnose the history of the Martian climate, which is now thought to have been warmer and wetter 2–3 billion years ago. The twin robots are also trying to parse the different contributions of wind, water, plate tectonics, volcanism, and cratering to the sculpting of the surface. Spirit and Opportunity provide "ground truth" data for calibrating the orbiters that continue to do remote geology and scout for future landing sites. A final goal is to prepare the stage for future astronauts by understanding the unique challenges of the Martian environment.

Meet the Rovers

Spirit and Opportunity are squat and sturdy, as large as golf carts and almost as tall as an adult (plate 3). Each weighs 400 pounds.[18] They have six wheels, each with its own, independent motor. Four-wheel steering allows them to make tight turns and swerve or turn on a dime. The design is based on the "rocker-bogie" system of the previous Sojourner rover.[19] The wheels can swivel in pairs and pivot vertically to maintain overall balance. The saddest thing that could happen to a very expensive rover millions of miles from home would be for it to tip over onto its side or back, wheels spinning uselessly. So Spirit and Opportunity have a suspension system that balances the load any time a wheel goes up or down. As a result, the bodies of the rovers experience half the range of motion of the wheels and legs. The rovers can tilt up to 45 degrees without overturning, but software is designed to sound the alarm any time the tilt exceeds 30 degrees. The wheels have cleats for gripping in soft sand and scrambling over rocks.

In case you had a mental image of a NASA engineer, cap turned sideways, slamming a joystick from left to right as the rover careens over the Martian dunes, the truth is a bit more sedate. The rovers have a top speed over smooth ground of 2 inches per second, or 1/10 of a mile per hour. But even at that speed, there's no danger of recklessness; hazard avoidance software is used that makes each rover stop and evaluate its progress every couple of seconds, reducing the true speed to a glacial 1/50 mph. On Earth, the rover might get overtaken by a snail.

The body of the rovers is the warm electronics box, a tough outer layer designed to protect the computer, electronics, and batteries—the rover's brains and heart. These vital organs must be buffered from the worst extremes of the Martian climate. Temperatures at the landing sites can plunge from a daytime high of 70°F to a nighttime low of –150°F, a range many times larger than we're accustomed to on Earth. Engineers use gold paint, heaters, and insulating aerogel to keep the rovers in a comfort zone. Within the warm electronic box, the "brain" of each rover is a computer that receives data from the scientific instruments and relays it to one of the Mars orbiters and then on to Earth. Spirit and Opportunity are exquisite machines, but they're far from state of the art in terms of processing power and data transmission rate. Moore's law—the near doubling of computational power every eighteen months—marches on, but the hardware on the rovers had to be tested for the severe conditions of space and "frozen" long before launch. So each rover's brain has only 128 Mb of RAM, and the data rate to the orbiters is 128 kb per second. That's the equivalent of a very low-end netbook and only twice as fast as a dial-up modem. If you have an iPhone, it easily eclipses the computational capabilities of both rovers.

The beating heart of the rovers is their solar panels. Mars receives half the intensity of sunlight that the Earth does, and it's a real challenge to gather enough energy to power the rover and its suit of instruments. Each rover can function on 140 Watts, the equivalent of a standard light bulb. The deck of each rover is tiled with solar panels which charge batteries within the warm electronic box. As the Martian winter approaches, there's insufficient power for driving, so each rover navigates to a north-facing slope and hunkers down for the months in which the temperature remains below freezing. This imperative has always been greater for Spirit, which is much farther from the equator than Opportunity.

Outfitting the Field Geologists

If you were a field geologist, your most valuable asset would be your eyes. Geologists are trained to recognize rocks and minerals and crystals, using archetypal images in textbooks and samples in the lab. But hand-picked samples, viewed in isolation, can never prepare a geologist for the complexity and apparent chaos of a real landscape (figure 3.1). Rocks are sometimes layered and sometimes jumbled, their color and texture can change according to lighting conditions and perspective, and each rock's story only makes sense in the light of the surrounding terrain. Experienced geologists must take all this in through their eyes. With little atmosphere on Mars to carry sound or smell, sight is the critical sense.

The rovers' eyes sit at the top of a mast, roughly five feet off the ground. Twin CCD cameras are used to form stereo images. The cameras can rotate 360 degrees to make a complete panorama, and pivot 18 degrees to scan the landscape or the sky. With 16 megapixels, they are similar to high-end digital cameras you could buy, and each one weighs nine ounces and could fit in the palm of your hand. But unlike a commercial device, the rover cameras have a large set of color filters designed to help diagnose the composition of rocks and minerals, and a set of solar filters for looking at the Sun. Like some new cars, the rovers also have "hazard avoidance" cameras, one pair on the front and one pair on the back,

Figure 3.1. A view down into Victoria Crater by the Opportunity rover as part of its seven-year exploration of the Martian surface. The dunes are similar to those seen on Earth, sculpted by wind in the thin atmosphere and shifting from season to season. In 2011, Opportunity reached Endeavor Crater, some 14 miles across (NASA/Jet Propulsion Laboratory).

each looking ten feet out to help avoid any unexpected collisions or obstacles.

Spirit and Opportunity may be one-armed geologists, but those arms are highly capable, with an elbow, a wrist, and four modes of motion. In the mechanical "fist" of each arm is a pint-sized imager that acts like the hand-held magnifying lens carried by many geologists. Just as no geologists would do any fieldwork without their trusty hammer, so each rover arm is equipped with a Rock Abrasion Tool (or RAT). The RAT is a muscular, diamond-studded grinder; in two hours it can carve out a hole 2 inches across and 1/5 inch deep into even the hardest volcanic rock.[20] Rock interiors can be quite different from the exteriors, which have been subject to weathering and radiation and may be coated in dust. After the RAT has finished its work, the microscopic imager can peer into the hole and two other science instruments can be pivoted into position to study the rock. One, called a Mössbauer spectrometer, is specifically designed to study iron-bearing minerals with high precision. It takes about twelve hours to do a single measurement. The other is called an Alpha Particle X-ray Spectrometer. It has an on-board source of alpha particles—high-energy helium nuclei—and it can bounce either those or X-rays off the rock sample. The results give the elemental composition, which is important in deciding how different chemicals came together to form minerals within the rock. These measurements take about ten hours per sample.

The rovers' suite of sophisticated instruments is completed by the Miniature Thermal Emission Spectrometer. All objects emit heat, and the spectrum of their heat can be used to deduce their composition. This instrument is particularly designed to look for minerals that can only be made in the presence of or by the action of water, such as carbonates and clays. It weighs five pounds and it sits in the rover's body, but it has a periscope so it can look out alongside the rover's eyes, and it can also gaze upward to gather measurements on the temperature, water vapor content, and presence of dust in the atmosphere.

Finally, like many good science fair projects, the Mars Exploration Rovers make use of magnets.

Magnetic dust grains are freeze-dried remnants of the planet's wetter past, where some types of molecules align with the Martian magnetic field when they are in solution, and then preserve that orientation when the mineral solidifies. Magnetic minerals give additional clues to the geological history. One pair of magnets sits at the end of the robotic arms, where it can gather magnetized particles ground out by the RAT. Another pair of magnets is at the front of the rover, tilted such that non-magnetic particles will fall off. Two of the spectrometers then analyze any particles that stick there.

The Mission

Although the rovers subsequently got into some amazing scrapes, there was nothing "seat of the pants" about NASA's planning for the mission. Each of the science instruments was exhaustively tested in the lab and in an environment designed to roughly mimic Mars.

Then, 150 potential landing sites were winnowed down to four and finally two. Site selection was an agonizing process, balancing science and safety, and taking into account any steep inclines and potentially hazardous rocks in the footprint of the landing area. To the scientists and engineers involved in the mission, the launch was a double-edged sword. On the one hand, after years of preparation and testing, nothing more could be done except watch the launch play out. On the other hand, $800 million of effort was on the line, and there were many opportunities for disaster. Everyone involved knew that over the previous thirty years, nearly half of the missions to Mars had been lost or had failed.

On January 4, 2003, after a journey of 300 million miles and seven months, Spirit entered the Martian atmosphere traveling at 12,000 mph, or twenty-five times the speed of sound. What followed was what team members called "six minutes of terror." The spacecraft heated up and was buffeted by winds in the upper atmosphere. Five miles up, the parachute deployed on schedule and then rockets fired to further slow the descent. Cocooned by airbags on all sides, Spirit hit the surface and bounced four stories high in a lazy arc in the weak gravity. After bouncing a dozen or so more times and traveling a quarter of a mile, it came to rest (figure 3.2). The "bouncing bag" method had been used for Mars Pathfinder; it removes the need for a completely soft landing, but at the expense of unpredictable ricochets from the surface. In the control room at JPL in Pasadena, over two hundred scientists, engineers, and NASA managers exhaled. And then let out a raucous cheer.

Spirit landed in the Gusev crater, on a rolling surface of red soil and small rocks. Opportunity landed three weeks later on the far side of Mars in the middle of a flat plain called *Meridiani Planum*, bouncing into a crater just 70 feet in diameter. Scientists were elated and called it a "hole in one" since the crater rim seemed geologically interesting, but the result was pure luck since the spacecraft could not have been targeted that accurately. This was the first example of serendipity or circumstance playing a role in the mission. Spirit and Opportunity have identical hardware "DNA," but their different environments began to play a role soon after landing. Both rovers went through a series of system checks before rolling gingerly off their platforms to begin exploration of the red planet.

Figure 3.2. The landing site, tracks, and location in September 1997 of the Spirit rover, as seen by the orbiting Mars Global Surveyor camera. Like its twin Opportunity, Spirit was designed to last for three months, yet it was active for over six years and drove nearly five miles before getting stuck in soft sand and being declared inactive by NASA in 2011 (NASA/Jet Propulsion Laboratory).

What We've Learned

Water, water everywhere, but not a drop to drink. The most stunning finding of the Mars rovers, proclaimed the "Breakthrough of the Year" for 2004 by *Science* magazine,[21] was the evidence for the prolonged presence of salty, acidic, and potentially life-supporting water on the surface. This water has long since disappeared, and climate change on Mars is the leitmotif of all the research done by Spirit and Opportunity. The Viking orbiter had provided suggestions of water, but the twin rovers provided indisputable evidence. Below the surface, there are large deposits of ice, with the possibility that some of it might be in aquifers kept liquid by pressure and a modest amount of natural radioactive heating from interior rocks.[22]

All life on Earth—from the tiniest bacterium to the mightiest redwood tree—needs water. The jury is still out on whether or not Mars has ever hosted life, but within weeks of arriving on the

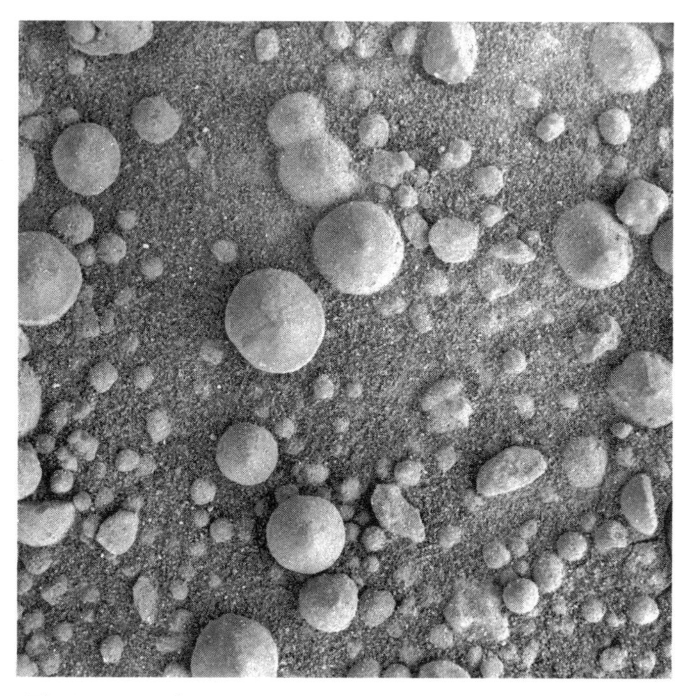

Figure 3.3. The Mars Exploration Rover Opportunity used its microscopic imager to take this picture of spherules a few millimeters in diameter, just north of Victoria Crater. Nicknamed "blueberries," the spherules are interpreted as iron-rich concretions formed inside deposits that had long ago been soaked in groundwater (NASA/JPL-Caltech/Cornell/USGS).

red planet, Opportunity showed that the *Meridiani Planum* had once been a water-soaked plain.[23] Pay dirt was a stone's throw from the landing site in the form of a rocky outcropping made of layers about as high as a street curb. The outcrop, nicknamed "El Capitan," contained numerous clues to a watery past. Opportunity found small hard spheres, called "blueberries" by the science team, which were sometimes loosely scattered on the surface and sometimes anchored into rock (figure 3.3). The blueberries were made of hematite, an iron-rich mineral that usually forms on Earth in the presence of water: oxygen atoms from the water bind to iron atoms in the mineral. The team speculated that groundwater carrying dissolved iron had percolated through the sandstone to form the spheres.[24] Later on, Opportunity discovered the mineral called jarosite, which only forms on Earth when acidic water is present.[25] Water that's acidic or rich in dissolved iron is quite capable of hosting microbial life, as we know from the ecosystems found in places like the runoff from the Rio Tinto mine in Spain.

Opportunity also found inch-high rock layers that overlapped and cut into each other. Geologists call such formations cross-beds, and their sizes and shapes indicated that they had been formed by flowing water. Some layers showed weathering by wind so the water must have been present intermittently. The minerals in the cross-beds were rich in sulfur, chlorine, and bromine, which had apparently settled to the bottom of a salty lake or shallow sea. Similar briny deposits are found in desert regions of the Earth. Opportunity also found small vugs, from the Cornish word for cave, and these inch-long or smaller cavities were probably left behind when concentrations of minerals were dispersed by groundwater. Mars minerals tell the tale of a watery past.

Meanwhile, Spirit wasn't spinning its wheels (yet). It had landed in a volcanic, rock-strewn plain with no obvious signs of sedimentation, but it soon roamed into its own discoveries. A volcanic rock called "Humphrey" had crevices filled with crystallized minerals that had most likely been dissolved in water. On another rocky target called "Clovis," Spirit found traces of the mineral goethite, which only forms in a terrestrial environment in the presence of water, and telltale enhancements of sulfur, chlorine, and bromine. Spirit also measured a soil sample with a very high concentration of salt, another indirect indication of water.[26] Science always deals with finite or incomplete information so it's important to rule out plausible alternative hypotheses. The alternative explanation for cross-bedding and other sculpting of the surface is wind erosion, but on Earth the signatures of wind and water erosion are quite distinct. Mars was volcanically active in the past, and volcanism can produce spheres from molten drops and cavities, and occasionally minerals like sulphates and jarosite. However, the web of chemical evidence from the rovers, plus the features that can only be explained by the action of water, present a compelling case for an ancient Mars with water on its surface.[27]

The finding of past water on Mars raises more questions than it answers. Mars has an obliquity—or tilt of its axis as it orbits the Sun—that varies much more than the Earth's. Orbital tilt causes seasons on any planet. Whereas our tilt is stabilized at close to

23.5 degrees by the Moon, the tilt of Mars is known to have varied from 10 to 60 degrees over the past 100,000 years, and dramatic climate variations were probably occurring in the distant past as well. The idea of a "warmer, wetter" Mars several billion years ago is appealing but is not well supported by climate models and evidence from Martian meteorites.[28] The problem is that the planet's modest gravity is incapable of retaining a thick atmosphere, and the early Sun was fainter than it is today. Some 65–70 degrees of greenhouse warming were needed to bring the surface up to the melting point of water, and while Mars did have early volcanism that generated carbon dioxide, which is a heat-trapping greenhouse gas, the early atmosphere was unlikely to have been thick enough to get the surface temperature above freezing. Mars climate change continues to be enigmatic and difficult to pin down.

The Little Rovers That Could

With all the human inspiration and perspiration that was poured into Spirit and Opportunity, it's hard not to anthropomorphize these little robots. Engineers and scientists on the project are clear-eyed and level-headed for the most part, but they've ridden an emotional roller coaster as their twin progeny encounter challenges and make discoveries. In their very different experiences on Mars, the rovers acquired distinct personalities. We can think of them as "lucky" Opportunity and "indomitable" Spirit.[29]

Opportunity's charmed life began when it landed 15 miles downrange of its intended landing site, careened into a crater, and found itself near rocks bearing signs of formation in water. By late 2009, Opportunity had roamed nearly 12 miles across the Martian landscape. First, it was sent on a kilometer jaunt to Endurance Crater. Entering the crater was a risky proposition; the team was resigned to the possibility that it might not get back out. Despite wheel slippage and tilting up to 30 degrees, Opportunity spent six productive months exploring the crater floor and got out again safely. Soon after leaving Endurance Crater, it had another stroke of luck. Sent to examine its own heat shield, it noticed a basketball-sized, iron-nickel meteorite, the first ever found on another planet.

On March 25, 2005, Opportunity set the one-day distance record for either rover: 720 feet. It then got stuck in the "Purgatory" sand dune for almost three months, extricating itself an inch at a time. The rover had always had a balky shoulder joint as a result of a failure of the switch that controlled the heating unit, and as a result the motors were in danger of wearing out. In 2008, as a last-ditch measure to save the arm, engineers stopped stowing it at night, so Opportunity now travels with its arm held out in front of it, like a blindfolded person afraid of stumbling into an obstacle. More good fortune came Opportunity's way in 2007. Choking dust storms reduced sunlight by 99 percent and power from the dust-cloaked solar panels was dangerously low. In a turn of events that was almost miraculous, the rover was restored to nearly full power by a series of "cleanings," most likely dust devils that whipped the dust off the solar panels.

Since then, Opportunity has continued its work with the endurance of a long-distance runner, pausing only to wait out the frigid Martian winters. It made the four-mile trek to Victoria Crater, and after cautiously exploring a quarter of the perimeter, it found a safe way down and worked there for nearly a year (plate 4). In August 2008, it set out on its most ambitious journey yet: a three-year, eight-mile journey to the huge Endeavor Crater, some 14 miles across. In keeping with its history, in April 2009 its power was boosted 40 percent by a cleaning event and it continued rolling purposefully toward a new set of adventures. After a detour in early 2011 to explore the quarter-mile diameter Santa Maria Crater, Opportunity trundled on toward Endeavor. In August 2011, it reached its goal and gained access to geological samples from an earlier period of Martian history than any it had found so far. At an outcrop on the lip of the crater, Opportunity found zinc concentrations that were indicative of water, and a bright vein of hydrated calcium sulfate that Steve Squyres called "the clearest evidence for liquid water on Mars that we have found in our eight years on the planet."[30] By the end of 2011, its odometer had clocked more than twenty-one miles, a Homeric distance for such a tiny traveler. Its metallic carapace becomes still and silent each Martian winter, but each spring, when the tepid Sun returns, it continues its epic journey.

Spirit's story is like a lesson in biology; the two rovers have identical genetic material, yet an unpredictable environment gave one of the rovers a much harder road. Spirit's first brush with disaster occurred less than three weeks after it landed, when it suddenly stopped talking to Mission Control. It got stuck in a "reboot loop," where a fault occurred during reboot, with the risk that the loop could continue forever. If it had been a hardware fault, it would have been fatal. However, the culprit turned out to be the flash memory, and careful management of that resource solved the problem.

Spirit had always had trouble with one of its wheels, but that didn't stop it from clambering up Husband Hill—taller than the Statue of Liberty—and working so diligently with its RAT that it wore out the diamond bits on the drill. In March 2005, it even got a taste of Opportunity's luck when a dust devil boosted power delivery from its solar panels by 50 percent. But a year later, JPL announced that Spirit's bad front wheel had stopped working altogether. After that, Spirit drove in reverse, dragging its wheel uselessly in the dirt. In December 2007, the dead wheel led to an important discovery. It scraped off a surface layer to reveal bright and pale material that was rich in silica. This find could only have come about in two ways: either hot water in a geyser dissolved silica in one place and deposited it somewhere else, or acidic steam percolating upward through rocks stripped them of their other minerals, leaving silica behind. Both of these scenarios are familiar on Earth and both are situations here that teem with microbial life.

After that, Spirit needed all its reserves of pluck and stoicism. Dust storms throughout 2007 and 2008 kept Spirit's power dangerously low, often on the point of total system failure. Early in 2009, beneficial winds increased the power level. That spring, at the base of the Columbia Hills, Spirit's left side broke through a dark crust and got embedded in pale, loose sand. This material is iron sulfate, which has almost no cohesion and so the rover wheels cannot get a grip in it. This time it was not a discovery; it was major trouble. While Spirit idled at this site, called "Troy," engineers tried to figure out a solution. At NASA's Jet Propulsion Laboratory and elsewhere, fans of the rover sported "Free Spirit" t-shirts. Progress was painfully slow. For example, on November 21, 2009, a series of commands to drive 16 feet produced just 1/10 inch of forward motion. As part of its aging, Spirit also experienced memory loss. After nine months of futile attempts to get out of the sand trap, NASA made "lemonade out of lemons" by declaring Spirit a stationary science platform. After nearly five miles of roving, it had come to the end of its road. JPL lost contact with Spirit on March 22, 2010. By May 25, 2011, after more than 1,300 commands without a response, the recovery effort was terminated. Spirit had finally succumbed to the vicissitudes of Mars.

Meet the Rover Drivers

Behind every successful robot rover there's a driver. Actually, fourteen drivers in the case of Spirit and Opportunity. For those people used to the gray, male gristle of the typical scene at Mission Control in Houston, the rover drivers are surprisingly young, and many are women. Drivers don't control the rovers in real time with a joystick; the reasons are that real-time control would be far too hazardous, and there's no immediate feedback. Depending on where the Earth and Mars are in their orbits, the distance can vary from 35 to 200 million miles, and the time delay for a signal can be as high as twenty minutes.

The drivers are part of a much larger team of engineers and scientists, numbering over two hundred, who are all involved at some level in what the rovers do. On a typical day, the results of the previous day—which are part of a larger strategy involving weeks or months of roving—are evaluated as quickly as possible, usually within an hour. Then the drivers work with the science team to map out the day's activities, which might involve measurements of an interesting rock or navigating around obstacles. This is turned into a set of commands that the rovers can execute. Commands are turned into a realistic animation and reviewed with the science team. Then they are picked apart for anything that could go wrong. All possible contingencies are considered. The final list of commands is reviewed twice and sent to the rovers to execute. Then the process starts again, as it has for over 2,500 days. The only break comes during each Martian winter when the rovers hibernate and conserve power.

Figure 3.4. Scott Maxwell works at NASA's Jet Propulsion Laboratory and is a senior driver for the Mars Exploration Rovers. Maxwell has a background in computer science and was a software engineer before becoming one of the rover drivers. He recounts his experiences on a blog called "Mars and Me," at http://marsandme.blogspot.com (Courtesy Scott Maxwell and NASA).

There's a catch. A Martian day is 40 minutes longer than a Terran day. So each day drivers begin their days 40 minutes later than the day before. As driver Scott Maxwell has said, "Pretty soon, you're starting your day at midnight, at 2 a.m., at 4 a.m. (figure 3.4). It's been called 'Martian jet lag'—it's tough on bodies, on brains, on relationships."[31] It leads to fatigue, which leads to mistakes. So many drivers watch their caffeine intake and keep to Mars time even on their days off, which puts further strain on relationships and adds to the "otherworldliness" of the job. As for what it feels like to control a robotic vehicle on another planet, listen to Ashley Stroupe, one of the most experienced drivers: "It's really just awe inspiring. Probably the closest I'll ever get to being an astronaut. Going to new places and being the first human eyes to see them is profound and hard to describe. It's the best job I could imagine."[32]

The different personalities of the rovers project into the driving experience, as Stroupe explains: "The rovers do behave differently!

Spirit and Opportunity are first in very different terrains, and so you have to drive them differently. Also, they have aged differently and have driven us to use very different strategies. We have to drive Spirit mostly backward to drag the broken right front wheel, and we have to drive Opportunity with the robotic arm out in front since one of the joints broke and we can't stow it anymore." There are also light moments, as Maxwell describes in his online blog: "Early in the mission, we nearly lost Spirit due to a problem with its flash file system. When we'd diagnosed and fixed the problem, cleaned up the flash drive, and knew that the danger was past, someone wrote this on one of our white boards: Spirit was willing, but the flash was weak." His greatest driving challenge was trying to get Spirit to a safe haven for the winter by driving across a dune with a balky wheel. As he said, "Imagine trying to cross a desert pushing a shopping cart with one stuck wheel."

Mars Is Kids' Stuff

Many of the rover drivers are younger than thirty-five years old. In general, planetary science is an older man's game; it takes more than a decade to plan and execute a space mission, and the proportion of women in the profession has been growing, but from a low base. However, NASA understands that the vitality of the space program depends on inspiring young people and broadening the participation of women. The Mars Exploration Rovers have set a strong example of engaging the next generation. It started with a third grader naming the rovers, as we saw in the opening vignette.

The trail was blazed by nine kids aged from ten to sixteen from around the world who won an even earlier essay contest. Their prize was to have guided a robotic rover on the Mars Surveyor mission, but that mission was cancelled. In March 2001, they came to the United States to work with the Mars Global Surveyor orbiter, where they became the first members of the public to ever command a NASA mission. The next set of eight students was selected from thousands of applicants who had to write a journal saying how they would use a rover to explore a hypothetical site on Mars. Aged eleven to seventeen, they came to JPL in Pasadena in 2002 to simulate two days of exploring Mars with a prototype of an advanced rover called Fido. They experienced the same training given to mission team scientists.

All this led to the selection of sixteen "Student Astronauts" from another international essay contest, sponsored by the Planetary Society. The eight boys and eight girls, ages thirteen to seventeen, came to JPL in early 2004 and were the first group of kids ever to participate in the daily operations of an ongoing Mars mission. They were in the thick of things as Spirit and Opportunity made some of their most interesting discoveries. Snippets from the children's online diaries give a sense of their experience. Courtney Dressing from the United States said, "Today was definitely the best day of my life! Spirit landed on Mars!" Saatvik Agarwal from India said, "It's really amazing how scientists just stop with whatever they are doing and explain it to us without feeling irritated!" Kristyn Rodzinyak from Canada commented, "Today has been a very exciting sol! I can't wait to start working on new images for these sols and on the other rover!" Camillia Zedan from Great Britain: "The overall message from all meetings is one of enthusiasm; just keep on truckin'. I must admit that I still can't believe that I'm actually here."[33]

A follow-up of these young people five years after the rovers landed showed that almost all of them are pursuing science degrees and heading for careers in science or the aerospace industry. Their passion for space is undiminished. Their dreams of other worlds were nurtured profoundly. They of course were lucky enough to have a singular experience, but the Mars rovers have also reached into the lives of a much larger number of people.

Mars Is Also Child's Play

To many people, participation in the space program seems completely unreachable. It requires too much training, too much specialized knowledge, and the hardware is too expensive for nonexperts to understand the issues, let alone be players. This mind-set ignores the great facility of the Net Generation with computers, games, and simulations, and it ignores the soaring aspirations of young people who have not yet tested their limits. To baby boomers who lived through the fallow years that followed the *Apollo* Moon shots, space travel is hard. To Millennials who are witnessing the opening of space to the commercial sector for the first time, space travel is natural and inevitable.

The Student Astronaut program and its precursors were spurred by an unusual collaboration between a nonprofit advocacy organization, the Planetary Society, an agency of the federal government, NASA, and a well-known corporation, the Lego Company. Their project was called "Red Rover Goes to Mars." It was preceded by a project called "Red Rover, Red Rover" which began in 1995. The executive director of the Planetary Society, Louis Friedman, saw a teacher at an educational workshop using a Lego product called Control Lab, which let students build motorized devices that could be controlled by a computer. He immediately saw a parallel to the robotic exploration of Mars. If kids could sit at a computer and control a rover in another room, or another country, they could experience the challenges of controlling a rover on another planet, a world that can only be explored through the limited senses of a robot. Starting in 2003, a network of "Mars Stations" was set up around the United States, and in Britain and Israel, each equipped with a different Mars-like terrain and a Lego rover with a web camera. Anyone who wants to can drive these rovers over the Internet.[34] Lego Education has products such that any kid can build and drive their own rover, assuming their parents don't object to the creation of an artificial Mars in the family home.

Another idea designed to appeal to kids was the inclusion of a Lego mini-figure on each of the Mars Exploration Rovers. Biff Starling and Sandy Moondust were named in another competition and they each "authored" freewheeling online diaries as the rovers explored Mars. Each Lego figure was attached to a mini-DVD, and before the rovers rolled off into the Martian dust, they took a few pictures looking back—it's a little incongruous to see Lego on Mars. Four million people have their names on the mini-DVDs; many are members of the Planetary Society and the rest signed up on the Planetary Society's special website.[35] This was only the second time privately developed hardware has flown on a planetary mission. Through these partnerships, the lure of space travel has been extended to a very wide audience. The Martian dreams of today's kids are very different from the dreams of kids four generations ago, which were fueled only by pulp fiction and fantasy.

A Planet with a Traveler's Guide

In the sleepy community of Redlands, California, on the local library shelves designated "Astronomy and Allied Sciences," one book stands out for its well-worn, bent, and torn cover. It's *A Traveler's Guide to Mars* by planetary scientist William K. Hartmann. Advertised as an "extraordinary Baedeker," the text is published in the format of the famous travel guide. Having sold so well the publisher reissued a second printing within the first two months of publication, the travel guide details features of the Martian surface. The adventures that await inside its pages include familiar and unknown craters, volcanoes, ancient river channels and flood plains, as well as the guide's foldout maps, dramatic color photographs of key geographical locations, and sidebar articles on featured terrain. This isn't a book for youth, but a serious guide for those interested in Mars. Readers are informed that they would be among the first to examine previously unpublished Mars Global Surveyor photos, as Hartmann served on the imaging team for that mission. His Martian travel guide offers serious analyses of geological formations in parallel with sidebars such as "What to Wear: A Look at Martian Weather." Readers learn that typical daily temperatures span from −13°F during the afternoon to −125°F at night and that the extremely thin carbon dioxide atmosphere and low barometric pressure are inhospitable for human survival without protective gear. To familiarize travelers with the Martian night skies, Hartmann explains: "The stars are brilliant at night after the glow of hazy sunsets fade, and the constellations are the same as the ones we see from Earth, with one exception: a blue-glowing 'evening star' with a faint companion 'star' is sometimes prominent for an hour or so after dusk."[36] One of Jim Bell's panoramas, almost certainly inspired by Carl Sagan, similarly captures from Gusev Crater a view of our pale blue dot on the Martian horizon.

Travelers dulled by the "been there, done that" aspect of some Earth-bound excursions might consider the many fascinating destinations available via Google Mars, which offers virtual tours of Mars and commentary about major landmarks excerpted from Hartmann's travel guide. Developed in collaboration with NASA by a Google team led by Noel Gorelick, and launched in 2009, Google's virtual Mars was designed so that planetary scientists and general users might have ready access to a rich photo archive of past and current missions.[37] Like Google Earth, click and zoom functions allow users to examine planetary features in 3D, as well as images from multiple NASA and ESA missions including Viking, Pathfinder, MER, Mars Global Surveyor, Mars Reconnaissance Orbiter, Mars Express, and Mars Odyssey Orbiter. Geographical and geological highlights are indicated with icons of two green mini-hikers, to reinforce Hartmann's Baedeker motif. Users might follow the tracks of Spirit and Opportunity, locate the Viking landers, peer over the canyon rim into *Valles Marineris*, or alter the perspective to the canyon floor, from which Hartmann notes its walls can soar upward for 13,000 feet.

Even more stunning are the spectacular landscapes produced at the University of California, San Diego in what is called the StarCAVE, a 3D virtual, immersive environment the size of a large closet that allows researchers to explore stretches of Martian terrain. The MER rover pancams, developed through a collaboration of the NASA Ames Research Center, Carnegie Mellon University, and Google, produce high-resolution photos that can be configured as highly detailed 360-degree panoramas. The StarCAVE panoramas extend across the floor and to the ceiling so that planetary scientists can *virtually* explore the Martian landscape in situ to search for clues regarding soil deposition, wind and water erosion, and other geological processes. Using a hand-held device to navigate the immersive environment, researchers can zoom in to rock or sediment layers or zoom out to survey the broader lay of the land. Larry Smarr, who heads the California Institute for Telecommunications and Information Technology (Calit2) that operates the StarCAVE, comments on the value of doing serious science in an immersive setting: "You can go into a room, and you're on Mars." He explains that the rendering is so fine that planetary scientists in effect can walk through the landscape, study rocks and geological features up close, as well as understand a site in relation to surrounding terrain.[38]

In the StarCAVE users must wear 3D glasses, but with the Personal Varrier technology developed at the University of Illinois, Chicago, researchers can work in immersive virtual environments without any headgear, somewhat like the fictional holodeck posited in the TV series *Star Trek: The Next Generation*. Even now researchers use this technology to engage with a variety of environments, such as walking through the temples at Luxor, or exploring, from the inside, a molecule or a segment of the human genome. The public impact of these emerging virtual learning environments will be profound. Both NASA and the U.S. Congress are interested in using similar technologies to make planetary science more accessible to everyone, so much so that the House of Representatives in its 2008 NASA Authorization Act invited the space agency to develop means by which general audiences can "experience missions to the Moon, Mars, and other bodies within our solar system" through technologies such as "high-definition video, stereo imagery, [and] 3-dimensional scene cameras."[39] For now, Google Mars and the immersive environments of the StarCAVE or Personal Varrier remind us that, whether in orbit or on the surface, our robotic partners precede us and increasingly unfold and make familiar nearby worlds.

Popular Discourse on Interplanetary Travel

Former NASA historian Steven Dick recounts the claims, first, by American physicist Nikola Tesla and a few decades later by Italian innovator of wireless telegraphy Guglielmo Marconi that they had received wireless signals from Mars. *Collier's Weekly* in 1901 reported that Tesla was convinced he was "the first to hear the greeting of one planet to another" and that the supposed radio signals were most likely from Mars.[40] By the early 1920s, Marconi apparently repeatedly attempted to receive short-wave radio messages from Mars. Dick writes: "Marconi's interest in interplanetary communication peaked during a trip from Southampton, England, to New York City aboard [his yacht] the *Electra* from May 23 to June 16, 1922. The *New York Times* noted that Marconi 'spent the time crossing the Atlantic performing many electrical experiments, principally by listening for signals from Mars.'"[41] William Sheehan and Steven O'Meara report that two years later, when Mars was at its closest since 1804, "radio stations around the world were urged to simultaneously cease transmissions at specified intervals, so as not to interfere with any attempts by Mars to radio the Earth."[42] Even the U.S. Navy, notes Dick, monitored radio transmissions for potential Martian messages.

News reportage of such events may be the reason so many assumed a Martian invasion in October 1938 when a radio broadcast, seemingly interrupting *The Mercury Theatre on the Air* programming, indicated Martians had landed at the Wilmarth farm in Grover's Mill, New Jersey. Orson Welles's radio presentation, discussed in chapter 1, is one of the most singular events in radio history. As Bruce Lenthall points out, broadcast radio was a major news source in the early twentieth century. For audiences in the United States, Lenthall claims: "Radio ownership more than doubled in the 1930's, from about 40% of families at the decade's start to nearly 90% ten years later. By 1940 more families had radios than had cars, telephones, electricity, or plumbing."[43] Lenthall estimates that 6 million people actually heard the radio play that Howard Koch adapted from H. G. Wells's *The War of the Worlds* (1898) and of that number, approximately 1 million listeners literally thought Martians had landed. Subsequent to Tesla and Marconi's independent claims of having detected radio messages from Mars, some contemplated whether Mars was inhabited by a sophisticated civilization whose radio signals the authorities had been listening for. T. S. Eliot noted in the poem "The Dry Salvages" (1941) that popular commentary on means to "communicate with Mars" was one of the "usual [p]astimes."[44]

Perhaps interplanetary travel from Mars didn't seem entirely fantastic given that the twentieth century emerged as an unprecedented era of global travel. In the first few decades of the century, train and ocean liner travel expanded exponentially, while motoring, and flight, first achieved in 1903, quickly developed as everyday experiences. Large numbers of people became mobile in ways that just a few years prior had been extremely arduous, or even unimaginable. With the serious development of rocketry in the 1920s and the genesis of the American Interplanetary Society in 1930, followed by the British Interplanetary Society in 1933, travel between Mars and Earth may have finally seemed possible and well within the realm of the imagination.

With large numbers for the first time traversing multiple time zones in a day, British author Virginia Woolf suggested that people began to internalize in finer detail Earth's global topography. She claimed that her own travel experiences afforded her a better sense of Earth's surface so that she could easily imagine and rehearse its large topographical contours. When she and her husband Leonard bought a used automobile from sales of her novel *To the Lighthouse* (1927), Woolf observed that motoring had been "a great opening up in our lives" that allowed her to "expand that curious thing, the map of the world in ones [sic] mind."[45] As early as 1909, while traveling through Italy, Woolf recorded in her diary: "It is strange how one begins to hold a globe in ones [sic] head; I can travel from Florence to Fitzroy Square [in London] on solid land all the time."[46] She meant that she could easily envision the contours of Earth's globe, as if she could turn the Earth around in her fingers and trace its continents, mountains, islands, and shorelines.

Coincident with the increase in global travel in the first few decades of the twentieth century was the construction of a new generation of large telescopes like the 100-inch at Mount Wilson Observatory. News reports covering astronomical discoveries frequently appeared in newspapers and weeklies, so much so that references to lunar and planetary landscapes were taken up in advertising and in the common parlance. In January 1926, while motoring through Persia, Woolf's close friend and celebrated author Vita Sackville-West wrote to Virginia describing the hills near Thebes in Egypt as a "mountains-of-the-moon landscape." A year later, in March 1927, Sackville-West was again touring in Persia and wrote to Woolf from Tehran, "[T]here is one little asteroid, called Ceres I think, only four miles across, the same size as the principality of Monaco, on which I have often thought I should like to live, revolving in lonely state round the Sun."[47] Presumably Sackville-West had read a news report regarding our largest asteroid, currently estimated at 975 kilometers or 606 miles in diameter. Prompted perhaps by Vita's evocation of other planetary landscapes, Woolf in September that year noted in her diary: "What I like . . . about motoring is the sense it gives one of lighting accidentally, like a voyager who touches another planet with the tip of his toe, upon scenes which would have gone on, have always gone on, will go on, unrecorded, save for this chance glimpse."[48] Decades later, Spirit and Opportunity offered our first extensive, close-up glimpse of Martian geological processes that have gone unrecorded for eons, and what we have found allows us to more fully internalize the red planet's landscapes and its geomorphology.

Travel Narratives for the Twenty-First Century

By the late 1960s and early 1970s, the Moon landings became the grand travel narratives of human history. Mars has subsequently emerged as the next great travel destination. There are multiple reasons why the planet so intrigues us. Through MER and other missions, Mars has become familiar and tangible. While our oceans obscure 70 percent of Earth's topography, on Mars we can visually survey the entire surface, so that planetary scientists to some extent have a better understanding of its surface than of our home planet. Global Surveyor's multi-colored elevation maps, for instance, reveal that Mars's northern hemisphere lies low in elevation compared to the southern highlands. Far fewer impact craters in the northern hemisphere may indicate that at one time this region was covered by a primordial sea. Prominent on the Tharsis bulge is the tallest volcano in the Solar System, Olympus Mons (figure 3.5), accompanied by three nearby shield volcanoes. Looking toward Mars's north pole, Arsia Mons is the most southern, then Pavonis Mons, and finally Ascraeus Mons. These adjacent volcanoes occur in a chain-like sequence reminiscent of the tectonic plate movement that produced the Hawaiian Islands, which raises the question of whether geothermal processes are still at work on Mars.

Moreover, extremophiles may thrive in possible underground water reserves, deep in the crust, or at the Martian polar caps. Re-

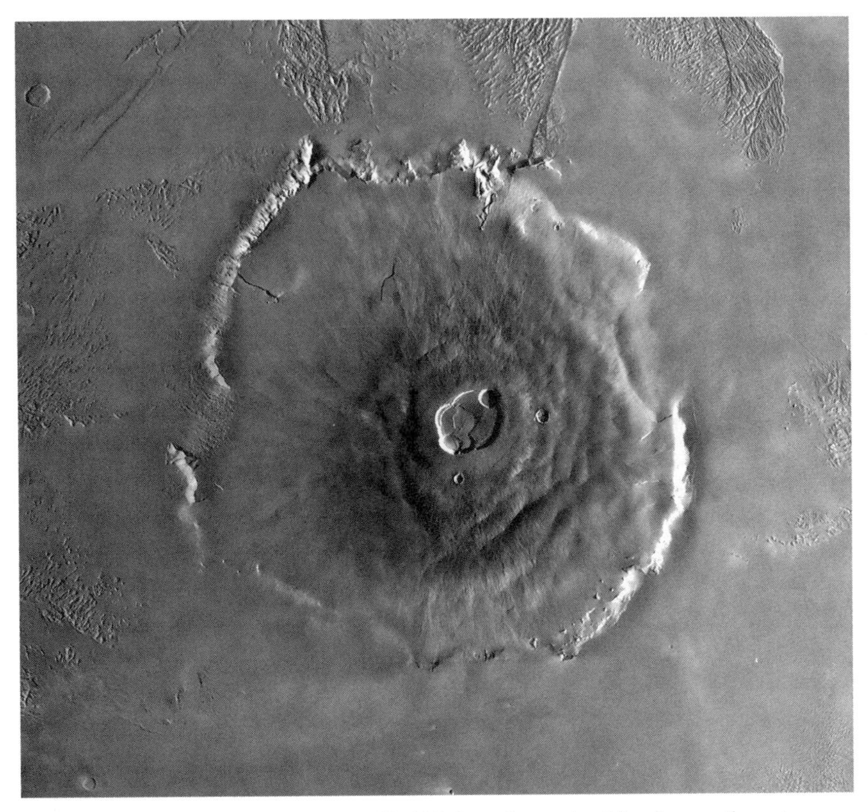

Figure 3.5. Olympus Mons rises to 69,480 feet above the Martian surface, is the largest volcano in the Solar System, and is a widely known and celebrated feature of the red planet. Nonetheless, Mars is so small that it only has mild tectonic activity and is almost dead, geologically speaking. Venus and Earth have much stronger volcanism. The width of the image is 1,000 km (NASA/Jet Propulsion Laboratory).

searchers have detected methane in the atmosphere, which could suggest active volcanism or a more exciting, possible microbial source.[49] Despite Mars's tenuous atmosphere, images by the Mars Reconnaissance Orbiter indicate groundwater sporadically breaks through the surface to briefly flow down steeply sloped terrain, and data from the 2008 Phoenix Lander indicate its soil is similar in acidity to Terran soil. Added to this are recent findings that ancient fossilized life on Earth, dating to more than 3.4 billion years ago, metabolized and lived on sulfur instead of oxygen, as Earth's atmosphere was not then oxygen-rich.[50] All of this suggests that Mars could harbor life deep within its ruddy dirt. If the Mars

Science Laboratory's rover Curiosity finds evidence of ancient or extant extremophiles, we might discover how life emerged on the early Earth.

The other tremendous appeal of Mars is that it is the only planet astronauts could travel to, and explore, in the near future. Alluding to the passion with which we dream of stepping foot on worlds beyond our Moon, Ray Bradbury has observed, "We know not why we thread an architecture of travel in a fiery path across a winter space and warm far worlds with our breath."[51] Perhaps we do so simply because we survive on our own relatively cold planet only as a result of technology. Anthropologist Ben Finney points out that we inhabit the globe, from pole to pole, solely with the aid of our technological ingenuity. Finney contends that the Moon or Mars represent a logical next step for human migration. He researched the ocean routes Polynesian peoples navigated via canoe over as much as 1,000 miles of open ocean, from South China and Southeast Asia to Hawaii and other islands, to demonstrate how humans might hopscotch to nearby planets. Finney has shown that Polynesian peoples sailed with all the supplies they needed to establish themselves on islands that they could have only speculated existed but had never charted or seen. This was an enormous feat without even the aid of a simple compass or sea charts. And yet, they accomplished it using horizon markers and the rising and setting of the Sun and stars.

The Polynesians succeeded, asserts Finney, precisely because "[t]hey did not regard the ocean as an alien environment, but one which was utterly natural—and essential to the spread of human life."[52] Even as ancient seafarers ventured far beyond known shores to settle uninhabited islands in the wastes of the Pacific, Finney argues that we could and should strike out across the Solar System. The extremely challenging act of humans venturing to Mars or an asteroid could mark the first step. Noted physicist Stephen Hawking would agree. Hawking contends that the future of our species depends on our becoming a truly space-faring species as a means of surviving natural or human-produced global catastrophes.

Archaeologist Peter Capelotti observes that extreme exploration, "begun more than two million years ago with a few rough stone tools, in what would become Africa's Olduvai Gorge, would culminate in *Apollo 11*." Our desire for travel and exploration will take us far beyond Earth, suggests Capelotti:

> By the time a human finally claimed to have stood at the North Pole, that icy coordinate had come to signify—and continues to represent—the human search for something larger in our culture and ourselves: the sense and the knowledge that our journey is a continuous one, and that the geographic "firsts" we credit to ourselves are merely waypoints on a continuum begun two million years ago. If our culture . . . (and, increasingly, our robots) allow it, this path may continue for millions of years more. We have set human feet upon only one North Pole; in space, there are millions of North Poles waiting to be discovered.[53]

As of this writing, Opportunity continues sampling the Martian terrain, and with each turn of its wheels, we take in more of the landscape, not just in images downloaded to NASA's archives or as attached links to Google Mars, but into the human imagination and into our dreams. As Capelotti indicates, there are many mountaintop vistas in our Solar System yet to be glimpsed, through the stereoscopic eyes of our robotic co-explorers and, eventually, through our own eyes. The sublime heights of Martian volcanoes entice the explorer in us. Mountain climbers and space enthusiasts dream of summiting Olympus Mons, rising 69,480 feet above the Martian surface, and imagine the sweeping vistas they would take in from the rim of its calderas, or from the top of nearby Ascraeus Mons reaching 59,699 feet, or Arsia Mons's summit of 57,085 feet.[54] Our tallest peaks above sea level pale in comparison, with Mount Everest reaching a mere 29,035 feet. Rock climbers likewise anticipate the grandeur of venturing into *Valles Marineris*, extending more than 2,500 miles in length and whose canyon walls soar from three to six miles high. But we don't have to wait for a brave explorer to rappel over the six-mile vertical scarp of Olympus Mons. Robotic eyes are even now taking us there.

The journey to, and across, Mars's cracked and cratered terrain is one of the great travel narratives of the twenty-first century. With our rovers, we have alighted upon and are motoring through its fascinating desertscapes. In July 1997, the Pathfinder Mission successfully landed the Sojourner Rover at *Ares Vallis*. Project scientist Matthew Golombek commented that, at the time, "an entire generation had never witnessed a landing on another planet" and reported that within the first month of operation, roughly 566 million people accessed the mission website.[55] The twin MER rovers captivated even larger audiences. As science studies scholar Robert Markley points out, "Sojourner had traveled a few yards to take readings of the mineral composition of several rocks; the 2004 rovers provided a photographic record that seemingly mimicked the experience of humans walking across a Martian landscape reminiscent of terrestrial deserts." Markley characterizes Spirit and Opportunity's images as "the first vacation photographs from another world" that in turn have become "part of the semiotics of human visualization."[56] Through these rovers, we've sampled the planet's sedimentary and volcanic rocks, as well as its ubiquitous dust. We've sniffed the air, measured barometric pressure, tasted the soil to determine its alkaline or saline qualities, and bored into rocks searching for fossilized evidence of water. On balmy afternoons, by Martian standards, we glimpsed the simultaneous dance of multiple dust devils. We've witnessed silent sunrises, seen morning fogs cling to the valleys and dissipate, and watched from Husband Hill through Spirit's cameras as the Sun sank below the horizon. In winter, the rovers weathered the cold, dim days and we too waited them out. Rather than our rovers exploring with us, it seems that we've tagged along with them as they traverse, photograph, and engage with this captivating and stark landscape.

Humans have a presence on Mars, even if for now it is limited to remote-controlled rovers ambling over its arid plains. Little could Steve Squyres know that his sturdy MER rovers would continue doing science for years beyond their ninety-day mission. Squyres wonders about the astronauts who might someday find the derelict rovers: "Above all, I simply hope that *someone* sees them again," writes Squyres. "*Spirit* and *Opportunity* have become more than just machines to me. The rovers are our surrogates, our robotic precursors" until astronauts someday leave their "boot prints in our wheel tracks at Eagle Crater."[57] Those astronauts are just as likely to find remnant tracks of Mars Science Laboratory's rover Curiosity. If W. H. Auden was right in his poem "Moon Landing" about the eventuality, from the first flaked flint, of humans walking on the Moon, then boot prints on Mars seem equally probable—even inevitable.[58]

4 Voyager

GRAND TOUR OF THE SOLAR SYSTEM

There's a NASA website where you can follow the two most distant human artifacts as they sail into the void of space. The real-time odometers for the Voyager 1 and Voyager 2 spacecraft flick silently upward. Single kilometers are a blur; even the tens of kilometers digit changes too fast to follow, while the hundreds of kilometers digit ratchets up by one every few seconds. These large and rapidly growing numbers are mesmerizing in the same way as counters of the national debt or the world's population; numbers this large are difficult to fathom. By late-2012, Voyager 1 was 18.4 billion kilometers or 11 billion miles from Earth, and its near-twin Voyager 2 was 15 billion kilometers or 9 billion miles from Earth. Their feeble radio signals take more than a day to reach the Earth as the probes streak through space at approximately 58,000 kilometers per hour, or roughly 36,000 miles per hour.[1]

To see why these spacecraft represented such a leap in our voyaging through space, consider a scale model of the Solar System where the Earth is the size of a golf ball. On this scale, the Moon is a grape where the two objects are held apart with outstretched arms. That gap is the farthest humans have ever traveled, and it took $150 billion at 2011 prices to get two dozen men there.[2] Mars on this scale is the size of a large marble at the distance of 1,100 feet at its closest approach. As we've seen, it took an arduous effort spanning more than a decade before NASA successfully landed a probe on our nearest neighbor. A very deep breath is needed to explore the outer Solar System. In our scale model, Jupiter and Saturn are large beach balls 1.5 and 3.5 miles away from Earth, respectively, and Uranus and Neptune are soccer balls 7 and 12 miles from the Earth. This large step up in distance was a great challenge for spacecraft designers and engineers. On this scale, the Voyager 1 and 2 spacecraft are metallic "motes of dust" 48 and 37 miles from home, respectively.

The great thirteenth-century polymath and Dominican friar Albertus Magnus, like many before him, wondered about other worlds. He framed the issue in a way that would be familiar to a modern scientist, saying it was " . . . one of the most noble and exalted questions in the study of Nature."[3] Before the Voyager spacecraft did their "Grand Tour" of the outer Solar System, the gas giant planets were ciphers, barely resolved by the largest ground-based telescopes. Imagine trying to see details on beach balls and soccer balls that are miles away. Appetites had been whetted by the Pioneer 10 and 11 probes, which flew by Jupiter in 1973 and 1974, with Pioneer 11 going onward to Saturn in 1979, but the twin Voyagers promised to send back much sharper pictures. In the 1970s, theory suggested that the gas giants were spheres of hydrogen and helium similar in composition to the Sun. If they had solid cores at all, the surfaces would be at temperatures of tens of thousands of degrees and pressures many millions of times that at the Earth's surface.[4] Their moons were assumed to be inert and uninteresting rocks like Mercury or the Moon. The word "world" comes from the Old English *woruld*, referring to human existence and the affairs of life. Yet the outer Solar System seems inhuman and inhospitable for life.

Or is it? Just a year before the launch of Voyager, Cornell University astronomers Carl Sagan and Ed Salpeter published a provocative paper in which they argued that free-floating life-forms might populate the temperate upper reaches of the gas giants.[5] The authors pushed the concept of life far beyond the bounds of terrestrial biology; aerial "gas bags" sounded like a conceit of science fiction, but at the time no one could prove them wrong. Like explorers venturing into terra incognita, nobody knew what the Voyagers might find.

Gravity's Helping Hand

Apollo 13 was crippled. Two hundred thousand miles from Earth, an oxygen tank had exploded. Damage to the Command Module was so extensive that the crew of three retreated to the Lunar Module, which they then had to use as a "lifeboat" to return home. Flight Director Gene Kranz aborted the Moon landing and considered the options. The simplest would have been to use the Command Module engines to reverse its direction and head for home, but it was unclear if the engines could be used safely, and *Apollo 13* was already in the grip of the Moon's gravity. So instead, Kranz authorized use of the Lunar Module engines to steer a course that would take the spacecraft behind the Moon and get a gravitational "assist" from the Moon that would send it on the correct trajectory for an Earth landing. It was a gutsy call. The world waited tensely as the spacecraft got a helping hand from gravity to bring the astronauts home safely.[6]

Traveling in the Solar System is a tussle with gravity, every aspect of which affects the amount of fuel required for a mission and so also its cost. The initial problem is leaving Earth's gravity. This is accomplished in two steps: first, by a large, expendable chemical rocket that can get the spacecraft into orbit, and then is jettisoned; second, by rockets onboard the spacecraft which give it the extra 40 percent of velocity needed to escape the Earth's pull. The escape velocity of the Earth is 11.2 kilometers per second, or 25,000 mph. Once achieved, the spacecraft is in the Sun's grip. Going to Venus or Mercury seems like the easy choice since the spacecraft can swoon into the Sun's gravity, which will pull it inward. But the problem emerges when the destination is reached. How do you slow down enough to carefully study or land on your target? Going to Mars and the outer planets means swimming against the tide of the Sun's gravity. The Sun is remote but extremely massive. So having left the Earth, it takes an extra 12.4 kilometers per second to reach Jupiter and an extra 17.3 kilometers per second to reach Saturn. Compared to what it took to leave the Earth, that's a similar amount of extra energy to get to Jupiter and twice as much extra energy to reach Saturn. Leaving the Solar System entirely requires a velocity of 42.1 kilometers per second, or a whopping 94,000 mph. Where is all this energy going to come from?

The answer to both questions lies in gravitational assist. Russian theorists Yuri Kondratyuk and Friedrich Zander pioneered the idea in the early twentieth century, and American Michael Minovitch added important refinements later.[7] Gravity assist was first used by the Russians to let Luna 3 photograph the far side of the Moon in 1959. NASA reprised the maneuver in 1970 to rescue *Apollo 13*. Here's how it works. Imagine standing beside a railroad track as a train approaches. You throw a tennis ball at the front of the train and it comes back in your direction at a higher speed because the train transfers some energy to the ball. Similarly, if the train was traveling away from you and you threw the ball, it would come back more slowly because the train took away some of the ball's energy. Notice that energy doesn't appear or disappear mysteriously. When the train is coming toward you, hitting it with the ball actually slows it down, but by an incredibly tiny amount because it's so massive. When the train is going away from you, hitting it with the ball speeds it up by a tiny amount. In the case of gravity, there's no physical contact as the force operates through a vacuum. When a spacecraft approaches a slower moving planet from behind, the spacecraft slows down by transferring some energy to the planet. Conversely, when a faster-moving planet approaches a spacecraft from behind, the spacecraft speeds up by gaining some energy from the planet. The idea works with moons as well as planets.[8]

Pioneer 10 was the first NASA spacecraft to benefit from gravity assist, using an encounter with Jupiter in 1973 to double its speed and send it someday out of the Solar System. Pioneer 11 followed suit a year later. Also in 1974, Mariner 10 passed close by Venus on its way to exploring Mercury. More recently, the MESSENGER probe needed one flyby of Earth, two flybys of Venus, and three flybys of Mercury to lose enough energy to be captured into an orbit of the innermost planet in 2011. For probes heading into the outer Solar System, it's worth going out of your way to get a boost from gravity. Galileo and Cassini both took inward detours to Venus to get a "kick" that helped them explore the gas giants. Energy truly is conserved in this gravitational ballet. When MES-

SENGER used Mercury to slow down and go into an orbit, it gave the planet some energy and nudged it a tiny bit farther from the Sun. And when Pioneer "robbed" Jupiter of some of its energy, it pushed the giant planet very slightly closer to the Sun.

The two Voyagers launched from Cape Canaveral in Florida in the summer of 1977. Their Titan III/Centaur launch vehicle only provided enough energy to get to the distance of Jupiter. Without Jupiter's help the spacecraft would have remained in elliptical orbits that never got closer to the Sun than Earth and never got farther from the Sun than Jupiter. But NASA engineers had planned for Jupiter to be coasting by at the right time to give each of the spacecraft a boost. Although Voyager 1 left second, it took a faster and more direct route that got it to Jupiter first, and then Saturn, but at the cost of not visiting the outermost planets. It could have visited Pluto, but this possibility was sacrificed for a close look at Titan. In 1998, Voyager 1 overtook the slower moving Pioneer 10 to become the most distant human artifact. Voyager 2 took a more circuitous route through the Solar System, flying by each of the four gas giants and gaining a modest gravity boost from each encounter (figure 4.1).[9] The Voyagers were originally conceived to take advantage of a once-in-176-year opportunity: the near-perfect alignment of all four gas giants and Pluto. Aerospace engineer Gary Flandro pushed for NASA to take advantage of the alignment with a planetary "Grand Tour," but the vision was compromised by budget cuts so the Voyager executed a scaled-back version of the concept.

The Tireless Twins

Here are the basics of these twin long-distance explorers. The Voyagers are nearly identical spacecraft, each weighing about 800 kilograms, the same as a very small car. The heart of each spacecraft is a ten-sided polygon which contains the electronics and which shelters a spherical tank holding the hydrazine fuel. The polygon is fronted by a ten-foot diameter high-gain antenna, which has kept communicating with the home planet even as it receded to a tiny

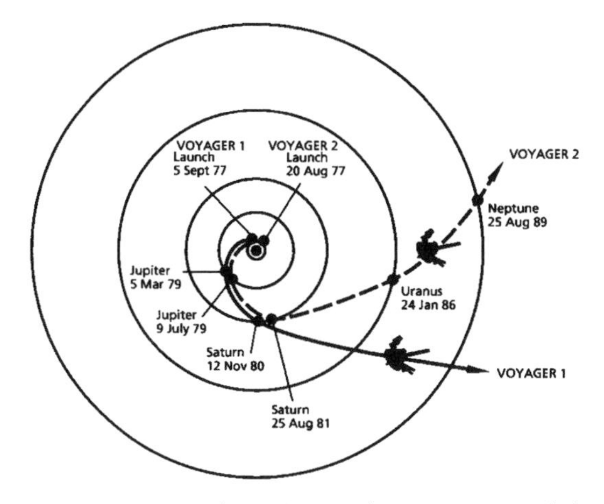

Figure 4.1. The twin Voyagers used gravitational assist to exceed the Sun's escape velocity and leave the Solar System. Despite its later launch, Voyager 1 reached Jupiter four months ahead of Voyager 2 due to a more direct path. Both spacecraft rendezvoused with Saturn, but only Voyager 2 had an encounter with Uranus and Neptune (NASA/Jet Propulsion Laboratory).

dot billions of miles away. Attached to one side of the polygon is a gold long-playing record; this apparent incongruity will be discussed later. Voyager's ten scientific instruments are attached to a couple of booms that extend away from the central polygon. The instruments include cameras and spectrometers, and there are devices for measuring magnetic fields, charged particles, cosmic rays, and radio waves. The video camera is based on a design developed by RCA in the 1950s. That seems hopelessly archaic, yet the Voyager vidicons were very robust and they worked flawlessly.[10] The mission cost $865 million through the Neptune encounter, and has been continuing on a dribble of funds as the spacecraft head beyond the Solar System. That works out to 5 cents per mile, cheaper than running a fuel-efficient car like a Toyota Prius. Actual fuel efficiency is even better. Each launch vehicle had 700 tons of fuel; so far the Voyagers are getting about 80,000 miles per gallon!

Solar power is inadequate in the outer Solar System; too little radiation would reach the spacecraft. Voyager used an onboard power system called a Radioisotope Thermoelectric Generator, or RTG. Three of these were attached to a boom coming off the spacecraft. RTG's had been used by the Vikings and the Pioneers previously, and subsequently they've been used by the Galileo, Ulysses, Cassini, and New Horizons spacecraft. Each RTG uses a fist-sized radiation source containing plutonium oxide to generate heat, which is continuously converted into electrical current.[11] Plutonum-238 (which is distinct from the Plutonium-239 that's used in nuclear weapons) decays with a half-life of eighty-eight years. The Voyagers were getting 470 Watts of power at launch, but now they're down to 250 Watts. That gentle decline in power is an inevitable consequence of nuclear physics, and as power has diminished, instruments have been switched off. In a few years, the gyroscopes will be shut down, and in the mid-2020s the Voyagers will go silent.

These are remarkable little spacecraft. Even though they are long in the tooth and obsolete by modern standards, they continue to generate scientific discoveries after thirty-five years. Think of it—they communicate at 160 bits per second, or a lower information rate than spoken speech and 25,000 times slower than "basic broadband" Internet service; they do their work with less than three light bulbs' worth of power; and they've barely used half of their 220 pounds of hydrazine fuel after traveling 10 billion miles! On the Earth, 8-track tape decks and LP records have been relegated to yard sales; at the edge of the Solar System they're still cutting edge. Moreover, the Voyagers are not too old to learn "new media" tricks. They have 25,000 followers on Twitter between them, with Voyager 2 being the more garrulous of the pair.

Ed Stone is the "father" of these middle-aged twins. Stone has had a storied career, serving as director of the Jet Propulsion Lab and chair of the Division of Physics, Mathematics and Astronomy at Caltech. He's a Fellow of the National Academy of Sciences and winner of the President's National Medal of Science and NASA's Distinguished Service Medal. He also has asteroid 5841 named after him. Stone has been principal investigator for nine NASA missions, but the Voyagers hold a special place in his affections. He started working on the project in 1972; he's now seventy-six, and he has no intention of retiring or slowing down. He intends to follow their progress all the way into the depths of interstellar space. In a 2011 interview, he allowed himself some introspection: "What a journey, what a thrill. . . . We were finding things we never imagined, gaining a clearer understanding of the environment the Earth was a part of. I can close my eyes and still remember every part of it."[12]

A Family Portrait

The Voyagers took a part of the Solar System that had been studied briefly and in little detail, and fleshed it out into a family portrait of four giant planets, their ring systems and magnetic fields, plus forty-eight of their moons. The additions and revisions have been enough to cause textbooks on astronomy to be rewritten.[13] Let's see what was learned about major members of the family.

First up was Jupiter. This mighty gas giant is three times more massive than any other planet and 320 times more massive than the Earth (figure 4.2). The Voyagers reached Jupiter in 1979, with a separation of four months. Even after centuries of telescopic observations, there were surprises. The Great Red Spot was revealed to be a huge anticyclone, large enough to swallow the Earth, with eddies and smaller storms around its periphery. It had changed color from orange to dark brown in the six years since the Pioneer flybys. As they passed behind the planet the Voyagers saw lightning illuminating the darkness of the night-side atmosphere.[14] Voyager 1 discovered a very faint ring system around Jupiter made of dust ejected from the inner moons after high-velocity impacts.[15] The rings are far less dramatic than Saturn's but are equally interesting scientifically. The main ring circles from 122,000 to 129,000 kilometers away from the center of the planet.

Voyager 1 discovered two substantial new moons of Jupiter: Thebe and Metis. Both are irregular in shape; Thebe is 70 miles in its largest dimension and Metis is only 35 miles long.[16] Voyager 2 got into the act by discovering Adrastea, which is no bigger than a small town, 15 miles across. However, the real excitement came from Io, Jupiter's closest moon and the fourth largest moon in the Solar System. This strange-looking rock, with its mottled yellow-

Figure 4.2. The Earth and Jupiter compared. The biggest atmospheric phenomena on Jupiter, like the Great Red Spot pictured here, rival the Earth in size and persist for centuries. The Voyagers provided new and unprecedented detail on the atmospheric properties of all four gas giants, and indirect evidence that they possess rocky cores (NASA Planetary Photojournal).

brown surface, looking like a moldy orange, is the most geologically active world in the Solar System. The Voyagers saw nine erupting volcanoes between them, marking the first time active volcanism had been seen anywhere other than the Earth. Plumes shoot out of the volcanoes at up to 2,000 mph and rise 300 miles above Io's surface.[17] There are more than 400 volcanoes dotting the moon's surface, not all of which are active.

Why is Io so lively? Normally, moons are geologically dead because there's not enough radioactive heating from their interior rocks to drive tectonic activity. But Io is in a gravitational "tug of war" with nearby Jupiter and the other three Galilean moons: Ganymede, Callisto, and Europa. This incessant pulling and pushing heats up the interior of the moon, like a racquetball heats up when it's flexed repeatedly. Io is distorted by the stretching force of Jupiter's gravity, and this departure from a sphere is called a tidal bulge. It bulges by up to 300 feet, which is enormous compared to the roughly one-foot stretching of the Earth's solid mass by the Moon. Sulfur and oxygen atoms ejected by the volcanoes create a torus of plasma around Jupiter, and these ionized atoms have been detected millions of miles away, right to the edge of Jupiter's magnetosphere. Lava flows episodically paint the surface red, orange, and yellow, and fizz off enough material to coat the entire surface with an inch of sulfur every year.

Galileo's other "children of Jupiter" turned out to have distinct personalities as well (plate 5). Ganymede is the largest moon in the Solar System, even larger than Mercury and more than twice as big as Pluto, which has seemed poor and misbegotten since astronomers demoted it to the status of a dwarf planet in 2006. Ganymede has terrain that's partially cratered and partially grooved, which is thought to indicate tectonic processes. Rocks mixed with ice comprise the top layer. More recent observations point to a liquid ocean under the icy crust.[18] The Voyagers saw huge impact craters on Callisto, with a couple indicating impacts almost large enough to blast the moon into fragments. However, the craters were strangely smooth, with almost no topographic relief, indicating that water ice at one point had flowed over them and filled them in. Europa attracted keen attention from the mission scientists when the first images came back. Low-resolution pictures from Voyager 1 showed linear features crisscrossing the surface. Higher resolution pictures from Voyager 2 increased the puzzle because the features were so flat they couldn't have been created by the familiar terrestrial process of slabs of crust sliding and colliding with each other. The only plausible explanation was that they were ice floes. Other observations pointed to a liquid ocean tens of kilometers deep under the few kilometers of ice. Europa vaulted into its current position as one of the most compelling targets for a future lander.[19]

The Voyager flybys of Saturn took place in 1980 and 1981. They got four times closer to Saturn's cloud-tops than they did to

Jupiter's upper atmosphere.[20] Saturn's magnificent rings were the source of many puzzles and a few surprises. Viewed up close, the rings had incredibly detailed structure. Not only were there concentric rings and gaps ranging from large and fuzzy to razor-sharp, but the cameras revealed radial spokes, kinks, and delicate braids. Time-lapse photography proved that some of these features came and went. The rings were made of mixed icy and rocky particles ranging from microscopic to house-sized, shepherded by Saturn's many small and irregularly shaped moons.

The rings of Saturn and the other gas giants are the result of an amazingly subtle gravitational dance. With no choreographer other than Newton's Law of Gravity, a disk of rocky and dusty material can naturally develop very complex structure. A resonance occurs for any pair of orbits where the periods are related by two small integers. In that case the two objects have a boosted interaction that can cause them to rearrange their position, cause one to be ejected from the system, or cause either one to clear out small particles in a ring at a particular radius. Some resonances are stable, such as the orbits of Jupiter's moons Ganymede, Europa, and Io, which have orbital periods related by the ratio 1:2:4. Think of a child on a swing; you can sustain or increase their motion not only by pushing them once per cycle of their motion, but also every second cycle or every third cycle, and so on. Although the principles of orbital resonance are understood, not all the subtle features in Saturn's rings have yet been explained. Voyager also earmarked Saturn's large moon Titan as a place to return to. Titan has a thick atmosphere of nitrogen and was inferred to have bodies of liquid ethane and methane on its surface. As we'll see in the next chapter, Cassini did give Titan the attention it deserved twenty years later.

Very little was known about dark and shadowy family members Uranus and Neptune before Voyager. The highlights of Voyager 2's solo flybys of the two outermost planets included the discovery that the magnetic fields are tilted far from their rotation axes. The new data trebled the number of moons of Uranus from 5 to 15 (27 are now known, most named after characters in the plays of William Shakespeare and a few named after characters in Alexander Pope's poem "The Rape of the Lock"), and it doubled the number of moons of Neptune from three to seven (thirteen are now known, named after Greek and Roman water gods). Miranda, the innermost of Uranus's five large moons, is a bizarre object. Even though it's only 300 miles in diameter, it has huge canyons and terraces ten miles high and mixed young and old surfaces. Voyager scientists thought it might have been the pieces of a smashed moon that came back together, but now it's thought that Miranda's topography arose from tidal heating at a time when it was in a much more eccentric orbit than it is now. Neptune rounds out the family portrait. This frigid and gloomy planet has howling winds of 1,200 mph and a Great Dark Spot similar in size to Jupiter's Great Red Spot.[21] Its large moon Triton might have been kept molten for a billion years after its capture by Neptune; geysers on its surface spew soot and nitrogen gas into its sparse atmosphere. Triton is the coldest place in the Solar System: –391°F, a temperature at which the air we breathe would freeze solid.

Beyond the Solar System

Mission "creep" is NASA jargon for the situation when new goals or capabilities (and usually extra associated costs) are added to a mission as it is being developed. The phrase has negative connotations, but the Voyager spacecraft wear it as a badge of honor. Their original two-planet mission was designed to last five years. With all the objectives for Jupiter and Saturn achieved, flybys of the two outermost planets—Uranus and Neptune—proved irresistible to mission planners. The five-year mission extended to twelve. As the spacecraft moved into uncharted territory, the time it took for a round-trip signal to be sent from JPL stretched to more than a day, so engineers figured out how to program the Voyagers for remote control, allowing them to be more autonomous. Now the mission has been running long enough that the Voyagers may outlive many of their designers. As the Voyagers recede from Earth, the modest power sent back with their high-gain antennas dilutes through an ever-larger volume of space. Currently, the deep-space tracking network detects a minuscule 10^{-16} Watts (equivalent to a billion billionth of a light bulb) from each spacecraft.[22]

What lies beyond the Solar System? Uranus is at 30 A.U., or thirty times the Earth-Sun distance. From 30 to 50 A.U. lies a vast ring of debris similar to the Asteroid Belt. Called the Kuiper Belt, it contains about 100,000 rocks larger than 30 miles across, including some dwarf planets.[23] Where the Solar System ends and interstellar space begins is a matter of argument. The Sun sends high-energy charged particles out from its upper atmosphere in all directions. This solar "wind" streaks past us at a speed of a million miles per hour and creates an evacuated region or bubble called the heliosphere. The heliosphere extends past all the planets and ends when the solar wind runs into the rarified gas of hydrogen and helium that permeates the regions between stars in the Milky Way. A shock wave is created as the rapidly moving wind slows to subsonic speeds. The location of this boundary layer is one of the big unanswered questions in astrophysics.

The Voyagers have several ways to diagnose the invisible processes that take place in the near vacuum of the outer Solar System. They can study the strength and orientation of the magnetic field of the Sun, the composition, direction, and energies of solar wind particles, and the strength of the radio emission from beyond the heliosphere. At the boundary is the place where the solar wind slows suddenly from a speed of 1 million miles per hour and becomes a lot denser and hotter. In December 2004, Voyager 1 crossed this shock.[24] A few years later, Voyager 2 followed its twin into the unknown. Recent results have been surprising. The edge of the Solar System isn't smooth but is filled with chaotic magnetic bubbles 100 million miles wide. These bubbles are formed when the Sun's distant magnetic field lines reorganize and form separate structures. Until this observation, it had been expected that the space far beyond the Sun would be smooth and featureless. In December 2010, Voyager 1 saw the outward speed of the solar wind particles slow to zero. Stagnation of the solar wind was unexpected and may mean that the spacecraft are on the verge of entering a new realm, the vast space between the stars.

Voyager is as ambitious as any project humans have undertaken. To date, the 13,000 work-years devoted to these spacecraft are half as much as all the labor summoned by King Cheops of

Egypt to build the Great Pyramid at Giza, with its slanted passageway so that the dead pharaoh's soul could reside with the stars.[25] Now, 4,500 years later, we have the means to go there directly as our robotic emissaries glide toward other worlds.

Strange New Worlds

Chesley Bonestell's lifelike paintings of Saturn as seen from the surface of Titan and other moons, published in 1944 in *LIFE* magazine, sparked the public imagination regarding the kind of geographies we would someday find in the outer Solar System. In characterizing the large gas planets and their moons, Voyager confirmed what astronomers had long suspected: the Earth to some degree serves as an analog for meteorological and geological processes on other worlds. Serge Brunier explains, "Fogs, clouds, glaciation, the cycle of the seasons, aerial erosion, and volcanism are universal phenomena."[26] But the geological processes occurring on worlds billions of miles from the Sun both surprised and mesmerized scientists and space enthusiasts across the globe. Close-up views of Jupiter's moons indicated internal heating as they flex in the tides of their host planet's gravity. A probable ocean of water captured under the fractured and icy surface of Europa, and Titan's seas of liquid methane, both suggested strange and captivating geographies.

Planet encounters garnered global attention and were keenly covered by the press. When Voyager 1 arrived at Saturn in November 1980, Henry Dethloff and Ronald Schorn report that approximately 100 million people tuned in to live television broadcasts from NASA's Jet Propulsion Laboratory while roughly five hundred reporters from across the globe provided news coverage "unprecedented in the history of unmanned space exploration."[27] It's no wonder public interest in the mission was so intense. Voyager literally beamed into our living rooms footage of worlds being charted for the first time. Of Voyager 1's encounter with the second largest planet in our Solar System, former JPL Director Bruce Murray writes: "Both *Time* and *Newsweek* ran Saturn cover stories. Live programming flowed daily from Pasadena to a network of viewers in countries ranging from Canada to Finland and, especially, Japan. Colorful images of Saturn surrounded by its magnificent rings rapidly became pop art cultural symbols." Saturn's weather turned out to be wilder than Jupiter's. A layer 1,200 miles thick has winds of over 1,000 mph. Murray recalls that President Jimmy Carter, curious about wind speeds and whether auroras had been detected on Saturn, phoned to say: "The Saturn pictures are fantastic. I watched two hours yesterday as well. Didn't expect to have so much time available." In the summer of 1989, when Voyager 2 arrived at Neptune and its moon Titan, black and white still images were broadcast live as they streamed from the spacecraft, and NASA, PBS, and CNN arranged for live reports at regular intervals throughout the encounter. Voyager's televised encounters allowed scientists and general audiences opportunity to ramble "through a cosmic Louvre," writes Murray. "For the millions viewing PBS's 'Jupiter Watch' telecasts, ABC's 'Nightline' programs, or Carl Sagan's 'Cosmos' series or watching the images appearing on screens in Japan, England, Mexico, and South America, Voyager revealed a treasure trove of abstract art."[28]

Carl Sagan was a member of Voyager's imaging science team and he incorporated into his PBS TV program *Cosmos* (1980) animations of planetary flybys and computer-generated graphics produced at JPL by the Voyager team.[29] Through the PBS series, his book of the same title, and late night talks with Johnny Carson, Sagan worked to capture and generate widespread public interest in planetary science and the Voyager mission. A celebrated senior astronomer at Cornell University, Sagan dedicated his life to popularizing the latest findings in astronomy and planetary science, at a time when most astronomers were unwilling to risk their scholarly reputations to do so. He fed a fascinated public exactly what they hoped for and wanted. Having researched the atmosphere of Venus as an example of runaway greenhouse effects, Sagan championed NASA's projects and delighted in articulating in descriptive imagery astronomical and planetary phenomena. It was Sagan who suggested that Voyager, once beyond the orbit of Neptune and receding from the ecliptic plane, should capture a photograph of Earth as it really appears in the scale of the Solar System—a seemingly unimpressive "pale blue dot" as he evocatively named it.

The Pale Blue Dot

Of the stunning images Voyager sent back, perhaps none have more profoundly shaped our understanding of humankind's place in space than Voyager 1's photo of Earth from the icy reaches of the outer Solar System. The "Pale Blue Dot" image represents one of the most significant of the mission and of our time. In February 1990, while 3.7 billion miles from Earth, Voyager 1 snapped sixty photographs looking toward the Sun, and over the next three months transmitted the images to Earth. Among the last six frames of the photo shoot, only one included humankind's most distant view of Earth. When Sagan suggested that Voyager's imaging team attempt the photographs, NASA administrators apparently were reluctant to authorize it, as programming the commands to take additional images was labor intensive and it was not clear whether the data would be scientifically useful. Candice Hansen, who helped design Voyager's imaging observations, and several others joined Sagan in arguing for the photographs. Hansen recalls how intently the imaging science team wanted to obtain what was later called the Family Portrait: "We were really after the picture of the Earth, so we had to look for the point in the Earth's orbit which would have it as far away as possible from the Sun. It turned out that we could also get Neptune, Uranus, Saturn, Jupiter, and Venus." Hansen, like Sagan, realized that an image of Earth viewed from billions of miles away would intrinsically "strike a chord . . . that might affect people's view of the world."[30] They couldn't have been more prescient.

By sheer chance, the Earth appears in the photo as though caught in the band of a rainbow. This artifact in the image is caused by sunlight reflecting off the spacecraft (figure 4.3). Sagan noted that it only accentuated how diminutive our world is in the empty wastes of outer space. Jurrie Van Der Woude, public infor-

Figure 4.3. The Pale Blue Dot is an image taken of the Earth by the Voyager 1 spacecraft in 1990 at a distance of 3.7 billion miles. Carl Sagan requested that the camera be pointed back toward Earth, seen projected behind a ray of scattered light from the Sun. The image became one of the most iconic in the history of the space program (NASA/JPL-Caltech).

mation officer for the mission, explained the difficulties in actually acquiring Voyager's final six photographs, including the singular image of Earth that so powerfully and visually characterized our minute place in space:

> They were stored onboard Voyager on the tape recorder, for the simple reason that the Deep Space Network—those big antennas in Australia, Madrid, and Goldstone, California—were too busy with Magellan and Galileo to listen to poor old Voyager. At the end of March, we played that sequence of photographs back, and the last series of photographs, five or six, were being received by the Madrid antenna when it rained. That blocked out the data, so in April we had to give it another shot. This time the antenna in use was the one at Goldstone, which turned out to have a hardware problem. As a result we still did not get those six photographs, so we had to do it again in May, when we finally got them.[31]

Wyn Wachhorst writes, "To stand on the moons of Saturn and see the Earth in perspective is to act out the unique identity of our species."[32] Wachhorst quite accurately points out that viewing Earth from the outer planets had radically altered our collective human consciousness. The Apollo missions afforded us the first opportunity to see the Earth from 240 million miles away. Apollo 8's photo of the gibbous Earth emerging from the limb of our desiccated and cratered Moon, *Earthrise*, along with *Apollo 17*'s *Whole Earth* photo are among the defining images of the manned lunar program and of the twentieth century. *Apollo 8* astronaut James Lovell, commenting on the view of Earth as the astronauts orbited the Moon in December 1968, noted: "The [E]arth from here is a grand oasis in the big vastness of space."[33] Voyager, conversely, had captured a chance glimpse of the dot of Earth forging into the empty abyss on its course around the Sun. In 4.5 billion years, our Solar System has never traversed the same wastes of space—nor will it ever. It is a staggering reorientation made visceral through the photo of the pale blue dot. With that single image, Voyager brought into focus the fractional blue world on which our lives, our narratives, and our art have mattered.

Voyager's Golden Record

Attached to each Voyager spacecraft is a gold-plated, copper phonograph record conveying greetings from Earth to civilizations potentially inhabiting nearby exoplanets in our galaxy (plate 6). Though not the first spacecraft to carry salutations from Earth, Voyager is the first to send our voices, music, and photographs. Pioneer 10 and 11 carry a plaque that includes a map of nearby pulsars indicating the Sun's location in the Milky Way, a schematic of our Solar System, their flight path from the third planet out, and an outline of the Pioneer spacecraft with the outlines of a man and woman standing in the foreground to demonstrate the humans' shape and comparable size. The male figure has his right hand raised in a sign of greeting. By contrast, Voyager carries much more detailed salutations to potential species that might happen upon the spacecraft.

The idea of attaching greetings from Earth to the Voyager spacecraft apparently was suggested by John Casani, who served as project manager for Voyager from 1975 to 1977. In October 1974, Casani had scribbled on a routine JPL concern and action report: "No plan for sending a message to our extrasolar system neighbors," as well as the phrase, "Send a message!"[34] Casani asked Carl Sagan to coordinate putting together a message. Sagan was a natural choice to head the effort; he and Cornell astronomer Frank Drake had organized development of the Pioneer 10 and 11 plaques. Sagan understood the powerful public impact of a greeting to possible galactic civilizations, eagerly accepted Casani's mandate, and recruited a group of scientists, professors, and business leaders for the NASA Voyager Record Committee.[35] Sagan additionally consulted with science fiction writers Arthur C. Clarke, Robert Heinlein, and Isaac Asimov and gathered a core group of collaborators including Drake, who in 1961 developed an equation for calculating the number of civilizations that might populate the Milky Way, writer Ann Druyan, science writer and professor Timothy Ferris, astronomy-inspired artist Jon Lomberg, and artist Linda Salzman Sagan, who drew the images for the Pioneer plaque. This core group selected the messages, sounds, music, photos, and diagrams now hurtling into the interstellar abyss.

Inscribed in the grooves of the Golden Record is an essay of sounds of Earth; spoken and written greetings; humpback whale song; 115 photographs and diagrams related to nature, science, and human activities; and ninety minutes of music. *Murmurs of Earth*, an account of the compilation of the interstellar record, reveals the intense deliberation that went into selecting its contents and determining the medium that would best preserve this special envoy from Earth. It was Drake who recommended using the equivalent of a phonograph record that could record sound as well as photographs that would render as television images. Since corrosion does not occur in the vacuum of space, it was calculated that, excepting a direct collision with micrometeoroids or other space debris in our Solar System, the records might remain intact for a billion years. "We felt like we were on the committee for Noah's Ark, that we were deciding which pieces of music and which sounds of Earth would be given eternal life, really a thousand million years. We took it as a kind of sacred and joyful task," recalls Druyan, who gathered the audio clips for the record's twelve-minute sound essay.[36]

The sound essay begins with a musical rendering of Kepler's *Harmonica Mundi*, a mathematical treatise transposed into sound at Bell Telephone Laboratories. "Each frequency represents a planet; the highest pitch represents the motion of Mercury around the Sun as seen from Earth; the lowest frequency represents Jupiter's orbital motion. . . . The particular segment that appears on the record corresponds to very roughly a century of planetary motion," explains Druyan. Following this are sounds of the primordial Earth: volcanoes, earthquakes, thunder, and mud pots. To illustrate the evolution of life, ocean surf, rain, and wind are followed by crickets and frogs, then vocalizations by birds, hyena and elephants, and a chimpanzee. Human footsteps, heartbeats, speech, and laughter come next and fade to the sound of fire and of flint being struck by rock to indicate the evolution of tool use. These are followed by a tame dog's bark, and the clamor produced by sheep herding, blacksmithing, sawing, and agriculture. Next is Morse code tapping out the phrase *Ad astra per aspera*, which translates as "to the stars through difficulties," a suggestion by Carl Sagan. To trace the evolution of travel technologies in the twentieth century, the clatter of horse-cart and noise of an automobile segue to an F-111 flyby and the cacophony of a Saturn V liftoff. Concluding the sound-essay are juxtaposed sound files of electromagnetic waves of human brain activity, actually those of Druyan, in case an advanced civilization could read human thoughts, and the radio output of a pulsar, a collapsed star in rapid rotation. As Druyan points out: "My recorded life signs sound a little like recorded radio static from the depths of space. The electrical signatures of a human being and a star seemed, in such recordings, not so different, and symbolized our relatedness and indebtedness to the cosmos."[37]

Frank Drake and artist Jon Lomberg were tasked with putting together the Golden Record's photo essay and selecting images that might make sense to a nonhuman species. Lomberg has sent more of his own art into space than perhaps anyone, having designed the sundial on the Spirit and Opportunity rovers and created the *Visions of Mars* DVD attached to NASA's Phoenix Lander as a message to future generations who explore the red planet. For Voyager's record, Drake and Lomberg compiled 115 photos and diagrams depicting insects, wildlife, ballet dancers, bushman hunters, various landscapes, architecture, an X-ray image of a hand, a radio telescope, the Earth in space, an astronaut space walking on orbit, and even a sunset, chosen in part to illustrate Earth's beauty and because "the reddening of the light contains information about our atmosphere."[38]

The elegantly etched aluminum cover of Voyager's record indicates how the record is to be played and is Lomberg's work as well. On the right of the cover are illustrations of the appropriate vertical to horizontal ratio for television images of the photographs and what the first image, a simple circle, should look like. On the left are top-down and side view graphics showing proper placement of the enclosed stylus and the correct rpm speed in binary. The Golden Record is not a conventional 33 1/3 rpm long playing record, as is often claimed. Wanting to include as much data as possible, the committee realized they could embed more music and images if they slowed the playtime to 16 2/3 revolutions per minute. Near the bottom right of the record cover is a diagram to indicate that the transition period of a hydrogen atom between its two lowest states, 0.7 billionths of a second, should be taken as equivalent to 1 in binary code. This is intended to give an accurate speed for spinning the record and for interpreting the pulsar map at the bottom left that depicts our Sun relative to 14 pulsars whose precise periods are notated in binary code.

The committee was adamant about sending music as a form of art, particularly since music expresses a potentially universal, mathematical language. But with only a ninety-minute segment dedicated to music, agonizing decisions had to be made in representing musical genres of the world. Hurtling into the depths of space are selections, among others, of Javanese gamelan, an initiation song from Zaire, Japanese shakuhachi, a raga from India, Melanesian panpipes, panpipes and drum from Peru, a Bulgarian shepherdess song, a Navajo Night Chant, tribal music from New Guinea, Louis Armstrong performing jazz, Chuck Berry playing rock and roll, and classical selections by Mozart, Bach, Stravinsky, and Beethoven. Also included is the Cavatina movement of Beethoven's String Quartet No. 13 in B Flat, *Opus 130*, along with an image of sheet music from this selection. It was a piece that Beethoven so cherished he once told a colleague he could cry when thinking of it.[39] In researching Beethoven's Cavatina, Druyan found on the score of an adjacent opus that the composer had written: "What will they think of my music on Uranus? How will they know me?" Beethoven apparently, comments Druyan, "toyed with the thought that his music might leave [our] planet."[40] His intuitive query seems indicative of an artist who saw far beyond his time. For that, and for his deafness, he was often misunderstood or considered eccentric. While reworking the finale rejected by his publisher for the String Quartet No. 13, Beethoven apparently strolled through fields waving his arms, shouting and likely singing, certainly in an attempt to feel since he could not hear, the music playing in his mind and that he found so compelling. So, it seems a perfectly harmonious outcome that Beethoven's music indeed traveled to Uranus and at this moment, aboard Voyager, is barreling into the pristine interstellar void.

Greetings from Earth

Linda Salzman Sagan coordinated the recording of greetings from the people of Earth. In an attempt to reflect the diversity of our species, the committee deliberately recruited volunteers from as many cultures, languages, and dialects as possible. The team decided to record greetings in the most commonly used languages as well as salutations in languages so ancient they are no longer spoken, such as Akkadian, Sumerian, and Hittite. To avoid complete anthropocentrism, the committee also included humpback whale song to give a voice, as Carl Sagan noted, to "another intelligent species from the planet Earth sending greetings to the stars."[41]

Actually, humpback whale songs are a shared form of communication and are now understood as cultural artifacts transmitted over hundreds or thousands of miles between humpback communities. Whale researcher Ellen Garland reports that specific songs are passed between humpback whale groups from west to east across the Pacific Ocean. The humpbacks learn the songs from each other and teach them to other whales, indicating "cultural change on a vast scale," asserts Garland, who along with her colleagues documented the same song, consisting of newly learned phrases, transmitted from one population to another.[42] Though researchers in the 1970s were only beginning to recognize whale song as purposeful communication, when Druyan inquired about a recording of humpback whales for the interstellar record, zoologist Roger Payne responded: "Proper respect! . . . Oh, at last! . . . The most beautiful whale greeting was once heard off the coast of Bermuda in 1970. . . . Please send that one." Druyan comments, "When we heard the tape, we were enchanted by its graceful exuberance, a series of expanding exultations so free and communicative. . . . We listened to it many times and always with a feeling of irony that our imagined extraterrestrials of a billion years hence might grasp a message from fellow earthlings that had been incomprehensible to us."[43]

Both fascinating and revealing are the fifty-five human salutations on the Voyager record. As might be expected, several people sent wishes of peace. Many simply said "hello." Carl Sagan's son, Nick, who was then about six years old, offered, "Hello from the children of planet Earth." Health and wellness marked a key refrain among the messages. A significant number of volunteers either wished presumed inhabitants of exoplanets good health or expressed concern regarding their wellness. For instance, Stella Fessler, speaking in Cantonese, sent this greeting: "Hi. How are you? Wish you peace, health and happiness." In Russian, Maria Rubinova said, "Be healthy—I greet you." Saul Moobola in Nyanja, a language of Zambia, asked, "How are all you people of other planets?" Maung Myo Lwin in Burmese queried, "Are you well?" Liang Ku commented in Mandarin Chinese: "Hope everyone's well. We are thinking about you all." Frederick Ahl stated in Welsh, "Good health to you now and forever." In Zulu, Fred Dube sent this message: "We greet you, great ones. We wish you longevity." Andrew Cehelsky in Ukranian stated, "We are sending greetings from our world, wishing you happiness, good health and many years." In Korean, Soon Hee Shin wondered, "How are you?" And Margaret Sook Ching and See Gebauer in the Amoy language of Eastern China asked: "Friends of space, how are you all? Have you eaten yet? Come visit us if you have time."[44] Slightly more than a quarter of the greetings explicitly mentioned wellness, good health, or longevity. Those salutations in particular seem to reflect a primal preoccupation with physical well-being. This is not surprising. For now, humans struggle with ubiquitous illnesses such as the common cold or influenza, severe and persistent back pain (a pervasive problem for upright walkers), or recovery following traumatic injury. The many wishes of good health to possible galactic neighbors on the Voyager Record suggest awareness of our own ephemerality.

Message in a Bottle

Thirty years after their launch and billions of miles from Earth, the Voyager spacecraft continue to transmit data on their way into interstellar space. NASA will track Voyager and listen for its transmissions until the spacecraft fall silent, so that we might learn something about what writer Stephen J. Pyne calls the "soft geography" of the outer Solar System and what lies beyond. Their secondary mission to carry messages of greeting beyond our Sun may, to some, seem futile. Pyne and others have noted the very low likelihood that even if another species came across the spacecraft, they could figure out how to play the phonograph record or even recognize it as a message. Today's teens, if presented with a copy of Voyager's record, might find the task difficult, if not impossible. "By the time Voyager reached Jupiter and Saturn," writes Pyne, "vinyl phonograph records were overtaken by magnetic tapes; by the time it reached Uranus and Neptune, tapes were fast fading before CDs; by the time it reached the heliosheath, CDs were passé compared with digital drives and iPods. The phonograph was hopelessly archaic just as the golden record reached the edge of the solar system—in technology years, barely beyond cuneiform tablets." As Pyne sees it, the Voyager spacecraft are in many ways ill-equipped for their journey into the unknown: "They were leaving the solar system with computer power inadequate to run a cell phone, and electrical power insufficient to animate a clock radio. Yet they had much yet to survey; the dynamics of the solar wind . . . reversals in the Sun's magnetic field, interstellar particles, radio emissions from various sources within and beyond the heliosphere, and of course the interstellar medium, if all went well." However, in spite of Voyager's technological limits, Pyne describes the spacecraft as the stuff of legend, whose tales will be told in future ages: "For now, it continued to send back reports from new settings. It was doing what no other spacecraft could. Its narrative simply defied closure from Earth."[45]

Voyager's ability to communicate with Earth is certainly limited by its plutonium supply, but like the species that launched it, the spacecraft and its record demonstrate their resilience precisely when faced with seemingly insurmountable limitations. Timothy Ferris, who coordinated the music selections, speculates that even if Voyager's record is someday retrieved but proves indecipherable, it nevertheless conveys a clear message: "However primitive we seem, however crude this spacecraft, we knew enough to envision ourselves citizens of the cosmos. . . . [W]e too once lived in this house of stars, and we thought of you."[46] Despite the possibility of becoming extinct, and precisely because we might, we sling these auspicious spacecraft into the unfathomable depths out of an innate optimism that runs deep in our species. The odds against another civilization retrieving and playing Voyager's record are astronomical, and yet we sent them in the recognition that understanding the universe and our place in it has mattered deeply from our earliest beginnings. Voyager's interstellar mission inadvertently and silently speaks of two intrinsic human traits: we are relatively physically fragile and, like other species, prone to extinction, and yet we possess an inexplicable capacity to hope even in the most dire circumstances. This instinctual, undaunted ability to hope against all odds must have evolved early on in Homo sapiens as a survival mechanism. So far, it has worked; we are the only extant hominid.

And, such unyielding expectation despite seemingly insurmountable circumstances has produced some of humankind's greatest accomplishments. It was that kind of resilience that compelled Beethoven, though completely deaf and fatally ill, to nevertheless compose some of his most acclaimed works. During *Apollo 13*, when it seemed the mission and possibly the crew were lost, NASA's engineers and astronauts refused to give up and brought the crew home via gravity assist. It was with similar abandon and hope that we sent Voyager's greetings to possible other galactic civilizations. Sagan eloquently illustrates this point: "Billions of years from now our Sun, then a distended red giant star, will have reduced Earth to a charred cinder. But the Voyager record will still be largely intact, in some other remote region of the Milky Way galaxy, preserving a murmur of an ancient civilization that once flourished—perhaps before moving on to greater deeds and other worlds—on the distant planet Earth."[47]

Envoy to the Galaxy . . . and to Ourselves

All human cultures communicate by signaling greetings, which serve as an opening that usually indicates a lack of hostility. Whether a presidential address, evening news program, a letter, telephone message, or a friend or stranger's passing acknowledgment on the street, we anticipate salutations.[48] Greetings also are an opening to what the ancient Greeks called *xenia*, hospitality to the unknown other. Biocultural theorist Brian Boyd contends that among ancient Greek cultures, the extension of hospitality, or *xenia*, initiated collaboration among strangers in a hostile time and region. Particularly in the *Iliad* and *Odyssey* texts, hospitality, asserts Boyd, is a core value:

> The word *xenos*, stranger-guest-host-friend, tells a whole exemplary tale in a single word. When a *stranger* arrives at my doorstep, I am obligated to welcome and feed him . . . even before asking him who he is. . . . The stranger becomes my *guest*, and to signify that he has therefore also become my *friend*, I should bestow on him a valuable gift at his departure, and help him on his onward journey. He is then obliged, should I arrive on *his* threshold, to become my *host*. . . . But more than that: the bond of *xenia* created by the initial act of welcome and cemented by the gift should endure between us for life and between our descendants.[49]

As an example of this, Boyd mentions a scene from the *Iliad* in which a Greek and a Trojan warrior refuse to fight as their relatives were *xenoi*.

From its inception, Voyager's Record was understood to be as much a message to ourselves as it was to those who may encounter it. Included are greetings from President Jimmy Carter, who wrote: "This is a present from a small distant world, a token of our sounds, our science, our images, our music, our thoughts and our feelings. We are attempting to survive our time so we may live into yours. We hope someday, having solved the problems we face, to join a community of galactic civilizations."[50] Even as Voyager extends hospitality to possible galactic civilizations, Carl Sagan often reiterated that we must extend to neighboring nations, those we know quite well, an equal largess if we wish to survive millions of years hence. That was the point of Sagan's talk given in 1988 on the 125th commemoration of the Battle of Gettysburg:

> Today there is an urgent, practical necessity to work together on arms control, on the world economy, on the global environment. It is clear that the nations of the world now can only rise and fall together. . . . The real triumph of Gettysburg was not, I think, in 1863, but in 1913, when the surviving veterans, the remnants of the adversary forces, the Blue and the Gray, met in celebration and solemn memorial. It had been the war that set brother against brother, and when the time came to remember, on the 50th anniversary of the battle, the survivors fell, sobbing, into one another's arms. They could not help themselves.[51]

Though seemingly ill-equipped for their mission, Voyager carries from a tiny blue dot of a planet into the unfathomable abyss a gift, a small repository of human artifacts, music, and greetings offering not just *xenia* but charity that neither expects nor requires reciprocity. That was the attribute that ensured our survival as a species and likewise has informed our dream of becoming members of an advanced galactic community. "In their exploratory intent," wrote Sagan, "in the lofty ambition of their objectives, in their utter lack of intent to do harm, and in the brilliance of their design and performance, these robots speak eloquently for us."[52]

Emissaries of Peaceful Exploration

Voyager's interstellar mission summons up the voice-over at the beginning of what began as a seemingly minor television series, initially aired between 1966 and 1969, titled *Star Trek*: "to boldly go where no man has gone before." Those words, immortalized by William Shatner in his role as Captain James T. Kirk of the starship *Enterprise*, have powerfully shaped popular discourse regarding space exploration. By the late 1960s, Americans were tuning their televisions to watch the space drama created by Gene Roddenberry and developed along with Herb Solow, Gene Coon, Matt Jeffries, and Bob Justman. The impact of *Star Trek* has been unprecedented and unparalleled. In decades of syndication, the series would inspire and captivate a global audience. Film historian Constance Penley points out that in the cultural discourse *Star Trek* became inextricably linked with NASA. Describing the conflation of NASA and the television series in its various iterations as a "powerful cultural icon," Penley contends that "NASA/TREK shapes our popular and institutional imaginings about space exploration." By 1976, NASA and *Star Trek* were so intertwined in the popular thinking that, at the request of *Star Trek* fans, President Gerald Ford was persuaded to change the name of the newly unveiled prototype space shuttle from *Constitution* to *Enterprise*. Penley writes, "Many of the show's cast members were there as the *Enterprise* . . . was rolled out onto the tarmac at the Edwards Air Force Base to the stirring sounds of Alexander Courage's theme from *Star Trek*."[53] Then JPL Director Bruce Murray recalls that a year later, in 1977 when the Voyager spacecraft were launched, funding for the planned Jupiter Orbiter with Probe (JOP) project had been cancelled. That summer, Gene Roddenberry happened to be speaking at a *Star Trek* convention in Philadelphia and encouraged the five thousand attendees to contact their congressional representatives to save the mission.[54] Whatever the reason was for the reversal, the Jupiter mission eventually was supported as the Galileo Orbiter.

That *Star Trek* touched a powerful chord in the public sphere is unquestioned. William Shatner has observed that the original *Star Trek* series went on to become the most successful television series ever produced, and has evolved into a huge industry comprising spun-off TV series, motion pictures including J. J. Abrams's *Star Trek* (2009) and *Star Trek Into Darkness* (2013), as well as novels, cartoons, action figures, Trek conventions, and many marketing products.[55] By aligning the space agency with the *Star Trek* franchise, NASA has realized even greater public interest. In 2011, the Kennedy Space Center (KSC) attracted visitors with "Summer of Sci Fi: Where Science Fiction Meets Science Fact." The event was designed to combine "the technology, innovation and exploration of NASA with the adventures of *Star Trek.*" Its website featured a retro image of Spock, modeled on the 1960s series, with his hand raised in the Vulcan greeting of "Live long and prosper."[56] Activities included a live theater production that posed the audience as new recruits for Starfleet Command, a shuttlecraft simulator, "Star Trek: The Exhibition," and an opportunity to win a suborbital flight with XCOR Aerospace, whose Lynx aircraft is to be piloted by former astronaut Rick Searfoss.

The Voyager mission was even featured as a plot device in *Star Trek: The Motion Picture* (1979) when Kirk reunites with his former crew to save the Earth from a sentient spacecraft that eradicates everything in its path as it searches for its creator. Unfortunately for Kirk and his crew, the entity views them as a carbon-based infestation of starships. It gains sentience after unknown aliens repair the old Earth spacecraft that forms its core, the name of which is V'Ger, a corruption of the word Voyager and likely derived from the acronyms given the Voyager spacecraft. A test model of Voyager was labeled VGR77-1, while the spacecraft actually launched were titled VGR77-2 and VGR77-3.[57] On a recent NPR *Science Friday* program, Ira Flatow asked Ed Stone about his reaction to the film, to which Stone replied: "I thought it was a really wonderful idea to take this spacecraft and somehow make it part of a sentient being. Of course, that's science fiction, but it really does illustrate the impact Voyager's had on [the] public imagination."[58]

The popular conflation of *NASA* and *Star Trek* produced a deep cultural narrative about the possibilities of exploring the universe through international and peaceful collaboration. Roddenberry's altruistic vision of human civilization four hundred years in the future is evidenced in the name chosen for his fictional starship. In deliberate counterpoint to the first nuclear-powered aircraft carrier, the *U.S.S. Enterprise*, Roddenberry christened his vessel the United Starship *Enterprise* and assigned its crew a mission for the peaceful exploration of space.[59] Currently, Richard Branson's *Virgin Spaceship Enterprise*, or *VSS Enterprise*, is poised to be among the first to offer commercial space tourism flights and is so named in recognition of *Star Trek*.

There's a lesson to be gleaned from the history of a low-budget television series, which was cancelled after its third season, that nevertheless has produced such an enduring vision for humankind's peaceful future. Jon Wagner and Jan Lundeen assert, "Myths are a people's deep stories—the narratives that structure their worldview." They point out that *Star Trek* and its spin-off series frequently drew upon ancient myths to rework them into a modern mythos about equality, regardless of ethnicity or species, and of a future time when humankind organized into a "Federation of Planets has eliminated intolerance, exploitation, greed, war, and materialism."[60] Penley likewise claims that "an astonishingly complex popular discourse about civic, social, moral, and political issues is filtered through the idiom and ideas of *Star Trek*" and this, in part, explains why the series has been such "a hugely popular story of things to come."[61]

In *Pale Blue Dot*, Sagan wrote: "The visions we offer our children shape the future. It matters what those visions are." Cultural narratives, even those spun from fiction, can powerfully shape a generation and a culture's vision for survival in ages long hence. Among the deeply resonant narratives of space exploration informing our generation is Voyager's epic journey, and Sagan's apt commentary on the spacecraft's view of Earth from the edge of our Solar System. "Look again at that dot," he admonished. "That's here. That's home. That's us. On it everyone you love, everyone you know, everyone you ever heard of, every human being who ever was, lived out their lives. . . . There is nowhere else, at least in the near future, to which our species could migrate. Visit, yes. Settle, not yet. Like it or not, for the moment the Earth is where we make our stand. . . . To me, it underscores our responsibility to deal more kindly with one another, and to preserve and cherish the pale blue dot, the only home we've ever known."[62]

Getting to the Worlds Next Door

To get a sense of the gulf of space that lies between us and the nearest star systems, let's return to the scale model from the start of the chapter. Visualize the Earth as a golf ball, a little less than half a mile from a glowing, 20-foot diameter globe that represents the Sun. Light speed shrinks as distance does in the model, slowing to just under three miles per hour, a steady walking pace. The Moon is a few feet away, a light travel time of just over a second. Mars is 1,100 feet away, a fifteen-minute stroll. The outer gas giant, Neptune, is 12 miles away, which is four hours of light travel, or walking in our scale model. We have shrunk space by a factor of 230 million; think of it as a map with a scale of 1:230,000,000. At the moment Voyager 1 is 48 miles away and Voyager 2 is 37 miles from Earth (figure 4.4). Proxima Centauri, the nearest star system, is 2,000 times farther away, just over 100,000 miles in this scale model. In the real universe, if it's taken the Voyagers thirty-five years to get where they are today, it will take them around 70,000 years to get as far as the nearest star. This daunting isolation makes other worlds seem out of reach. The trajectories of the Voyagers were never intended to aim at a particular star. They will each drift past other stars as the eons pass. In approximately 40,000 years, Voyager 1 will come within 1.5 light-years, or 9 trillion miles, of an anonymous star in the constellation Camelopardalis. Meanwhile, Voyager 2 is heading toward Sirius, the brightest star in the sky, and will pass within 25 trillion miles of it in 300,000 years.[63] Long before then, in less than 25 years, both spacecraft will lose power for all of their instruments and become silent sentinels gliding through the Milky Way.

The stars seem as far away as ever. With the Space Shuttle program over and the International Space Station expensive and unpopular among most scientists, who don't see it as a cost-effective or compelling platform for research, NASA is struggling to recapture the vision that fueled the Apollo program and the "Golden Age" of planetary exploration epitomized by the Voyager probes. Reaching the stars will take much greater speeds than are currently possible. Traveling at the highway speed limit of 55 mph, Proxima Centauri is a 50-million-year trip. At the speed of the

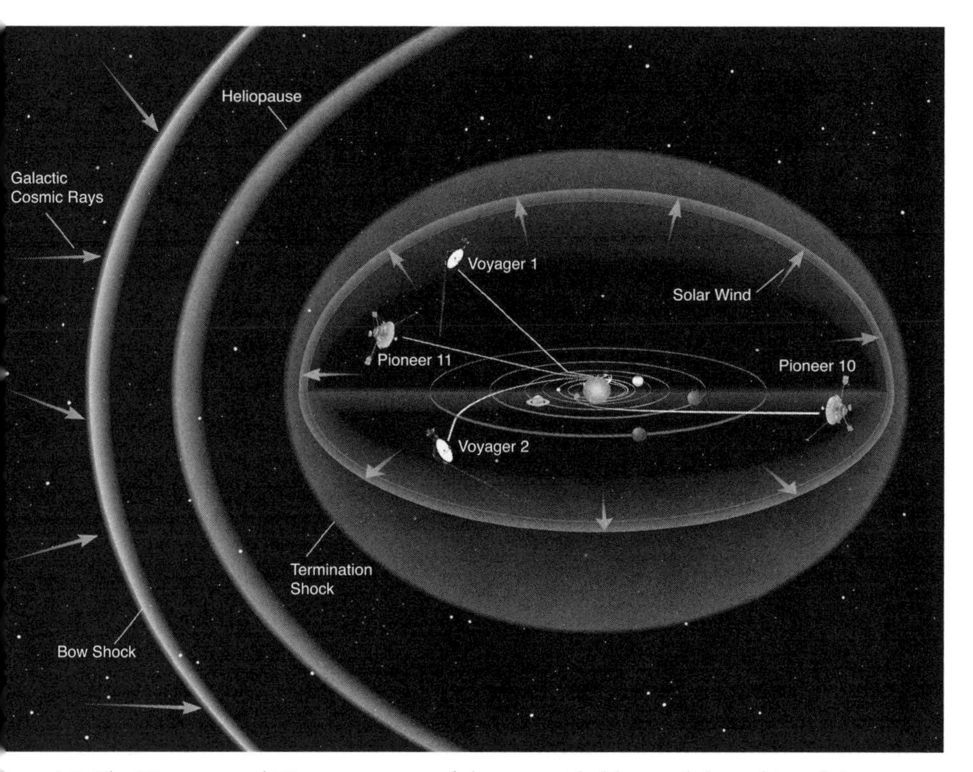

igure 4.4. The Voyager and Pioneer spacecraft have traveled beyond the orbits of the out-rmost planets and are now in the uncharted territory of the heliopause, where the solar vind meets the diffuse medium between stars, which is a hot and nearly perfect vacuum. At ıeir speed of travel, it will take tens of millennia for the Voyagers to reach the nearest stars NASA Science News).

Apollo spacecraft, it's a million-year trip. And Voyager traveling at 37,000 mph would take 80,000 years to get there. Sending a probe to a nearby star in a human lifetime runs into the obdurate principles of physics. Reaching a speed a thousand times faster than Voyager requires a million times the energy, since kinetic energy is proportional to velocity squared. The space program so far has been powered exclusively by chemical rockets, which are inefficient because they get their energy from chemical bonds. Fusion releases energy from atomic nuclei and is 10 million times more efficient.[64] Sending a Shuttle-sized craft to Proxima Centauri in fifty years would take about 10^{20} Joules, which is the amount of energy consumed in the United States in a year. Unfortunately, that would require 500,000 kg of hydrogen fuel, and fusion technology is nowhere near being able to put that capability into a Shuttle-sized package.[65] Ideally, there would be no need to carry and accelerate all that fuel, but solar sails don't work efficiently when far from a star, and there's not enough interstellar hydrogen to scoop up and use along the way. It sounds like a bridge too far.

At the speed of light, traversing the span of our galaxy would take 100,000 years. Even so, we dream of plying the eternal wastes beyond that. It is a remarkable aspiration for such a fragile species. Some engineers consider interstellar travel an impossible goal. That may be, but in launching spacecraft in the direction of Earth's rotation, we use in a very simple way a natural energy resource of our planet. Similarly, Voyager harnessed the rotational energy of the large outer planets to sling itself from one planet to the next. In essence, through the gravity assist deployed in Voyager and other missions, we tap into a resource of the Solar System itself and, however minutely, borrow energy from, and leave our mark on, the rotation of nearby planets. Sagan, who characterized walking and driving on the Moon or using gravity assist as natural steps in human evolution, wrote of Voyager: "They are the ships that first explored what may be homelands of our remote descendants. . . . [U]nless we destroy ourselves first we will be inventing new technologies as strange to us as Voyager might be to our hunter-gatherer ancestors."[66]

Research on propulsion concepts for interstellar travel was undertaken at NASA's Glenn Research Center from 1985 to 1992, and then the seed funding of $1.5 million ran out. One tangible result was a 740-page volume that quickly became the "bible" of propulsion science.[67] Project leader Marc Millis offers a cautionary note to NASA's Breakthrough Propulsion Physics website:

> On a topic this visionary and whose implications are profound, there is a risk of encountering, premature conclusions in the literature, driven by overzealous enthusiasts as well as pedantic pessimists. The most productive path is to seek out and build upon publications that focus on the critical make-break issues and lingering unknowns, both from the innovators' perspective and their skeptical challengers. Avoid works with broad-sweeping and unsubstantiated claims, either supportive or dismissive.[68]

Yet the visionaries are alive and well, and planning for a future when we will slip the bonds of the Solar System.[69] In a move that mirrors the U.S. government's encouragement of the private sector to develop new launch capabilities, in 2011 the Defense Advanced Research Projects Agency (DARPA) initiated an annual strategic planning workshop and symposium to bring together a wide range of technologists, engineers, members of space agencies, entrepreneurs, space advocates, science fiction writers, those working in medicine, education, and the arts, as well as the general public. Organized to foster collaboration among these groups as well as academicians and financiers, the "100 Year Starship" project isn't immediately planning to design a starship that could travel to the stars, but is hoping to kick-start private-public partnerships that will address the technical problems of interstellar travel in the next one hundred years. Colloquia for the project focus on propulsion at light speed, medicine and space travel, possible destinations, the economic, legal, and philosophical implications, as well as the need to communicate a vision for interstellar travel through narrative, or storytelling.[70]

The 100 Year Starship initiative exemplifies the potential of an idea, suggested in mere fiction, to become reality. Flip cell phones modeled on the *Star Trek* communicator, desktop computers, downloadable mp3 files, iPods, and iPads were all inspired by the fiction of *Star Trek*. Chief engineer and mission manager for NASA's Dawn spacecraft, Marc D. Rayman, attributes his design of ion propulsion for interplanetary spacecraft to a *Star Trek* episode titled "Spock's Brain" in which the term was used (figure 4.5). The ion propulsion powering the Dawn mission allows continuous firing of its engine so the spacecraft can attain speeds surpassing chemical propulsion.[71] Similarly, the Qualcomm Tricorder X Prize competition, which was launched in 2012 by the X Prize Foundation for the first team to develop an inexpensive device to readily diagnose illness, is yet another *Trek*-inspired potential innovation.[72]

Futurist Ray Kurzweil thinks we are facing a critical stage in human evolution due to exponential advances in information technologies, genetics, nanotechnology, and robotics. Kurzweil and others anticipate that emerging technologies in three-dimensional

Figure 4.5. Gene Roddenberry's *Starship Enterprise* has been a cultural icon for more than fifty years and inspired the 100 Year Starship project. There has been an intriguing interplay between the science fiction and fictional characters of *Star Trek* and the astronauts of NASA's space program. This replica of the *Starship Enterprise* (NCC-1701-E) is on display at the Famous Players Colossus Theater in Langley, British Columbia (Wikimedia Commons/Despayre).

printing will radically change the world as we know it. Myriad items can be produced via 3D printers, including parts for flyable aircraft printed out of plastics or complete components for buildings that could be printed from liquid concrete. Made in Space, a fledgling company inspired by Singularity University, based at the NASA-Ames Research Center in California and co-founded by Kurzweil and Peter Diamandis, has proposed using 3D printers on the International Space Station or for establishing outposts on the Moon or Mars. In 2011, NASA funded research focused on printable spacecraft, as well as 3D printing technologies to construct planetary surface habitats. When astronauts return to the Moon or someday walk on the surface of Mars, they may carry 3D printers with them and computer files for printing the tools and habitat components needed to build livable spacehabs using nothing more than lunar regolith or the ruddy dirt of Mars. The process of 3D printing involves a sequential layering of powdered or liquid plastics or metals, even potentially human cells. The medical community is interested in using human stem cells to print three-dimensional vertebral discs for repairing spine injury, or a human heart or liver as organ transplants.[73]

In the next thirty years, Kurzweil predicts, disease will be eradicated through this and other medical advances in biochemistry, gene therapy, and bio and nanotechnologies. He may be on target, given the success of researchers at MIT who are in the developmental stage of a drug that can destroy human cells infected with viruses while leaving normal cells untouched. So far, trials with mice eradicated H1N1, the most common flu virus in humans, and the drug looks very promising in effectively treating stomach viruses and the common cold.[74] Kurzweil anticipates that in the next few decades we will begin to incorporate artificial intelligence the size of a blood cell into the human body to enhance our health as well as our intellectual and computational capacities so that it will be possible to extend our lives for hundreds or thousands of years, possibly even indefinitely.[75]

That is not an unprecedented idea. Consider *Turritopsis Dohrnii*, the immortal jellyfish discovered in 1988 by marine biology student Christian Sommer and now being researched by marine biologist Shin Kubota. Upon reaching maturity, the jellyfish reverts in age to its nascent state and then begins life again. The organism's natural life cycle doesn't ever end. Kubota believes that in unlocking the jellyfish's genome humans too might become immortal. In such a scenario, human missions to nearby stars seem nearly feasible.

If we're not attached to sending humans, with our present need for expensive life support, nanotech might propel us into interstellar space even sooner. Exoplanet hunters like Geoff Marcy and Debra Fischer are studying the two brightest stars in the nearby Alpha Centauri system, the nearest solar systems to Earth and comprised of three suns. In October 2012, a graduate student at Geneva Observatory, Xavier Dumusque, detected a planet comparable in mass to Earth orbiting the star Alpha Centauri B. Though this exoplanet orbits too close to its star to be habitable, the general consensus among astronomers is that where there is one planet there are likely to be more. Fischer, though not associated with the discovery, has called the finding "the story of the decade."[76] Progress in the detection of exoplanets has been stunning. Starting with a first detection of a Jupiter-mass object in 1995, Earth-mass exoplanets are now routinely detected both with ground-based and orbiting telescopes.

Unfortunately, studying these planets in detail from afar will be extremely difficult; think of a golf ball seen at a distance of 100,000 miles. Listen to what Marcy hopes will happen next: "NASA will immediately convene a committee of its most thoughtful space propulsion experts, and they'll attempt to ascertain whether they can get a probe there, something scarcely more than a digital camera, at let's say a tenth the speed of light. They'll plan the first-ever mission to the stars."[77] If there is an "Earth" next door, Alpha Centauri will become the compelling destination that the Moon was fifty years ago. Assuming the tricky issue of miniaturized propulsion can be resolved, nanobot probes can be small enough that the energy requirements are tractable. They'll take pictures of the habitable worlds and we'll see them back on Earth within a generation. These pint-sized emissaries will be able to carry far more information than the quaint phonograph records of the Voyagers; their digital storage will contain the sum of all human knowledge. With the Voyagers, we took our first tentative baby steps beyond the Solar System. The future beckons.

5 Cassini

BRIGHT RINGS AND ICY WORLDS

"It is a drama as ancient as the Sun, as unflinching as time . . . a never-ending whirl of celestial movements, scripted and precise, in a silent show of cosmic force, played out in light and shadow. It is a drama called equinox," writes Cassini Imaging Team Leader Carolyn Porco in her "Captain's Log," an online diary of the Cassini spacecraft's observations of the ringed world of Saturn and its moons. It takes approximately thirty years for Saturn to orbit the Sun, so the planet only experiences an equinox, when the Sun shines equally on its northern and southern hemispheres, every fifteen years. In August 2009, equinox returned once again for Saturn as Cassini explored Saturn and its moons. "To its operators at significant remove, a billion miles away, it has been a long and gripping wait for this special season about to unfold . . . when the Sun passing overhead from south to north begins to set on the rings," writes Porco. Observing Saturn in equinox from on site, she reminds us, is "a solemn celestial phenomenon no human has beheld before."[1]

All of the large outer planets of our Solar System have rings, though none as magnificent as Saturn's. The rings, no more than tens of meters thick yet spanning nearly 155,000 miles in diameter, come into sharp and rare relief during equinox when the angle of the Sun's rays is lowered relative to the ring plane and casts long shadows across the rings (figure 5.1).[2] In eloquent and poetic prose Porco comments, "Like the seas of Earth, this wide icy expanse

Figure 5.1. Cassini spent twelve hours in Saturn's shadow in 2006 and took this image looking back toward the eclipsed Sun. Saturn's night side is partially lit by light reflected from the rings and the rings appear dark where silhouetted against the planet (NASA/JPL/ESA/CICLOPS).

[. . .] froths and churns, not by wind but by the convulsive forces of Saturnian moons. This famous adornment, impressed deep in the human mind for four centuries as a pure, two-dimensional form, has now, as if by trickery, sprung into the third dimension."[3] Equinox on Saturn has since faded to northern summer and we along with Cassini have observed the clear, deep blue skies of Saturn's northern hemisphere cloud over to reflect Saturn's signature peach or faded orange hue. Robotic explorers like Cassini have given us entirely new perspectives of other worlds in the outer Solar System.

The Cultural Significance of Saturn and Its Rings

Saturn's ring system became etched in the human imagination some time after the mid-1600s when Dutch mathematician and astronomer Christiaan Huygens first illustrated the rings.[4] He also was first to identify its moon Titan.[5] However, in recent memory, it was Chesley Bonestell's renderings of Saturn that brought the planet into the public purview. By training Bonestell was an architect, but he's probably best known for the stunning paintings of Saturn published in May 1944 in *LIFE* magazine. The most famous and striking of these is titled "Saturn as Seen from Titan." Readers were amazed by the suite of paintings that appeared in the May 29 issue, perhaps in part because the issue was largely dedicated to news and advertisements related to the war effort. Among 130 pages of news reports on American soldiers in Europe, war-themed ads, and largely black and white photos of troops, Bonestell's realistic and full color renderings of Saturn lumbering in the sky of Titan transported readers into exotic and delightful planetary vistas.

Bonestell recounted how in 1905 he was inspired to paint the spectacularly ringed planet: "When I was seventeen, an important event occurred in determining my future career, although I little suspected it then." Having once hiked with a friend to Lick Observatory at the summit of Mount Hamilton, Bonestell recalled, "That night I saw for the first time the Moon through the 36-inch refractor, but most impressive and beautiful was Saturn through the 12-inch refractor. As soon as I got home I painted a picture of Saturn."[6] Although that painting was lost in the fires caused by the Great Earthquake of 1906 in San Francisco, Saturn had made a lasting impression on the young artist.

When nearly forty years later Bonestell submitted the series of paintings that included "Saturn as Seen from Titan" to the editors of *LIFE*, they quickly agreed to publish them. Ron Miller and Frederick Durant explain: "No one had ever before seen such paintings—they looked exactly like snapshots taken by a *National Geographic* photographer (figure 5.2). For the first time, renderings of the planets made them look like real places and not mere 'artists' impressions."[7] Miller and Durant write that Carl Sagan maintained "he didn't know what other worlds looked like until he saw Bonestell's paintings," while science fiction writer Arthur C. Clarke suggested that in a sense Bonestell had walked on the Moon long before Neil Armstrong and reportedly quipped, "Tranquility Base was established over Bonestell's tracks and discarded squeezed-out paint tubes."[8]

Wyn Wachhorst has explored why Bonestell's famous painting "Saturn as Seen from Titan" is so compelling: "Since Titan is the only satellite in the Solar System with an atmosphere, the giant Saturn looms low in a dark blue sky like an alien ship, a thin,

Figure 5.2. Chesley Bonestell's "Saturn as Seen from Titan" made tangible planetary vistas human eyes had not yet seen. This iconic and influential image was an early example of the school of "realistic" space art that often informed and inspired the planetary scientists working to learn more about distant worlds (Chesley Bonestell).

gleaming crescent bisected by the glowing edge of its rings, afloat between jagged cliffs that jut from a frozen sea. . . . A hint of dawn lights the far horizon; and beyond a lofty pinnacle, out under the glow of the great crescent, lies a distant patch of noonday plain." Among the other Bonestell paintings in the *LIFE* layout were imagined scenes of Saturn from its moons Phoebe, Iapetus, Mimas, and Dione. One depicted Saturn's rings passing overhead from the perspective of the planet's cloud tops. Wachhorst explains that the suite of paintings was intended to offer varying views of Saturn on approach from its outer moons.[9]

Though he painted numerous panoramas of planetary landscapes ranging from Mercury to Pluto, Bonestell was aesthetically captivated by Saturn, a subject he repeatedly returned to throughout his life. He painted numerous iterations of Saturn from Titan and its other moons. In 1949, for instance, he completed paintings of Saturn from Dione, in which the full body of Saturn is glimpsed from the mouth of a cave. His panorama for the Griffith Observatory, completed in 1959, featured a prescient vision of the frozen landscape of Titan with Saturn low on the horizon. Throughout the 1960s, Bonestell reworked different views of Saturn from Titan, changing the lighting or subtly altering Titan's landscape. In 1972, he completed two separate paintings of Saturn from Iapetus, as well as a painting of Saturn from Enceladus for Arthur C. Clarke's book *Beyond Jupiter*. Bonestell returned to the subject of Saturn again and again, in various configurations, settings, and lighting.

All this from an architect whose work included contributions to the design of San Francisco's Golden Gate Bridge, the layout of the well-known Seventeen Mile Drive at Pebble Beach in Monterey, the Eagle gargoyles and art deco façade of the iconic Chrysler building in New York City, and the design of buildings for the California Institute of Technology in Pasadena.[10] He became the highest paid special effects artist in Hollywood, working on films like *The Hunchback of Notre Dame* (1939) and *Citizen Kane* (1941). Bonestell's turn to space art played out in popular magazines such as *LIFE* and *Collier's*, and in films like *Destination Moon* (1950), *When Worlds Collide* (1951), and *Conquest of Space* (1955). His work inspired generations to imagine the stark and beautiful planetary landscapes in our Solar System and in far-flung star systems of the galaxy. "Bonestell brought the edge of infinity out of the abstract and into the realm of direct experience," comments Wachhorst.[11] His paintings suggested planetary vistas human eyes had not yet seen, and sometimes included figures of astronauts dwarfed by a vast surrounding terrain. This was true of his painting of Saturn from Mimas in the *LIFE* layout, and of an iteration of Saturn from Titan completed in 1969, which situates three tiny astronauts on a cliff, looking out at a fully lit Saturn as one astronaut points to the rings.

Bonestell apparently learned the technique for rendering his realistic paintings from science illustrator Scriven Bolton while working in the 1920s at the *Illustrated London News*. Bolton, a Fellow of the Royal Astronomical Society, constructed plaster cast models of planetary landscapes, photographed them, and then painted in planets and stars.[12] Working from this technique, Bonestell would project light onto his plaster landscapes to get a sense of how sunlight and shadows might fall across terrain, and then painted based on photos of these lit scenes. This resulted in landscapes that seemed reachable and tangible. His widely celebrated renderings "invited viewers into the possible planetary landscapes that exist on moons of the outer solar system. In Bonestell's depiction, Titan's landscape resembles that of the American Southwest or perhaps the craggy cliffs of the Rocky Mountains in winter. The deep blue of the sky recalls that of Earth."[13] Though such spectacular views from Titan may be unlikely given the moon's hazy, methane-rich atmosphere, Bonestell was prescient in suggesting the sublime experience of standing on the shore of one world to view another in close proximity.

Wyn Wachhorst contends that Bonestell's art purposely evokes "a kind of cosmic shoreline, a composite of stark and eerie beaches on the near edge of the starry deep," and that the seashore is the "root metaphor" of Bonestell's art, meant to evoke Earth's horizon as the shoreline between Earth and outer space. Bonestell's art reminds us that from Earth we stand "on the shore of the cosmic ocean, riding our wisp of blue and white like mites on a floating leaf, in the whorls and eddies of a great galactic reef."[14] Carl Sagan wrote in *Pale Blue Dot* that humans have from time immemorial been innately drawn to the horizon. Ancient Egyptians identified their god Horus with the Sun on the horizon and with the planet Saturn, thought to represent Horus the Bull.[15] The Great Sphinx in Giza apparently was associated with Horus and specifically is oriented toward, and draws the eye to, the Eastern horizon.[16] Without question, Bonestell's work inspired Sagan, whose first episode of the PBS series *Cosmos* was titled "The Shores of the Cosmic Ocean."

The Lure of the Shoreline

The seashore marks a liminal space and appeals to us in part due to the sheer tenacity of life to survive there despite its harsh environment. Marine zoologist Rachel Carson characterized the seashore as a harsh but vibrant biome. Known primarily for her book

Silent Spring, Carson's area of expertise and real passion was the ocean. She published three best-selling books on the topic, but *The Edge of the Sea* specifically focused on the rigor necessary for species to survive in the hostile terrain of the shoreline.[17] Describing the seashore as "the place of our dim ancestral beginnings," Carson writes:

> Only the most hardy and adaptable can survive in a region so mutable, yet the area between the tide lines is crowded with plants and animals. In this difficult world of the shore, life displays its enormous toughness and vitality by occupying almost every conceivable niche. Visibly, it carpets the intertidal rocks; or half hidden, it descends into fissures and crevices, or hides under boulders, or lurks in the wet gloom of sea caves. Invisibly, where the casual observer would say there is no life, it lies deep in the sand, in burrows and tubes and passageways. It tunnels into solid rock and bores into peat and clay. It encrusts weeds or drifting spars or the hard, chitinous shell of a lobster. It exists minutely, as the film of bacteria that spreads over a rock surface or a wharf piling; as spheres of protozoa small as pinpricks, sparkling at the surface of the sea; and as Lilliputian beings swimming through dark pools that lie between the grains of sand.[18]

Celebrated anthropologist and naturalist Loren Eiseley, in his essay "The Star Thrower," similarly contemplated the formidable environment of the shoreline in Sanibel, Florida, which he describes as "littered with the debris of life": "Shells are cast up in windrows; a hermit crab, fumbling for a new home in the depths, is tossed naked ashore, where the waiting gulls cut him to pieces. Along the strip of wet sand that marks the ebbing and flowing of the tide, death walks hugely and in many forms. Even the torn fragments of a green sponge yield bits of scrambling life striving to return to the great mother that has nourished and protected them."[19]

Eiseley recounts walking along the beach and encountering a man picking up starfish washed ashore by the rugged waves. Eiseley pauses and together the men notice a starfish that "thrust its arms up stiffly and was holding its body away from the stifling mud." The man picks up the star and tosses it beyond the breakers back into the sea. At some point, Eiseley joins in the effort to save at least a few of the beached starfish. He describes lifting up one star, "whose tube feet ventured timidly among my fingers while . . . it cried soundlessly for life."[20] In such prose Eiseley offers a snapshot of life's tenacity to survive in a threshold landscape. Evolved approximately 500 million years ago, starfish are ancestors to more complex organisms, including us. In that momentary embrace between Eiseley's fingers and the tube feet of the starfish, life primordial touches the present as two beings communicate a simple and mutually understood message, each wishing only to cling to life. Such vivid depictions of species surviving by a bare foothold, or tubed foothold, inspire us to imagine the distant shores on Titan, with its carved river channels and methane lakes, and wonder what microbes or variants of terrestrial mollusks might subsist on its frozen shorelines. While Bonestell painted his spacescapes at the dawn of the Space Age, we have since sent our spacecraft billions of miles from home, and in time we'll seek out the answers to what must have been Bonestell's questions regarding life-forms that might cling to the rocks and crevasses on Saturn's icy moons.

A Complex Mission

In October 1997, a six-ton spacecraft the size of a school bus set off on a billion-mile journey to Saturn. It was named after the seventeenth-century Italian astronomer Giovanni Domenico Cassini, who discovered four moons of Saturn—Tethys, Dione, Iapetus, and Rhea—along with co-discovering Jupiter's Great Red Spot and first spotting the gap in Saturn's rings that bears his name. Cassini, along with its deployable probe Huygens, is one of the most complex and ambitious missions in NASA's history.[21] Scientists can spend their entire careers working on a planetary probe like Cassini. The first concept was floated in 1982 and it got a boost in the late 1980s when it was conceived of as a joint mission with the European Space Agency. That helped it survive budget-cutting by Congress in the early 1990s. More than thirty years after it was first conceived it's still going strong. Cassini is an exemplar of international collaboration in space. More than five thousand scientists and engineers in seventeen countries have worked on the mission. It's still the heaviest spacecraft NASA has ever launched to a destination beyond the Moon.

To date, Cassini-Huygens has cost about $3.5 billion, and the very high cost of complex planetary probes makes many people flinch. It was Cassini's ballooning budget in part that made incoming NASA Administrator Dan Goldin embrace a faster, better, and cheaper approach in the early 1990s.[22] During the Goldin years, NASA launched nearly 150 payloads at an average cost of $100 million, with a failure rate of less than 10 percent. But the move to crank out more frequent, stripped-down missions at lower cost was not universally popular; a NASA report in 2001 argued that the strategy had cut too many corners and produced an unacceptably high failure rate.[23] The public took notice when the high profile Mars Climate Orbiter and the Polar Lander missions failed. The former was notoriously lost due to a failure to convert English to metric units, but the latter was successful in a later incarnation as the 2007 Phoenix mission to Mars.

The debate may be a false dichotomy. Special-purpose missions such as the recent LCROSS probe to the Moon and Mars Global Surveyor are only designed to do one thing. Billion-dollar missions typically have a dozen instruments and are very versatile; they're like the Swiss army knives of the space program. Among them, the two Voyagers and Galileo lasted twelve years beyond their design lifetimes and have fulfilled all their scientific goals. Cassini has finished its primary mission and in 2008 it was approved for a two-year extended phase called the Cassini Equinox Mission. In 2010, it was extended until at least 2017 and renamed the Cassini Solstice Mission. These names reflect the fact that the spacecraft will have witnessed an entire cycle of Saturn's seasons by late 2017. So far, more than 1,500 research papers have been published based on Cassini and Huygens data, making this the most productive planetary probe ever. The mission will end with a dramatic flourish. Current plans call for Cassini to dive inside Saturn's rings on September 15, 2017, orbit Saturn twenty-two times, and then plunge to its death in the atmosphere. One hopes a musician will be inspired to write a suitably operatic theme for this planetary finale.

Flying Rings around Saturn

At its closest approach to the Earth, Saturn is about eight times the Earth-Sun distance, or 800 million miles away. Yet, by the time Cassini reached Saturn it had traveled 2.2 billion miles. NASA launched Cassini with the best rocket available, but it wasn't powerful enough to get the spacecraft to Saturn directly since it would be fighting against the Sun's gravity all the way. So mission designers used gravity assist, colloquially called the gravitational "slingshot" mechanism, to get it to its target. As we saw in the last chapter with Voyager, this technique has been used since the 1970s to get nature's help in hefting planetary probes away from the Sun's gravity.[24] The assist is provided by bringing the spacecraft alongside a planet from behind and letting it get a "kick" from the orbital angular momentum of the planet. In principle the spacecraft can get a speed boost of up to twice the planet's orbital velocity.

Cassini passed by Venus twice, then the Earth, and finally Jupiter before heading to Saturn. The terrestrial flyby was controversial because of the nature of Cassini's power source. Solar panels aren't feasible for a mission so far from the Sun's rays. So Cassini has three generators where radioactive decay of plutonium-238 generates electricity via a thermocouple. The power source had raised congressional eyebrows before launch, but as the flyby approached, NASA was told to do an environmental impact assessment on the possibility of Cassini impacting the Earth. For a worst case scenario of a shallow angle of entry to the atmosphere and slow dispersal of the radioactive materials, the odds were one in 10 million.[25] NASA was allowed to proceed. There were demonstrators and a few lawsuits, but the launch and flyby went off without a hitch and now the plutonium is at a safe distance of a billion miles.

These flybys were just a warm-up for the amazing series of gravitational dances that Cassini engaged in when it got to Saturn (figure 5.3). Over its core mission, Cassini orbited Saturn 140 times. To see Saturn, its rings, its largest moons, and its magnetosphere from all conceivable angles, Cassini is using its rockets and seventy gravity-assist flybys of Titan to tweak its orbit size, period, velocity, and inclination from Saturn. As the largest moon, Titan is

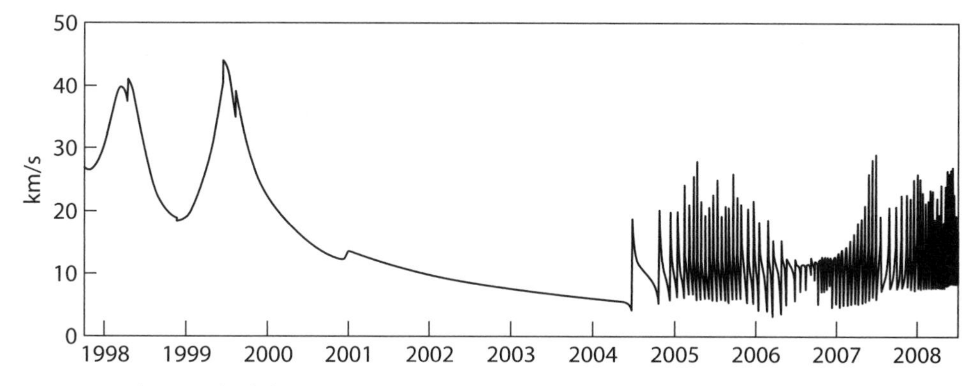

Figure 5.3. The speed of the Cassini spacecraft relative to the Sun during its first decade of travel. The early peaks are gravitational slingshot maneuvers to enable it to reach Saturn with the appropriate velocity for insertion into orbit around Saturn. The later variations are cleverly arranged flybys of Saturn's moons (Wikimedia Commons/YaoHua2000).

the most useful in "steering" Cassini around the Saturnian system. Each Titan flyby is engineered to return Cassini into the proper trajectory for its next Titan flyby. Encounters with other moons are performed opportunistically with what's called a targeted flyby. About fifteen are planned by the end of the mission, half to the intriguing small moon Enceladus. From 2004 through 2011, Cassini did a dizzying hundred flybys, with another dozen completed in 2012. NASA hosts a clock counting down the time until the next swooping visit to a moon and coyly calls these "Tour Dates" to appeal to a younger generation.[26] By clever planning, NASA engineers have doubled the length of the mission even though just a quarter tank of fuel remains.

Titan's gravity has also been used to gradually tilt the inclination of Cassini's orbit, allowing it to see the rings from above and below, and to see atmospheric phenomena of Saturn's poles for the first time. Flybys of Titan can't get closer than 600 miles, or the large moon's thick atmosphere would slow the spacecraft. Some flybys to small moons have been as close as 15 miles. At a distance of a billion miles, that's like hitting a golf ball coast to coast and dropping it within an inch of the hole! There's another problem with operating such a remote probe. Depending on where Earth and Saturn are in their orbits, the distance between them can vary from eight to ten astronomical units or Earth-Sun distances. It therefore takes 70 to 85 minutes for radio commands to travel from mission control to the spacecraft, and the same for the reverse journey. Controllers can't give "real-time" commands. Even if they responded immediately to a problem, nearly three hours would pass before Cassini would get the response. This casts the flybys in an entirely new light, since the margin of error is less than a second.

Tools of the Swiss Army Knife

The Cassini orbiter has twelve scientific instruments and the Huygens probe had six. Tools on this "Swiss army knife" fall into three categories: remote sensing using visible light, remote sensing using microwaves, and studies of the environment near the spacecraft (figure 5.4). Despite using 13,000 electronic components and 10 miles of cable, each of the instruments has worked as planned. Many scientific studies use data from more than one instrument. The optical remote sensing instruments are all mounted on a palette so that they face the same way. The main camera on Cassini is the Imaging Science Subsystem and it has been most people's entrée into the exotic world of Saturn and its moons. Carolyn Porco, the principal investigator of the instrument team, maintains a website loaded with amazing images and evocative descriptions.[27] As noted earlier, she uses a Captain's Log motif with affectionate allusions to *Star Trek*, and its clear she's having the time of her life as Cassini images new worlds with unprecedented levels of detail. She's compared a leadership role in a big space mission with child rearing—a journey of thrills and heartache, excitement and occasional disappointments, lasting twenty years or more. Even the most jaded cynic about big science would be entranced by the best pictures from the main imager.

Two mapping spectrometers look at the properties of Saturn, its moons, and its rings at optical and infrared wavelengths. This is the key to determining their composition and temperature. An ultraviolet spectrometer does the same thing at shorter wavelengths than the eye can see, as another guide to chemical composition. Cassini can't capture samples of the atmospheres or rings, so remote sensing with spectrometers is the best guide to chemistry. The

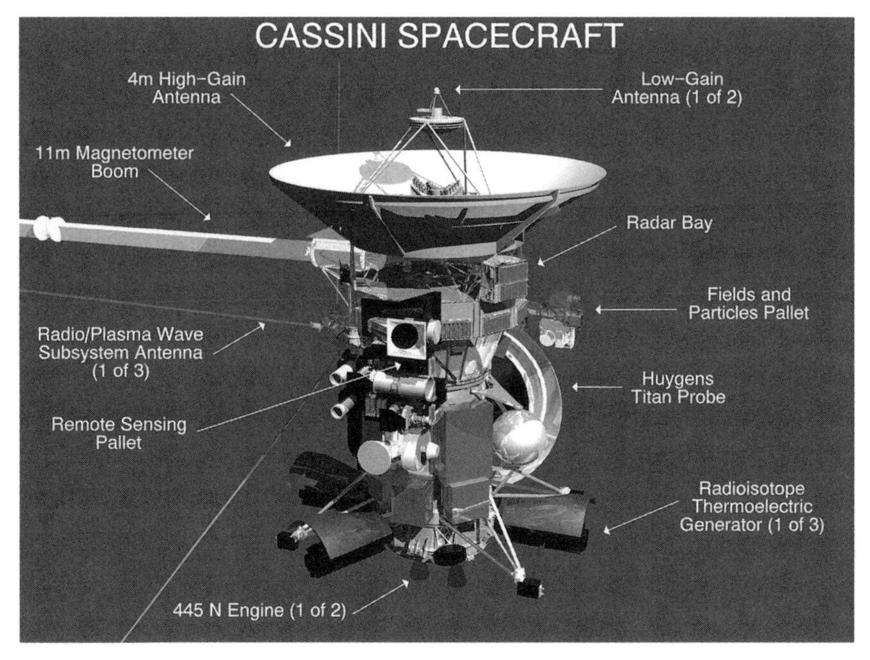

Figure 5.4. Cassini is a large and complex spacecraft about the size of a bus, with twelve scientific instruments, plus six on the Huygens lander. The instruments include cameras and spectrometers and others to measure magnetic fields, high energy particles, radio waves, microwaves. The power source is radioactive plutonium (NASA/Jet Propulsion Laboratory).

spacecraft's microwave remote sensing instruments function a bit differently. Optical remote sensing just uses available light or radiation reflected from the planet, moon, or ring particle. Cassini has to generate its own radio waves and microwaves, send them to the target with a 4-meter high gain antenna at one end of the spacecraft, then "listen" for a weak echo signal. The radar instrument can penetrate Titan's atmosphere and so make topographic and compositional maps. It can also see deeper into the atmosphere of Saturn than any other instrument. The radio instrument looks for fine structure. It also makes Doppler measurements that allow the masses of Jupiter's moons to be precisely calculated.[28]

A set of onboard instruments measure energetic particles, ions, and magnetic field strengths at the position of the spacecraft at any time. Here the complex orbital gymnastics are essential; to make a map of these properties the onboard instruments need to sample data from as many locations within the system as possible.

Several of the six instruments are devoted to understanding Saturn's magnetic field via the charged particles that it often accelerates. To avoid overspecialization, NASA selected six teams of interdisciplinary scientists. Their job is to use Cassini's instruments in concert to maximize the learning and perhaps answer questions that hadn't been anticipated. All this instrumentation is power-hungry and like other probes to the outer Solar System, there isn't enough sunlight for solar panels, so Cassini uses 72 pounds of Plutonium.

What Cassini Discovered

Cassini has rewritten the book on Saturn, making many scientific discoveries; the first six years of study were summarized in 2010 in two papers in the journal *Science*.[29] Two of the greatest advances were conceptual. The first was to paint a sharply etched portrait of a gas giant planet and its moons as intriguing worlds with "personalities" and quirks that made them noteworthy. Voyager paved the way, but Cassini got closer to its targets and spent a larger amount of time in their neighborhood. Another realization was the plausibility of the outer Solar System for biology. Although far from the Sun's warming rays, the largest of Saturn's moons get internal energy from radioactive decay of their rocky material, and many of the moons get extra heat by being tidally "squeezed" by the gravity of the massive planet. If life's minimum requirements are energy, liquid water, and organic material, those conditions may be met on half a dozen of the moons of Saturn. This animates the search for life elsewhere in the universe, since giant planets with attendant moon systems are expected to be commonplace.

Not all scientific data are equally digestible to public audiences. Spectra or measures of magnetic field strengths and charged particle fluxes are abstract and esoteric without a lot of background information and context. So the general awareness of Cassini has been based on its imaging. In that respect the Imaging Science Subsystem is preeminent, and ISS team leader Carolyn Porco has been an eloquent spokesperson for why we should care about the outer planets and their moons. The portraits made of the Saturn system, best seen at a glance in a poster mosaic of sixty-four scenes from Saturn on the ciclops.org website, are an eloquent testimony to the interplay of shadows and light in the realm where there's no air to diffuse and scatter light. The shadows of moons fall on rings, or on each other, and shallow-angled light casts deep shadows on the hills and pock-marked surfaces of the moons. The chiaroscuro is worthy of Caravaggio. In fact, there's an unbroken lineage in the natural depiction of strong light and shadow that connects Leonardo and Galileo's sketches of the Moon, Bonestell's paintings, and the actual images from Cassini.

The science started flowing long before Cassini got to Saturn. As it passed Jupiter to get a gravity assist, the spacecraft took 26,000 pictures and studied the circulation patterns that produce counter-rotating atmospheric bands on the giant planet. It also provided evidence that Jupiter's faint ring originates from micro-meteorite impacts on the smaller moons.[30] Cassini gave a nod to Albert Einstein with a new test of his general theory of relativity. Gravity very slightly bends and slows down light and any other form of radiation passing a massive object. Cassini sent radio signals past the Sun to the Earth so this delay could be measured. The results agreed with the predictions of general relativity theory with a precision of one part in 50,000, improving on the precision of previous tests by a factor of fifty.[31]

In 2004, Cassini discovered three new moons of Saturn, bringing the total to sixty-one. They're small, between three and four kilometers across, like free-floating mountains in space. A year later it found a slightly larger moon in one of the gaps in the rings. The moons and rings engage in a complex gravitational waltz. Some gaps in the rings are caused by a moon clearing out particles at that distance, but others are caused by a more distant moon driving a resonance. If orbits have periods that are related by the ratio of two small integers, then the outer object can influence the inner one, and in the case of an outer moon, it clears out a gap in the rings, as described in the last chapter. Since there are many moons, there are many ratios that can cause resonance. Harmonic effects like this cause much of the complexity of the ring system.[32]

They orbit silently in the vacuum of space, but ring and moon systems have the timbre of a beautiful musical instrument, each one unique.

Cassini's arrival at Saturn in July 2004 involved a daring and risky maneuver. It shot the gap between the F and G rings, equivalent to threading the eye of a needle. The High-Gain Antenna had to be pivoted away from Earth and along the flight path to protect it from hits by small particles, and then the rockets were used for a very precise deceleration that allowed Saturn to capture the spacecraft. At its closest approach it skimmed just 13,000 miles above Saturn's cloud tops. Imagine the majestic view if humans had been along for the ride, watching the towering cloudscapes of a planet so big 760 Earths would fit inside. Cassini provided fascinating new details on the ring system.[33] To a casual observer it seems miraculous that such subtle and complex patterns could arise by unguided natural processes, but the gravitational mechanisms behind the rings have been known for decades. Cassini measured the size distribution of ring particles more accurately than before, and its data showed how ring particles can join into loose aggregations or "piles of rubble." And, it found instances of moons stealing particles from rings as well as times when moons eject particles. New data showed that the rings are 90 percent water, so they're most similar to a huge bumper car ride with careening chunks of ice of all sizes—from microscopic to the size of a house. Surprisingly, when seen up close the rings are tinged red, which scientists speculate is due to rust or small organic molecules mixed in with the water ice (plate 7). To the connoisseur of gravitational dynamics, there's a life's work in understanding the shifting spokes, the spiral density waves, the embedded moonlets, and the features that look like waves and straw and rope.[34]

Much of the time thereafter has been spent planning and executing flybys. An early passage near Phoebe was the only one that will be possible for this curious little moon. Phoebe is one of the outermost moons and only 150 miles across, but parts of its heavily cratered surface appear very bright and scientists think there's ice under the surface. The lion's share of the flybys have been used to explore Titan. Titan is the second largest moon in the Solar System.

Only Jupiter's Ganymede is larger. Titan is 50 percent larger than the Earth's Moon and nearly twice the mass. It is unprecedented in having an atmosphere thicker than Earth, made of the same primary ingredient, nitrogen. Setting aside episodic lava flows on Io, it's the only object beyond Earth where stable bodies of liquid have been seen on the surface. Cassini has used radar to penetrate the thick murk of the atmosphere and reveal surface details as small as a kilometer.[35]

Titan's orange haze is naturally produced photochemical smog. Methane and ethane are mixed in with the nitrogen and they form clouds and rain, which falls on the surface.[36] There's weather here but it's a completely alien chemical environment. Other trace ingredients include propane, acetylene, argon, and hydrogen cyanide. Add some oxygen and you'd have the recipe for a real conflagration, and it's not anything humans would want to breathe. Hydrocarbons break up and recombine in the upper atmosphere under the action of sunlight. The methane present would be converted into more complex molecules in only 50 million years, suggesting that it must be replenished from Titan's interior. Robert Zubrin pointed out that the base of Titan's atmosphere is so dense and its gravity is so gentle that astronauts could potentially fly by wearing powered wings attached to their arms.[37]

The real prize on Titan is not the chemical haze in the atmosphere but the glittering liquid on the surface. In late 2009, NASA released a gorgeous picture of the northern polar region, back-lit by the Sun, with a glint of specular reflection from a body of liquid.[38] Other images show that the liquid levels do not vary by more than 3 millimeters—there's little surface wind on Titan. The north and south polar regions are dotted with lakes varying from a mile across to larger than the largest lake on Earth. The main ingredients of the lakes are likely to be methane and ethane, with smaller contributions from ammonia and water.[39] Evaporation from the lakes is not enough to supply the methane seen in the atmosphere, implying even larger reservoirs of liquid methane underground. Titan is literally swimming in organic materials. There's hundreds of times more mass of liquid hydrocarbons on Titan than the sum of all the oil and gas reserves on Earth.

Huygens Pays a Visit to Titan

The centerpiece of Cassini's symphonic mission, a brief aria of great excitement and beauty, was the descent of the Huygens probe to the surface of Titan in 2005.[40] Titan is the most Earth-like world in the Solar System. It has weather, erosion, active geology, and a complex topography of lakes and rivers and flood plains. Three billion years ago, when the Sun was dimmer and no oxygen had been pumped into Earth's atmosphere by microbes, the two cold worlds had strong similarities.

Huygens represented the main contribution of the European Space Agency to the Cassini mission and even though it returned a modest amount of data, it was not a disappointment. On December 25, 2004, the probe separated from the main craft and began its perilous descent. Buffeted by winds in the upper atmosphere and unable to get a navigational lock on the Sun due to the thick smog, it slowed by parachute and landed on January 14, 2005, on what appeared to be a flood plain, scattered with cobbles of water ice (plate 8). Since the surface conditions of Titan were unknown, Huygens was never designed to be a lander. Rather, it was designed to survive landing on any surface from rocks to ocean, and transmit a small amount of data before expiring. This was dictated by the limitations of the batteries, which only had three hours of life, much of which was taken up by the descent. There was one minor disaster in the mission, when a software error prevented some of the lander's images from being uploaded. While 350 images were returned, a similar number were lost. There was also a major disaster averted. Long after launch, some dedicated and persistent engineers discovered that Cassini's communication equipment had a major design flaw which would have caused the loss of all Huygens's data. The probe had to send its data by radio to Cassini's 4-meter antenna and then on to Earth. However, the acceleration of the probe would have Doppler-shifted its data out of range of the Cassini hardware, and the hardware could not be reprogrammed. To salvage this situation, flight engineers changed the landing and flyby trajectory so that the Doppler shift was greatly reduced.

Huygens weighed 700 pounds and carried six scientific instruments, most of which had been designed to study the atmosphere.

A microphone on one instrument captured the first sounds ever recorded on any planetary body apart from Earth. Another instrument mapped wind speeds at all elevations down to the ground. A third carried a lamp to illuminate the surface, which was useful, since Titan is a very murky moon. Huygens team member Martin Tomasko recalled: "We had great difficulty obtaining these pictures. We had only one percent of the illumination from the Sun, we're going into a very thick atmosphere with lots of haze that blocks light from penetrating to low levels, and we're taking pictures of an asphalt parking lot at dusk."[41] A fourth instrument did the clever trick of heating itself up just before the impact so it could analyze the vapor that came off the surface. The surface was a frigid –290°F or –179°C, cold enough that ice is brittle and methane is a liquid. Scientists clustered around monitors in the mission control room got to stare at a bleak and remarkable scene for several hours, longer than the expected thirty minutes, before the batteries finally died.

Big Surprises from Tiny Enceladus

Before Cassini arrived at Saturn, astronomers had paid little attention to Enceladus, a moon one-tenth the size of Titan. The Voyagers had shown in the 1970s that Enceladus was like an icy billiard ball, reflecting almost 100 percent of the Sun's light. Its surface gave indications of activity since some parts were old and heavily cratered while others seemed to have been altered by volcanism in the last hundred million years. But nothing in earlier data prepared scientists for what Cassini would reveal.

In 2005, plumes were seen rising from the fractured, icy surface (figure 5.5). It took several years and numerous observations by Cassini's instruments to build up a picture of what was going on, but this is what we know. Enceladus emits geysers of tiny ice particles from a number of hot spots on its surface, near the southern polar region. The plumes are ejected at over 1,000 mph, greater than the escape velocity. They rise thousands of miles above the surface and form Saturn's E ring.[42] The geysers arise from geological features called "tiger stripes" that are 100–200°F warmer than

Figure 5.5. Saturn's tiny moon Enceladus has all the ingredients for life: liquid water, energy, and organic material. Evidence for subsurface water came in the form of plumes visible above the moon's sunlit edge. The plumes are composed of tiny ice crystals, ejected at hot spots on the surface from a salty underground ocean (NASA/JPL/SSI).

other areas of the moon. Geologists think there's spreading and tectonic activity in the tiger stripes, similar to what happens near deep-sea ocean ridges on Earth. Tidal heating must play a role, but the cause of the active geology on Enceladus is still a mystery, since the neighboring and similarly sized moon Mimas is inactive. Cassini has swooped through the plumes on several occasions, three times approaching the moon within 30 miles, "tasting" the material with its instruments to determine the chemical composition. The plumes are made of tiny ice particles and vapor that includes methane, ethane, propane, acetylene, and other organic molecules. The chemical composition may be like a comet. Most excitingly, the plumes contain sodium chloride—common salt. That's the best indication so far that Enceladus has a subsurface ocean that occasionally erupts through the surface.

Much of this information was gathered in a series of swooping flybys, the closest of which zoomed within 15 miles of the surface. The imaging team calls these maneuvers "skeet shoots." The spacecraft is moving so fast and is so close to the moon that the camera can't track or lock onto any particular geological feature. Some images resolve features as small as 10 meters across, about the size of a living room. In late November 2009, Cassini made its eighth flyby of the tiny moon, the last before Enceladus entered the shadows of the long, cold Saturnian winter. With no new mission slated to return to the outer Solar System for at least fifteen years, it will be a while before we see images like this again. Meanwhile, the presence of liquids on Titan and inside Enceladus naturally leads to speculations about biology. Our dreams turn to the possibility of creatures floating in the dark and frigid depths of lunar seas far from their sheltering stars.

Earth's Oceans as Analogs for Extraterrestrial Seas

In *Alien Ocean: Anthropological Voyages in Microbial Seas*, Stefan Helmreich argues that the discovery of life deep in Earth's oceans provoked increased speculation about the possibility of life in extraterrestrial seas. Prolific communities of tubeworms, krill, and other hypothermophiles were discovered near deep-sea hydrothermal vents in 1977 in the Pacific Ocean.[43] These organisms live in complete darkness—no sunlight can penetrate to these depths. Scientists speculate whether life in primordial times thrived in the sulfur-rich heat plumes that spew nutrients from the ocean floor. Besides the fact that we believe all life on Earth originated in the sea, we have long imagined, notes Helmreich, that extant primeval life-forms might somehow survive in the deep, even today. Life's origins may best be represented by extremophiles currently living in extreme environments. He writes, "Some marine microbiologists maintain that vent hyperthermophiles are the most conserved life forms on the planet, direct lines back to the origin of life."[44]

What we are learning about extremophiles or methanotrophes—bacteria that metabolize methane—may indicate something of the possibilities for life on other worlds. "Astrobiologists treat unusual environments on Earth, such as methane seeps and hydrothermal vents, as models for extraterrestrial ecologies," writes Helmreich.[45] Cassini has revealed that several of Saturn's moons have liquid or frozen oceans and briny geysers. Enceladus, with its outgassing plumes of ice water and salt, and Titan, are of particular interest.[46] As the planetary body in the Solar System most similar to

Earth in atmospheric composition, Titan could reveal how life on Earth emerged, suggests astrobiologist Chris McKay.[47] Since Titan and Enceladus are likely heated internally by Saturn's gravitational squeezing, astrobiologists imagine that extremophiles might huddle near hydrothermal vents on the floors of their arctic seas. On Earth, Lake Vida in the Antarctic has been capped under a 50-foot thick ice sheet for 2,800 years and yet researchers found ancient microbial life thriving there.[48] Helmreich comments, "For astrobiologists, life, extremophilic or no, will exist in a liquid medium. It is for this reason that extraterrestrial seas—alien oceans—are such objects of fascination."[49] In these discoveries alone, the implications of the Cassini mission are as deeply cultural as they are scientific.

Robotic and Biological Symbionts

In the introduction to the stunning coffee table volume *Saturn: A New View*, Kim Stanley Robinson comments on the amazing photographs Cassini has archived in its ongoing exploration of Saturn. Noting that "the gorgeous concentricities of Saturn's rings look like gravitation itself made visible," Robinson is wistful that astronauts have not yet journeyed to Enceladus or Titan. "Eventually, we might even go to Saturn ourselves," writes Robinson, "It would be a kind of pilgrimage: it would be a sublime experience."[50] However, roboticist Rodney Brooks would likely argue that we have already journeyed to Saturn and landed on one of its moons. NASA's planetary missions are extensions of ourselves. These little machines, with the ability to travel billions of miles across the chasm of interplanetary space, enhance our vision—like a pair of contacts or glasses—and extend our sense of touch, our ability to sample the atmosphere of another world. Analogous to cochlear implants, pacemakers, or titanium prosthetic legs that allow paraplegic athletes to run faster than Olympians, Cassini has taken us to the far shores of the outer Solar System and continues to record, in fine detail, the state of affairs at Saturn and its moons. Cassini, like our other planetary science missions, serves as a highly technical extension of humankind. These robotic explorers not only extend our fingertips into the frigid outer Solar System, but Brooks argues that our machines are "us," and that biotechnology of the future will reconfigure what we think of as human. "Our machines will become much more like us, and we will become much more like our machines," predicts Brooks. "The distinction between us and robots is going to disappear."[51]

Futurist Ray Kurzweil couldn't agree more. Kurzweil predicts that in the next thirty years we will use biochemistry, biotechnology, and nanotechnology to reconfigure the human body, in part, by readily incorporating technology into our bodies to enhance longevity and our intellectual capacity. Kurzweil points to the evolution of sight to illustrate how technology has exponentially enhanced our biological capabilities:

> There are many ramifications of the increasing order and complexity that have resulted from biological evolution and its continuation through technology. Consider the boundaries of observation. Early biological life could observe local events several millimeters away, using chemical gradients. When sighted animals evolved, they were able to observe events that were miles away. With the invention of the telescope, humans could see other galaxies millions of light-years away. Conversely, using microscopes, they could see cellular-sized structures. Today humans armed with contemporary technology can see to the edge of the observable universe, a distance of more than thirteen billion light-years, and down to quantum-scale subatomic particles.[52]

Kurzweil compares this exponential advance in visual observation to the evolution of information technology. He notes that microorganisms can respond to and communicate events in their immediate environment, but with the evolution of humans, language, and the technology of writing, we have recorded information that persists for thousands of years. The simple technology of writing, whether in cuneiform or in modern languages, has exponentially expanded our scientific knowledge and reach. Our robotic partners in space are no less an extension of ourselves than a telescope or the technology of writing and have powerfully shaped what we know about our planet and the Solar System, and the billions of worlds we have yet to explore.

Even now we are joint explorers with our smart machines. Anthropologist Stefan Helmreich comments, "What it means to do oceanography and ethnography is changing. In an age of remotely operated robots, Internet ocean observatories, multi-sited fieldwork, and online ethnography, presence in 'the field' is increasingly simultaneously partial, fractionated, and prosthetic; it is not just distributed across spaces—multi-sited—but cobbled together from different genres of experience, apprehension, and data collection."[53] This collaborative scientific exploration, already being undertaken between humans and machines, affords us a kind of distributed intelligence across the Solar System.

Helmreich, Brooks, and Kurzweil suggest that we think of our machines as symbionts, without whose help we could not explore Earth's ocean depths, much less the depths of lakes and oceans on icy moons orbiting Jupiter or Saturn. Our collaboration with smart machines incites Helmreich to consider one other order of unsuspected collaboration—that between humans and microorganisms. He suggests that alien microorganisms, if such exist in the frozen ocean on Enceladus or in Titan's hydrocarbon lakes, may be more akin to life on Earth than we imagine. As microbiologist Jo Handelsman points out, "We have ten times more bacterial cells in our bodies than human cells, so we're 90 percent bacteria."[54] Of the microbes coexisting in our bodies, scientists explain that we have "evolved with them in a symbiotic relationship, which raises the question of who is occupying whom."[55]

In fact, instead of thinking of microorganisms as alien to us, doctors have begun to recruit them in fighting cancer. Researchers at the University of Pennsylvania are relying on our symbiotic relationship with viruses and other microorganisms to attack and kill cancer cells. They're using viruses to insert DNA into patients' T-cells that in turn causes the T-cells to selectively attack and kill cancer cells. As Stefan Helmreich makes clear:

> Microbes are not simple echoes of a left-behind origin for humans, orphaned from all evolutionary association. Microbes are historical and contemporary partners, part of our bodies "microbiomes." "The" human genome is full of their stories. . . . The bacteria that inhabit our bodies do not simply mirror the bacteria that inhabit the sea—as

> might brine in our blood. This is not human nature reflecting ocean nature. It is an entanglement of natures, an intimacy with the alien. Such dynamics shift the grounds upon which *anthropos* might be figured, perhaps transforming humanity into *Homo alienus*."[56]

Evidence of this is the fact that people in different regions of the world have different genetic makeup partly due to local microorganisms. People of Japanese descent have "acquired a gene for a seaweed-digesting enzyme from a marine bacteria. The gene, not found in the guts of North Americans, may aid in the digestion of sushi wrappers."[57]

As noted in the chapter on the Viking mission, Lynn Margulis's contribution to Gaia Theory was to highlight the extent to which our existence is intimately bound up with the Earth's microorganisms. Having proposed the theory of symbiogenesis, which claims that the mechanism for evolution is the symbiotic sharing of genetic material, Margulis demonstrated that bacteria invading single-celled organisms became their mitochondria and chloroplasts. We do not know whether extremophiles exist on Titan, Enceladus, or other worlds such as Europa, one of Jupiter's moons. What drives our continued exploration of those distant shores is that our beginnings may be entangled with theirs.

Is There Life on Saturn's Moons?

Huygens was a pinnacle among many high peaks for the Cassini mission, marking the first time humans had landed a spacecraft on any body in the outer Solar System. It saw a terrain sculpted by wind and liquid, with vast hydrocarbon dunes and sinuous channels carved into the shoreline. Volcanoes in the distance are likely to emit water instead of lava. Meanwhile, a tepid ocean sits under the icy crust of little Enceladus. These two moons are worlds at once alien and familiar, and if they host life, it will be unlike any form of life known on Earth. In 2011, the sum of all the evidence so far caused scientists at a major meeting to elevate Enceladus to the status of the "most habitable spot in the Solar System" beyond Earth, above Titan, and even above Mars.[58]

One of the conundrums of the search for life beyond Earth is the difficulty of moving away from anthropocentric thinking. All life on this planet is one thing, derived from a single common ancestor and a single implementation of information storage in genetic material. There's no way to inductively formulate a "general theory" of biology or know how life might have evolved if conditions had been slightly different, or if it would have evolved at all. Since the hypotheticals cannot be answered, we don't know whether the origin of life on Earth was a historical accident or the nearly inevitable outcome of physical and chemical conditions plus the passage of time. These two scenarios project very different roles for life elsewhere in the universe: sparseness or abundance. We do know that life has adapted to almost every conceivable ecological niche on the Earth, including environments below water's freezing point and places where the source of energy is not sunlight or photosynthesis. Life grips the planet like a fever, thriving in almost all places we can imagine and in some that are nearly unimaginable.

If Titan and Enceladus host life, we'll not just have to expand the envelope of our traditional thinking about life processes, we'll have to throw away the box completely. If the moon of a giant planet is hospitable to biology, then by focusing only on finding Earth-like terrestrial planets we'll be missing a large part of the story. If a tiny moon far from the Sun can be habitable, the number of potential living worlds throughout the Milky Way galaxy rises dramatically to several billion. Biochemistry based on ethane and methane, augmented by water and ammonia, would be unlike anything we've ever seen on Earth. Computer and lab simulations might help, but we still know too little about physical and chemical conditions on these moons for them to be reliable. The next time window to get a gravity assist and revisit the Saturn system is 2015 to 2017, which adds a sense of urgency since the opportunity following that one is not until 2030. Cassini and Huygens have given us just a taste of the potential for life beyond Earth; if we want to address this profound question we'll have to go back.

6 Stardust

CATCHING A COMET BY THE TAIL

The story of life in the universe is a story of stars. As the first clouds of gas formed stars in the infant universe, more than 13 billion years ago, the universe contained only hydrogen, helium, and a few other trace light elements. The nuclei of these light elements were forged in the intense heat a few minutes after the big bang, when the entire universe was as hot as the core of the Sun is now. As the universe rapidly expanded, radiation eased its grip and a scant half million years after the big bang, it had cooled enough for electrons to mate with nuclei and for hydrogen and helium atoms to form. Chemistry was now possible, but a universe made of the two simplest elements is singularly dull—hydrogen atoms can only join to form a hydrogen molecule, while helium is inert.

As the first stars congealed out of the expanding gas, there were no planets because there was nothing to make them out of. There was no life because there was no carbon and no nitrogen and no oxygen.[1] Our existence on a rocky planet depends on generation after generation of stars fusing heavy elements in their cores and ejecting them into space to become the raw material for solar systems.[2] The fireworks couldn't start until gravity had used its long reach to gather matter into concentrations dense enough to counter the omnipresent cosmic expansion. This took several hundred million years. But then the pockets of spherical collapse ignited legions of stars that could slam atomic nuclei together hard enough for them to fuse and populate the periodic table for the first time.

Every carbon atom in our bodies was once in a star in a remote region of space more than 4.5 billion years ago. Some atoms have cycled through multiple generations of stars; their myriad stories played out over eons until they were co-opted and incorporated into our fleeting human story. We are made of stardust.

Understanding the way in which the products of stellar fusion enriched the nebula that formed the Sun and planets requires finding primordial material in the Solar System. The most pristine samples available are certain types of meteorites and comets. There may be as many as a trillion comets and they spend most of their time far from the Sun and Earth in the deep freeze of space. Material from the outer Solar System has been radioactively dated back to 4.567 billion years, which is taken to be the formation epoch. The spherical comet cloud extends to 100,000 Earth-Sun distances and it's a tenuous relic of the time when the Sun switched on for the first time. In the outer part of their orbits, comets are dark and dead, but they become lively and visible when they approach the Sun. This diaphanous shroud of frozen worlds holds important clues to our origins.

Catching a Comet by the Tail

To gather stardust, first you have to catch a comet by the tail. That's quite a trick, and the mission designed to capture material from a comet was one of the most ambitious NASA had ever undertaken. The Stardust spacecraft was launched on February 7, 1999, from Cape Canaveral on a Delta 2 rocket. It was about the size and weight of a fully laden refrigerator and it consumed a similar amount of power, 350 Watts.[3] Stardust had to survive the rigors of deep space so it was built around a coffin-sized aluminum frame called a spacecraft bus that provided the basis for a number of interplanetary probes. It was sheathed in a material called Kapton to protect it from micrometeorites; Kapton is the same material as used for the outer layer of space suits since the Apollo era.[4] Stardust's destination was an Earth-approaching comet, Wild 2 (figure 6.1).

Figure 6.1. Visualization of the Stardust spacecraft approaching comet Wild 2. This comet represents pristine material since it has only been traveling in the inner Solar System for a few decades, after billions of years in deep space far from the Sun. With each close passage of the Sun, frozen gases are boiled off into a glowing tail (NASA Planetary Photojournal).

Why Wild 2? The vast majority of comets not only spend more of their time in the very distant Solar System, but they never even range within the orbits of the planets.[5] Another set of familiar comets like Halley's make periodic passages into the inner Solar System and close to the Sun, where they grow spectacular tails vaporized from ice mixed in with the rocks. A third and very special class of comets is those that are brand new visitors to the inner Solar System. Wild 2 had a fairly uneventful life for 4.5 billion years until 1974, when it passed within a million kilometers of Jupiter and the gas giant's strong gravity changed its orbit. As a result its orbital period shrank from forty-three years to six years and it began to visit the inner Solar System, approaching to the distance of Mars from the Sun.[6] By the time Stardust encountered it, Wild 2 had only made five trips around the Sun so it was almost unaltered from its original composition. Imagine being a naturalist when an exotic and poorly understood animal that normally roams in remote forests is suddenly standing in your backyard where you can study it in detail. By contrast, the more famous Halley's Comet has made more than a hundred passes close to the Earth and Sun, so it has been heavily altered by solar radiation.

After a comet has passed close to the Sun a thousand or more times, it has been baked dry and is little more than a stone. Wild 2 contains frozen liquids and gases like water, carbon monoxide, carbon dioxide, ammonia, and methane—called volatiles—in abundance.[7] As it was launched, Stardust didn't have enough energy to reach its target. So the spacecraft was sent back toward Earth for a "gravity-assist" maneuver.[8] As we've seen, a gravity assist is the clever way mission planners "steal" energy from a more massive object to boost a spacecraft's velocity and overcome the Sun's gravity. Typically, the spacecraft comes up on the larger object from behind and gains some momentum as it whips by. Since a planet is so much more massive than a spacecraft, the slight slowing of the planet is undetectable. So nobody in the United States was any the wiser or looked up from their breakfast on the morning of January 15, 2001, when Stardust flew by Earth and grabbed some of its energy to get out to Wild 2. On the way, it even managed an opportunistic flyby of the small asteroid Annefrank.[9]

Mission controllers tried to sneak up behind Wild 2 to minimize the relative speed of the two objects. Even so Stardust was moving 13,000 mph, or five times the speed of a rifle bullet, as it flew through the glowing coma of the comet. It took seventy-two close-up photographs. That may not seem like many, but keeping the relatively small comet in the camera field of view during such a fleeting and high-speed encounter was a major feat.[10] The images showed a surface riddled with depressions with flat bottoms and sheer walls, ranging in size from dozens of meters to several kilometers. The comet itself is irregular in shape and five kilometers in diameter. The features are impact craters and gas vents; ten vents were active when Stardust flew by.

The neatest trick Stardust had up its sleeve was gathering material from the comet tail. Ashes to ashes; dust to dust. Behind this somber incantation is astronomical fact. All of the solid objects in the universe were built from microscopic dust particles—stardust. The probe was designed to capture material too small to see in its eight-minute ride through the comet's tail and then its long ride home. The basic goal was to bring back two kinds of material that might speak to our origins. One was comet particles that would pepper the aerogel as it swept through the tail—particles dating back to the birth of the Solar System. The other was samples from much rarer impacts due to interstellar dust—particles that might be older than 4.5 billion years and so would have to come from previous generations of stars. To do this, it deployed an instrument that looked like an outstretched tennis racket in a cover. Inside the racket was an extraordinary substance called an aerogel.

Solid Smoke

Aerogel was created in 1931 as the result of a bet. Steven Kistler wagered a colleague that he could replace the liquid inside a jam jar without producing any shrinkage. In other words, he thought he could fill the volume with something that had the same structural properties but was completely dry. The material that won the bet was 99.8 percent air and has the lowest density of any known solid. Aerogel is made by extracting the liquid from a gel by supercritical drying. This allows the liquid to be slowly drawn off without the solid matrix of the gel collapsing under its capillary action, as would occur during evaporation. The first aerogels were made of silica; more recent ingredients include alumina and carbon.[11]

If you touch an aerogel, it feels like Styrofoam or the green foam that flowers are often pressed into. Pressing on it softly doesn't leave a mark and pressing on it firmly leaves a slight depression. But pressing down sharply enough will cause a catastrophic breakdown of the dendritic structure, making it shatter like glass. It's light and strong, supporting four thousand times its own weight. Aerogel is a thousand times less dense than glass, and the myriad tiny cells of air trapped inside the material make it one of the best insulators known. Engineers at NASA's Jet Propulsion Lab learned how to make extremely pure aerogel and it was used as an insulator on the Mars Pathfinder mission. Peter Tsou is the wizard at JPL who fabricates aerogels; because of the importance of his skills, he was named deputy principal investigator on Stardust.

With Stardust, the challenge was to capture small particles moving at six times the speed of a rifle bullet without vaporizing them or altering them chemically. Aerogel is perfect for this job; the rigid foam that's not much denser than air slows the particles down and brings them to a relatively gentle halt, each one leaving a carrot-shaped wake two hundred times its size. Imagine firing bullets into a swimming pool filled with Jello. Stardust's aerogel was fitted into a module the size and shape of a tennis racket that swung out when the spacecraft approached the comet. One side was turned to face Wild 2, and the other side was turned to face interstellar dust encountered on the journey. Before and after use, the module was stored in its protective Sample Return Capsule (plate 9).[12]

Stardust flew within 150 miles of the comet on January 2, 2004 and headed back to Earth with its precious cargo trapped like tiny flies in a silica spider web. On January 15, 2006, Stardust returned home after seven years and nearly 3 billion miles of traveling. First, the mission controllers did a short rocket burn to divert the spacecraft from hitting the Earth, leaving it with just 20 kg of fuel. Then they fired two cable-cutters and three retention bolts to release the 46-kg return capsule and watched as springs on the spacecraft pushed the capsule away. The capsule streaked into the pre-dawn California sky at 29,000 mph, faster than any man-made object had ever been returned to Earth. The heat shield and parachutes worked flawlessly and the capsule landed in the Utah desert at 5:10 a.m. The few people up and outside that morning saw a fireball and heard a sonic boom.

Within two days, the package containing the aerogel was opened in a clean room at the Johnson Space Center in Houston. Stardust was subject to the maximum contamination restrictions, since it returned material from an extraterrestrial object with the potential to host life. In practice, the risk of "infecting" the Earth with alien life was low, since any known organism would almost certainly be destroyed by the high impact speeds in the aerogel, but NASA took no chances. The mission was carried out under a Category 5 planetary protection policy, which is even more stringent than Biosafety Level 4, the protocol used to deal with hemorrhagic fevers like Ebola and Marburg.[13] That means sterilization by heat, chemicals, and radiation before the spacecraft is launched, and a requirement that the returned samples are handled in a secure facility and never come into direct contact with humans.

Members of the team opened the sample return package in a clean room just down the hall from where hundreds of kilos of Moon rocks are kept, brought back by the Apollo astronauts.[14] The room was a hundred times cleaner than a hospital operating theater. They were delighted to see the aerogel segments littered with particle tracks, looking like burrows left behind by microscopic creatures. The mission had clearly been a success.

Stardust at Home

Scientists all along anticipated that tracing our primordial origins with interstellar dust would be like looking for a needle in a haystack (or a piece of grit in a gel block). They estimated that, while the side of the collector that faced the comet debris might collect a million particles, they'd be lucky to gather a few dozen from the side trawling for dust from interstellar space. Such was the case. One side of the aerogel was peppered with trails from comet particles, but interstellar dust was very rare, and difficult to spot among all the blemishes and markings that the other side of the aerogel had suffered after seven years in space. Imagine searching for a few dozen ants nestled deep in the grass of a football field. So the two hundred members of the international science team decided to get some help.

Stardust@home has engaged nearly 30,000 members of the public around the world in the search for interstellar dust. The archetypal citizen science project was SETI@home, where the "spare" CPU cycles of millions of PCs were harnessed to analyze chunks of radio data in order to search for transmissions from intelligent aliens.[15] SETI@home had distributed computing as a model and no thought or intervention was required by people who participated. Stardust@home is more like Galaxy Zoo, where human eyes and brains are harnessed in pursuit of science goals and participants must undergo training.[16] Citizen science is one of the exciting recent developments in outreach and the "democratization" of research, where interested members of the public get online training in categorizing and sifting through large amounts of data, and then are able to contribute to the creation of new knowledge. Occasionally, these very attentive amateurs make important discoveries.[17]

The raw material for Stardust@home is a huge number of images made with an optical microscope which can automatically focus at different depths in the aerogel. A set of forty images of a small area are taken with the focus ranging from just above the surface to 100 microns into the aerogel. These images are turned into an animated sequence or "movie" so the viewer seems to move through the aerogel. Altogether, 1.6 million movies were needed to cover the 1,000-square-centimeter surface of the collector. This huge number is part of the reason help was needed. Starting in August 2006, Stardust "movies" were made available to the general public. Each eager participant first had to undergo a short training session and take a test to show that they could indeed recognize particle tracks. Then they were unleashed on the "haystack." The signature of a cosmic dust particle is a hollow wake that ends in a tiny particle, often no bigger than a micron in size (figure 6.2). A million such particles ploughed into the aerogel. Of these, only

ure 6.2. In this image from the Stardust mission, a particle entered from the bottom of the ıme, penetrating the aluminum foil protecting the aerogel block, and leaving a spray of ·cta in the aerogel. The image is of a region a millimeter across. Hundreds of dedicated ıtizen scientists" were involved in characterizing the traces left by particles in the aerogel 'ASA News Archives).

ten were large enough to see by eye—a tenth of a millimeter or larger—and only one was as big as a millimeter across. Computer programs are unable to reliably identify telltale signs of a particle impact, and they can't be trained since such detections haven't yet been made! Additional information has come from the aluminum foil detectors, which were also peppered with dust impacts.[18]

Citizen scientists can't get instant gratification from the project. They have to use the "Virtual Microscope" program in a web browser and report their results to Stardust @home headquarters in Berkeley. Each movie is sent to four users who each scan it independently. Only if a majority of users claim a particle detection does it go to the Stardust science team for confirmation. What do the volunteers get in return for their labors? Mostly online certificates, and the knowledge that they're contributing directly to an important science mission. Bruce Hudson from Ontario in Canada did a bit better. He had suffered a stroke and turned to the Stardust mission as a good way to pass the large amount of time he had on his hands. Working up to fifteen hours a day for over a year, he not only found the first confirmed interstellar dust particle in the aerogel, he then found a second, named them (Orion and Sirius), and he'll be a co-author on the paper that results. Hudson might be amused by the irony that astronomers rarely give names to asteroids or craters less than a kilometer across, yet he put names on objects a billion times smaller. Interstellar dust is distinguished from comet dust by chemical analysis. Particles from deep space are glassy and contain lots of aluminum, along with manganese, nickel, chromium, iron, and gallium. Researchers take particular care not to drop or lose these particles—it would cost $300 million to replace them.

If that seems rather too high-tech and difficult, you can take on the somewhat easier task of gathering comet (and asteroid) dust in the comfort of your own home. Or at least on your roof. Each year 10,000 tons of micrometeorites and 100 tons of space dust land on Earth and a little of that material will also land on the roof of a house. (They're not recoverable from the ground because they're too similar to particles in the dirt.) The best scenario is a sloping metal, tin, or slate roof, with no overhanging trees. Collect the runoff from a day or more and filter it sequentially with a window screen and then a finer mesh, to remove all leaves, paint flakes, and other artificial materials. The next step involves using a very strong magnet (such as a Neodymium or rare earth magnet, easily obtained by mail order) to gather metallic morsels from the sludge that remains.[19]

This will isolate the primarily metallic particles, but many terrestrial forms of debris can be magnetic so the last step involves a hand-held magnifier or cheap microscope. With a magnified view, the rounded, melted, and pitted shape of micrometeorites readily distinguishes them from more mundane terrestrial metal particles. Following this method patiently and carefully will net you a number of particles from deep space, without leaving home, and for a much lower price tag than several hundred million dollars.

Stardust Surprises

Stardust had some surprises in store for scientists working on the mission.[20] The first came during the flyby. Team members and comet experts expected Wild 2 to be bland-looking, like a large black potato. Instead, the seventy-two images sent back to Earth revealed a dramatic scene (figure 6.3). There were kilometer-sized holes bounded by vertical and sometimes overhanging cliffs, spiky pinnacles hundreds of meters high, and numerous jets of gas and dust surging into space. Some of the holes didn't look like the kind of impact crater found on Mercury, the Moon, and every other airless surface exposed to space. The ravines and chasms indicated that the surface is very young and is constantly being created and altered by dynamic processes within the comet. Some of the jets were on the dark side of the comet, proving that they are not generated by the action of the Sun. Wild 2 doesn't look like any other comet or asteroid that's been imaged by spacecraft (plate 10).

A related surprise came as the spacecraft flew through debris escaping from the comet. It was expected that the rate of particle hits would increase with time, reach a smooth peak, and then decline as the comet rapidly disappeared in the rearview mirror. Instead, the impact rate surged and fell in bursts, presumably as the spacecraft passed through jets of dust escaping from the surface and the breakup of cometary "clods" as the ice holding them together evaporated.[21]

The most exciting discovery came in 2009, when a team at Goddard Space Flight Center reported the discovery of the amino acid glycine in tiny particles they were analyzing.[22] It's not a total surprise that comets contain molecules like amino acids, but their survival intact in material traveling at thousands of miles per hour was unexpected. An isotopic study of the glycine showed it was not a contaminant from Earth. This discovery shows that at least

Figure 6.3. An artist's concept of the comet Wild 2, as seen by the Stardust spacecraft during its close approach. As it nears the Sun, the comet becomes active and releases jets of hot gases and clumps of material into space. Its surface is etched with steep canyons and dramatic valleys, all on an object the size of a small town on Earth (NASA/Jet Propulsion Laboratory).

one of the essential ingredients for life can be delivered by comets, and since most stars are thought to have comet clouds, comets will be an important delivery system for delivering pre-biotic molecules to Earth-like planets, so "setting the table" for life. Another unexpected result was the discovery of iron and copper sulfide minerals in Wild 2 that could only have formed in the presence of water.[23] This means that the comet has spent some of its time at balmy temperatures in the range 50–200°C, where water is liquid or steam, so it's not the "frozen snowball" that everyone expected.

Often in science, a simple model or theory formed based on limited data has to be modified or even discarded when better data are gathered. Nature is messier than our wishful thinking. The Stardust mission has played an important role in retooling the conventional ideas of comet structure and formation mechanisms. It was long thought that short-period comets formed from materials that condensed beyond the orbit of Neptune in the cold Kuiper Belt. Long-period comets, by contrast, were presumed to have formed at higher temperatures closer to the Sun, among the giant planets, and then subsequently ejected (primarily by the gravity of Jupiter) to orbits that extend halfway to the nearest star. The discovery of trans-Neptunian objects that feed both the short- and the long-period comet population complicates this simple picture. Data from Stardust have complicated it further.

Our direct information on comets is still modest. The first probe to reach a comet was launched in 1978 and approached Comet Giacobini-Zinner, providing support for the "dirty snowball" theory. Comet Halley's first visit to the inner Solar System in the age of spacecraft was in 1986, and received visits from an armada of spacecraft—Giotto, Suisei, Sakigake, Vega 1, and Vega 2. Fifteen years later, Deep Space 1 flew by comet Borrelly. Stardust came next, and in 2005, Deep Impact excavated 10,000 tons of material from comet Tempel 1 by smashing into it at 23,000 mph. When Stardust returned its payload in 2006, it was the first time we had gathered physical samples from any other place than the Moon. As the only human relic to go on a multi-billion-mile journey and return home, Stardust earned a very auspicious resting place. In 2008, on NASA's fiftieth anniversary, the sample return capsule went on display in the Smithsonian Museum's "Milestones of Flight" gallery, alongside iconic artifacts such as the *Apollo 11* command module, Charles Lindbergh's "Spirit of St. Louis," and the Wright brothers' 1903 Flyer.

Rethinking Comets

The comet cloud defines the outer edge of the Solar System, extending a thousand times further than the planets from the Sun, and a significant fraction of the distance to the nearest stars. It had always been presumed that these small bodies were frozen relics from the formation epoch, essentially unchanged in 4.5 billion years. The expectation was that comets would be made of dust from a previous generation of stars, or pre-solar grains. That's why the mission was named Stardust.

Instead, something quite unexpected was found: fire and ice.[24] Comets contain ice that formed in the frigid zone beyond Neptune, but the bulk of a comet's mass is rock, and that rocky material seems to have formed under conditions hot enough to vaporize bricks. The comet particles embedded in Stardust's aerogel included two ingredients that are found in meteorites, debris from asteroids that formed between the orbits of Mars and Jupiter. One is chondrules, rounded droplets of rock found in many primitive meteorites that melted and quickly cooled as they orbited the Sun. The other is a much rarer mineral called a Calcium Aluminum Inclusion, irregular white particles of very unusual chemical composition that can only form at very high temperature. The Stardust team puzzled over this discovery, but the implication seemed clear. Matter that formed readily in the inner Solar System was somehow transported to the edge of the young Solar System where the comets formed. Wild 2 does contain grains of pre-solar stardust material, but they're very rare. Comets are not made of material left over from other stars; they're made mostly from material that formed close to the Sun. As such, they provide insights into how planets and moons were built 4.5 billion years ago.

Stardust's primary contribution to planetary science is verification of the idea that there was extensive radial mixing of material in the solar nebula just as the Solar System was forming. Also, comets are more physically diverse and variegated than anyone imagined. Each of the comets where we've got up close and personal—Halley, Borrelly, Tempel 1, and Wild 2—have different shapes, surface textures and features, and levels of activity. Comets may be modest in size compared to planets and moons, but their role in planetary systems is anything but modest. They deliver water and organic building blocks of life to terrestrial planets, and as we know from studies of our geological record, when they hit a planet they can decisively alter the history of ecosystems and even the entire biosphere. It's appropriate that all human cultures have been fascinated by comets, and viewed them as harbingers of life and death.

To learn more requires an increase in our modest amount of direct information. So it was very exciting when Stardust was given a new lease on life after its sample return. In 2007, NASA approved the New Exploration of Tempel 1 (NExT) mission, a return to the site of the Deep Impact mission in 2005.[25] Stardust had conserved just enough hydrazine fuel to make a second journey to a comet, and the goal was to observe the impact made by the earlier spacecraft and see any changes on Tempel 1 caused by its last close approach to the Sun (Deep Impact's cameras were blinded by dust released from the impact). In early 2010, a controlled burn ensured Stardust would approach at the optimal speed. The flyby was challenging because the camera had to catch the correct side of the tumbling comet nucleus while coming up on it at 7 miles per second. Appropriately, our first second date with a comet took place on Valentine's Day, 2011. Stardust showed that the 500-foot wide crater created in 2005 was indistinct and had a mound in the center, indicating that much of the ejected material had fallen straight back down in the weak comet gravity. The results indicated that the comet nucleus was fragile and only weakly held together.

All good things must come to an end. For Stardust the end came on March 24, 2011, when mission controllers deliberately ordered a "burn to completion." They told the spacecraft, which was already running on fumes, to fire its main engines until there was no fuel left. NASA used the final burn to refine fuel consumption models for these kinds of engines, ensuring that Stardust gave useful data right up to its last gasp. The next day, Project Manager Tim Larson put the spacecraft into a state called safe mode, turned off the transmitter, and walked away from the console. Mission accomplished.

We Are Made from Stars

In August 1929, the *New York Times* science section ran an article titled "The Star Stuff that is Man." Astronomer Harlow Shapley had been popularizing the point that humans are the mere byproducts of stars. In a radio talk series a few years earlier produced by the Harvard College Observatory, Shapley pointed out that "we are made out of the same materials that constitute the stars."[26] In the *Times* article, Shapley similarly observed: "We are made of the same stuff as the stars, so when we study astronomy we are in a way only investigating our remote ancestry and our place in the universe of star stuff. Our very bodies consist of the same chemical elements found in the most distant nebulae."[27] At the time, some were disconcerted to think that humans might be little more than the product of fission and fusion occurring in stars.[28]

Cosmologists have good evidence that all the hydrogen and most of the helium in the universe have existed since close to the time of the big bang. Hydrogen, the simplest element, is truly primordial, while most of the helium was formed soon after in a process known as primordial nucleosynthesis or big bang nucleosynthesis. "The primordial nuclei of the matter constituting the universe were formed in the first three minutes," explains CERN theoretical physicist Luis Alvarez Gaumé. "The cosmic oven produced a number of nuclei, made up of about 75% hydrogen and 24% helium. Small amounts of deuterium, tritium, lithium and beryllium were also produced, but hardly any of the other atoms that make up our bodies and the matter around us: carbon, nitrogen, oxygen, silicon, phosphorus, calcium, magnesium, iron, etc. All of these were formed in the cosmic ovens of subsequent generations of stars. As Carl Sagan put it, we are just stardust, remains of dead stars."[29]

The *Oxford English Dictionary* lists a definition for the term stardust as "fine particles supposed to fall upon the Earth from space; 'cosmic dust.'" The *OED* likewise cites one published commentary from 1879 that claimed "the very star-dust which falls from outer space forms an appreciable part" of the mud accumulated on the ocean floor.[30] In fact, the estimate for interplanetary dust particles swept up each year by the surface of the Earth as it churns through space is a substantial 10,000 tons, but that's a tiny fraction of the material added to the seafloor. The most pristine reservoirs of stardust or interplanetary dust are comets. Having formed early in the life of our Solar System approximately 10 million years after the Sun's protoplanetary disc stabilized, comets spend most of their time in the cold extreme outskirts of the Solar System. As a result, comet nuclei are largely preserved from heating, melting, and collisions with other planetary bodies. The ice composition and dust grains of comets reveal elements present in the Solar System's primordial past.

Nuclear fusion reactions in stars create heavier elements by processing or fusing together lighter atomic nuclei. The base of the fusion "pyramid" is the fusion of hydrogen into helium, which occurs in the Sun and low-mass stars. Stars approximately the mass of our Sun can also synthesize helium, carbon, and oxygen, but thereafter, they never reach a dense enough or hot enough state to fuse yet heavier elements. Massive stars have the ability to fuse their hydrogen to produce helium, carbon, and oxygen, but then progress in successive stages of evolution to neon, magnesium, silicon, sulfur, nickel, and iron. Iron is the most stable element, so typically the fusion chain stops there. Still heavier elements are forged two ways: by the slow capture of neutrons in the atmospheres of massive stars and more rapidly by stellar explosions known as novae and supernovae.[31] Since rarer, larger mass stars are required to generate the heaviest elements, these elements are cosmically scarce compared to the light elements hydrogen and helium.

The first generation of stars probably formed approximately 200–300 million years after the big bang.[32] According to current theoretical ideas, none of those stars still exist and certainly none have yet been found. They have since exploded in novae or massive supernovae, collapsed into black holes, or have otherwise expended their fuel and burned out. The oldest stars currently are generations removed from the first stars that shone in the universe.[33] Astronomers estimate that approximately twenty-five novae occur each year in an average galaxy generating interstellar dust enriched with heavier metals. From the clouds of gas, dust, carbon, silicon, and metals produced by repeated generations of novae and supernovae, new stars and their attendant planets are born.

A nova is a sudden brightening of a star believed to occur when the outer layers of a star are pulled by gravity onto the surface of a companion white dwarf. Pressure due to the mass that accumulates on the white dwarf causes nuclear fusion of hydrogen into helium at its surface that in turn blows material off the surface of the white dwarf. One type of supernova is an extreme form of a nova, where mass acquired from a young companion star causes a white dwarf to begin carbon fusion and explosively detonate. Another type, a core collapse supernova, occurs when a single massive star exhausts its nuclear fuel and suffers a core collapse followed by explosive detonation. In comparison to a nova, a supernova will produce a million times the energy and can for weeks shine as brightly as all the stars of an entire galaxy. In the case of the Milky Way, that's equivalent to about 400 billion Suns. A supernova is a prodigious alchemical event; one that detonated in a nearby galaxy in 1987 generated enough stardust to build 200,000 Earths.[34]

The Milky Way formed approximately 10 billion years ago, but our Sun is only 4.5 billion years old. The Sun and its neighboring stars were likely born in a nearby region of dense, coalescing interstellar gas and dust. Isotopic studies of meteorite samples indicate that our Sun formed from the detritus of a massive supernova explosion in nearby space roughly 1 million years prior to the formation of the protoplanetary disc that became our Solar System.[35] We know this simply by the presence of metals that our Sun could not have produced, such as the iron that makes up the Earth's core. Other than hydrogen, which is primordial and dates back to cosmic genesis, almost all the other atoms in the universe have been recycled, silent witnesses to amazing trips through fiery cauldrons and into frigid space, some experiencing multiple adventures. Unfortunately, no atom bears the imprint of its particular passages through the core of a star; astronomers can only describe the origins statistically.

Just Stardust in Your Eyes

By number of atoms, humans and all other living creatures are roughly 63 percent hydrogen, 26 percent oxygen, 10 percent carbon, and 1 percent nitrogen.[36] Not only the iron in your blood, or the calcium in your teeth and bones, but every living organism is made from recycled stardust. John and Mary Gribbin emphasize that "the nitrogen in the air you breathe and in your DNA (along with most of the carbon in your body) had a previous existence as part of a planetary nebula, and was expelled from one or more red giant stars." The lead in a pencil, potassium in bananas, zinc and selenium in vitamins were all forged in stars. Nickel in coins, oxygen in Earth's atmosphere, mercury in a thermometer, as well as precious metals such as gold and silver are examples of the heavy elements forged as stars blow off their outer layers and die. "This is where neon in neon lighting, the sodium in common salt, and the magnesium used (appropriately) in fireworks comes from—carbon burning inside stars."[37] And, ultraviolet starlight irradiating the envelope of gases surrounding nearby dying stars can produce large volumes of water, essential to the formation of Earth's atmosphere and oceans and a key marker for life as we know it.[38]

The composition of a comet is the frozen record of the chemical environment of our region of the Milky Way as the Sun formed, the sum of generation upon generation of star birth and death. When a comet comes near the Sun, the energy reanimates chemicals that have been interred from 4.5 billion years. Fittingly, the comet lights to form a feast for our eyes, in a pale echo of the blinding stellar light in which those atoms formed, long before there were eyes to see them.

Comet Impacts as Vectors for Life

Robert Burnham points out that it was not until the Space Age that astronomers realized just how pervasive cometary and asteroid impacts are throughout the Solar System. Beyond the Moon's cratered and craggy surface, NASA's planetary science missions have charted numerous bodies in our star system riddled with impact craters. Those missions reveal, explains Burnham, "that by an overwhelming margin the most common feature in the Solar System is the impact crater. Entire planets, moons, and asteroids turned out to be covered with them, at all scales ranging from global to microscopic. Craters or their traces can be found on every solid surface from Mercury to Pluto and surely beyond."[39] It's likely the object that devastated the Siberian taiga near the Stony Tunguska River at approximately 7:15 a.m. on June 30, 1908, was a meteorite. However, the unknown projectile was not made of iron like the meteorite that excavated Meteor Crater in Arizona approximately 50,000 years ago. It is estimated the impactor exploded above ground with the force of 1,000 atomic bombs, more than adequate to destroy a modern city. The explosion reportedly flattened 830 square miles of forest, leveling 30 million trees, and was heard 500 miles away. Multiple remarkable witness accounts of this event included a man, 40 miles from the impact site, who reported being knocked off his feet by a searing blast wave.[40] Atmospheric pressure waves registered on barographs after racing around the world at hundreds of miles per hour. The explosion apparently caused a local geomagnetic storm that lasted for days. People in London wondered about the strange, glowing atmospheric effects and pink-fringed clouds appearing in evening skies. One British meteor specialist published his observations in *Nature* in 1908, saying that the night sky over Bristol, England, was so bright few stars could be seen.[41]

In July 1994, Comet Shoemaker-Levy 9 broke into fragments that subsequently slammed into Jupiter with an estimated explosive force of six hundred times humanity's entire nuclear arsenal, and spewed plumes of material thousands of miles above Jupiter's clouds. Millions globally, including astronomers, world leaders and politicians, and a fascinated public, found themselves captivated by television news reports and images streaming from JPL's designated website. That single comet impact raised global awareness of the very real danger such events pose for Earth's populations (figure 6.4). The comet began to impact Jupiter on July 16. Within three days, the U.S. House Committee on Science and Technology voted to establish a NASA program to track comets and asteroids that could threaten the Earth. As co-discoverer of the comet David Levy recalls, "This vote reflected the nation's fascination, and its growing awareness that such a trail of destruction could have been headed at Earth. The initiative was designed to protect future generations of people." Should the threat of a comet or an asteroid impact be successfully deflected, Levy contends the global attention focused on Comet Shoemaker-Levy will have in effect "rescued our planet."[42]

Though the devastation of comet impacts is obvious, these primordial snowballs may very likely be the reason for life on Earth. Given their ubiquity, it's believed that cometary impacts are a common vector by which life may have emerged on Earth. As noted previously, researchers working with the Stardust data reported finding the amino acid glycine in Comet Wild 2 samples. Amino acids are comprised of what is commonly referred to as CHON, or carbon, hydrogen, oxygen, and nitrogen atoms, and the molecules that CHON, along with phosphorus and sulfur, can combine to

Figure 6.4. Comets are carriers of both life and death. This artist's impression shows a large comet impact on the early Earth. Cometary material includes water that led to the Earth's oceans and some basic building blocks of life like amino acids, but occasionally impacts by comets have caused devastation and destruction over the Earth's history, including several mass extinctions in the past half billion years (NASA/Don Davis).

make are all chemical ingredients of DNA. Comet and meteorite impacts on Earth could have dispersed such organics or even synthesized them.[43]

Science writer Connie Barlow claims: "Our ancestors include ancient stars. Stars are part of our genealogy."[44] She means that if we want to discover our origins, we can't simply peer into a microscope at possible progenitor organisms. Instead, we need to look to the stars, where heavy elements necessary for life, like carbon, are forged. But we also can attribute our genetic makeup to comets and asteroids that may have deposited on Earth the chemistry for life. All known organisms are the products of their DNA, and some components of DNA have been detected in meteorites. In 2011, NASA scientists reported finding in meteorites "adenine and guanine, two of the four so-called nucleobases that, along with cytosine and thymine, form the rungs of DNA's ladder-like structure."[45] DNA is the coded information in our cells that determines biologically who and what we are. "At the center of the ladder-like DNA molecule lie ring-like structures called nucleobases. It's these tiny rings that scientists at NASA and the Carnegie Institution for Science in Washington found in 11 of 12 meteorites they scrutinized."[46] Meteorites, if not jettisoned from a neighboring planet, are remnant chunks of rocky debris left over from the Sun's formation. Given that such meteorites deliver organic molecules to Earth, molecules from our star's earliest days are inevitably in our DNA. Scientists reported in the journal *Nature* having found fossils of multicellular life dating back 2.1 billion years, roughly 1.5 billion years prior to the Cambrian explosion. These fossils in black shales in West Africa rewrite the history of when multicellular organisms first emerged.[47] Microbial life on Earth predates even this. Biologists, astronomers, and astrobiologists now faced with recalibrating the timescales on which life began on Earth are reworking our appreciation for comets and their critical role in dispersing stardust throughout the Solar System.

Stardust's mission to capture the interplanetary dust grains that pervade our star system is an attempt to understand not only how our Sun originated and evolved, but also the characteristics of the early Solar System that led to the origin of life. What we learned from the Stardust mission is that comets and asteroids harbor the building blocks of life. Twenty different amino acids, occurring naturally in a wide number of arrangements, produce the proteins that create all living organisms on Earth. John and Mary Gribbin report, "Formic acid (the stuff some ants squirt out as a defensive weapon, and the stinging ingredient in stinging nettles) and methanimine are two of the polyatomic organic molecules that have been identified in dense interstellar clouds. Together, they combine to form an amino acid, glycine."[48] As James Lovelock, originator of Gaia Theory, has observed, "It seems almost as if our Galaxy were a giant warehouse containing the spare parts needed for life."[49] Perhaps the universe is built for life. The Gribbins suggest that the chemistry for life is endemic to the entire universe. They assert that "one of the most profound discoveries made by science in the twentieth century" is that the universe comprises "the raw materials for life, and that these raw materials are the inevitable product of the processes of star birth and star death."[50]

Though the normal matter that makes up stars, planets, and stellar dust clouds constitutes only a small fraction of the total matter in the universe, most of which is dark and still mysterious, the stuff of stars nevertheless provides all the components necessary for life. "The raw material from which the first living molecules were assembled on Earth was brought down to the surface of the Earth in tiny grains of interplanetary material, preserved in the frozen hearts of comets," write Gribbin and Gribbin, who eloquently observe, "Those grains themselves literally, not metaphorically, formed from material ejected by stars. The 'manna from heaven' that carried the precursors of life down to the surface of the Earth was literally, not metaphorically, stardust. And so are we."[51] That we are made of stardust is not to be disparaged. It's likely that life has emerged on planets orbiting other stars in galaxies far, far away. If so, it's certain that the blood in their veins and the calcium in their bones, should they have them, were forged in the fires of their suns.

A World by Any Other Name

Comets and asteroids provide evidence of the formation epoch of our Solar System. They represent the pristine stardust of which we are made. They're implicated in life and death in the evolution of life on Earth. But they're minor bodies of the Solar System. The three targets of Stardust's study are a motley, misshapen collection. There's 5535 Annefrank, three miles across and shaped like a right-angled prism. There's Wild 2, similar in size, with a pitted and coruscated surface. And there's 9P/Tempel, slightly larger and shaped like a potato. Are they big enough to be called worlds? Are they substantial enough to have a place in our dreams?

Yes, absolutely. They may be too small to hold an atmosphere, but comets and asteroids have unique shapes and topographies that imbue them with personality. Our technology is good enough to throw a gravity "lasso" around a space rock that ventures near us and steer it into Earth orbit. Comets contain water ice and organic material needed to sustain life, and the hydrogen and oxygen in the water are all that's needed to make rocket fuel. A comet retained in high Earth orbit would be a perfect jumping-off point for manned exploration of the outer Solar System and beyond. Asteroids also contain valuable metals and minerals; according to one estimate, the mineral wealth in the asteroid belt amounts to $100 billion for every person on Earth.[52] Wild 2 probably contains around $20 trillion worth of precious and industrial metals, and over $100 billion worth of platinum alone.[53] In fact, the precious metals we do have were gifts from comets in the early days of the Solar System. Metals present when the Earth formed settled directly to the core while the planet was molten; those we can reach in the crust were "dusted" onto the planet by later waves of comets.[54]

Space mining aside, a small, barren rock might not be worth calling your own. The 1967 United Nations Outer Space Treaty prohibits any State owning or controlling the Moon, an asteroid, or a comet, but the treaty has a loophole since it doesn't specifically exclude the ownership rights of an individual.[55] A small world might seem limiting, but think of the pleasure in owning a world the size of a small town and surveying the domain like a colossus. The gravity of Wild 2 is so weak you would literally be as light as a feather. A small push and you could escape your world and sail into deep space. And think of the glittering minerals—a hoard magnificent enough to power all the dreams ever dreamed.

7 SOHO

LIVING WITH A RESTLESS STAR

Imagine you woke up one morning and your planet had been engulfed by the atmosphere of a star. High-energy particles slam into the atmosphere, creating shimmering auroras. Sunspots flicker, each one releasing more energy than the largest atomic bomb. Great loops of hot gas uncoil from the star's surface, extending millions of miles. Each whip crack of activity causes mayhem on orbiting satellites, frying their circuit boards and wreaking havoc on their guidance systems. Your planet is assailed by high-energy particles traveling at nearly the speed of light. Looking at the star, its emission doesn't change detectably from day to day or year to year, but its invisible short wavelength radiation fluctuates wildly and unpredictably.

Where is this place? It's the Earth, and the star is the Sun. This activity isn't because the Sun has run out of nuclear fuel and has turned into a red giant; our star is in the dull middle age of its life. But the clean, crisp edge the Sun seems to have in the daytime sky is an illusion. That curved surface—the edge of the photosphere—corresponds to a place where the gas of the Sun has thinned out to the point where photons no longer careen into particles and they can travel freely.[1] From that point they travel in straight lines without interruption, streaking to the Earth in eight minutes. Inside the photosphere, which is a slender sheath thinner than the skin on an apple compared to its radius, the Sun is opaque and our view is blocked. We see the boundary layer as having a sharp edge, just like a cloud has a well-defined edge.[2]

The edge is an illusion because there's no physical surface or barrier. If you could penetrate the Sun in a spaceship, you'd feel no bump. The material of the Sun extends much further into space than its visible edge. In its physical quantities of temperature, pressure, and density, the Sun varies continuously and smoothly from its fusion core outward. It possesses a surprising multi-million-degree corona, the influence of which extends past the Earth's orbit almost to the edge of the Solar System. Invisible forms of high-energy radiation and subatomic particles—ultraviolet waves, X-rays, gamma rays, and relativistic cosmic rays—streak toward us and past us at the speed of light. In a very real sense, we live "inside" the atmosphere of an active star, with profound and surprising consequences for life on Earth.

Living with a Star

Humans have always been aware that the Sun's rays sustain life on Earth. But until the invention of the telescope, the Sun was thought to be smooth and pristine. There's some evidence of naked-eye observations of sunspots in China as early as 28 BC, and possibly several centuries earlier.[3] Aristotle's strong supposition that the heavens were perfect meant that sunspots probably went undiscovered in Europe for nearly two thousand years. Galileo saw sunspots in 1610 and noticed that they rotated with the Sun. Jesuit astronomer Christoph Scheiner also observed them soon after, and he believed them to be orbiting satellites.[4] But by 1630, Scheiner had abandoned his Aristotelian view and accepted that sunspots rotated at different rates, faster at the equator and slower at the poles. This proved that the Sun couldn't be a solid object but was more probably composed of gas or liquid. Through the seventeenth century, the number of sunspots declined almost to zero, giving a first hint of the Sun's effect on our weather, as a period of low sunspot numbers called the Maunder minimum corresponded to decades when Europe went into a deep freeze, with southern England ice-bound and people in Holland skating on the frozen canals all summer. It wasn't until the 1840s that Heinrich Schwabe discovered that sunspot numbers varied with an eleven-year cycle (figure 7.1).

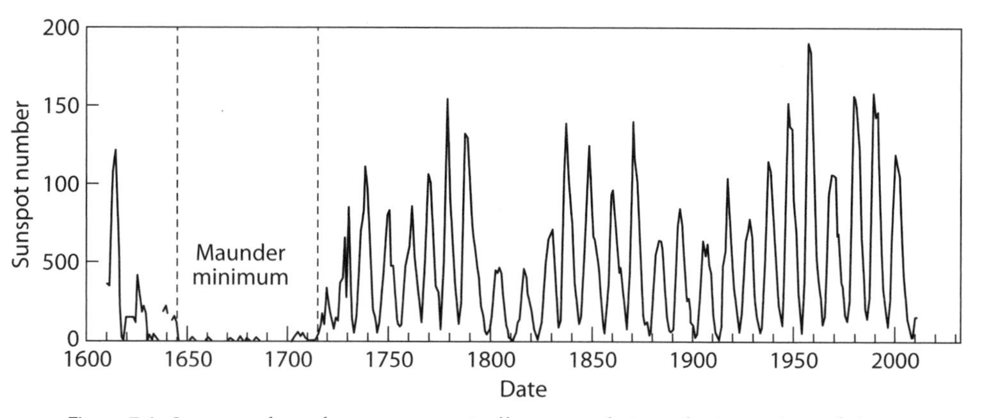

Figure 7.1. Sunspots have been systematically counted since the invention of the telescope. There was an extended "Maunder minimum" of sunspot numbers soon thereafter, which corresponded to a period of very cold climate in northern Europe. This was an early indication of the possible connection between Sun variation and climate (NASA/David Hathaway).

Around the same time that Galileo was doing his pioneering work with the telescope, William Gilbert recognized that the Earth had a magnetic field similar to the dipole field of magnetized rocks called lodestones. Over the next several hundred years, scientists explored the Earth's invisible magnetic field and showed that it fluctuated in ways that seemed to depend on the Sun. In 1777, Johann Wilcke noticed that auroral rays in the polar regions lay along the field lines of Gilbert's dipole. Then, in 1859, two scientists observed a white light flare on the Sun, followed two days later by a geomagnetic storm on the Earth.[5] This once-in-a-millennium event caused aurorae all around the world, including places that had never experienced such a phenomenon in human memory, such as the Caribbean. The northern lights were so bright in the North American Rockies that miners ate breakfast, convinced that the day had started. Telegraph systems failed worldwide and operators reported that they received severe shocks by sparks leaping from the equipment.

Coincidences like this were provocative, but they didn't amount to an explanation. In 1896, Norwegian scientist Kristian Birkeland devised an elegant experiment where the Sun was represented by a cathode and the Earth by a spherical, magnetized anode. With both situated in a large vacuum chamber, he watched as the apparatus behaved like a miniature Earth-Sun system, with an electrical discharge like an aurora occurring at high geomagnetic latitudes. Not long afterward, George Ellery Hale built the first instrument to study solar flares in detail, on Mount Wilson in California, and he confirmed the two-day lag between flares and geomagnetic storms on Earth. Hale proved that sunspots were magnetized—the first detection of magnetic fields beyond the Earth. He also showed that the polarity of sunspot pairs is the same within a cycle, but reverses from one sunspot cycle to the next.[6]

Many researchers have looked for correlations between sunspots and weather on the Earth. Sunspots are actually cooler than the area around them, because the tight bundle of magnetic field lines that they contain restricts the convection that carries heat up from the solar interior. How, therefore, could a sunspot minimum correspond to colder temperatures on the Earth? In part it's because the area around a sunspot is brighter than an average patch of the Sun's surface (figure 7.2). So the Sun is brighter when there are more sunspots. But that still leaves a mystery. When scientists developed digital detectors in the 1970s, they found that the full variation of the Sun's brightness over a sunspot cycle was only one part in a thousand, or 0.1 percent, not nearly big enough to explain significant temperature variations on Earth. The Maunder Minimum aside, there's no convincing evidence that the sunspot cycle drives non-seasonal climate change, especially the rise in global temperatures over the past half century.[7]

However, the variations of the Sun in invisible forms of high-energy radiation are much more extreme than its variation in visible light. Moreover, space isn't the simple and absolute vacuum that most scientists once imagined. Fifty years ago it was shown that the Sun is surrounded by a vast region of plasma, a diffuse, magnetized gas so hot that the particles are traveling near light speed. The stream of particles emerging from the Sun—the solar wind—travels millions of miles to us, where it hits and distorts the Earth's magnetic field.[8] Even though the nature of the effects is complex and subtle, the Sun and Earth are one connected system.

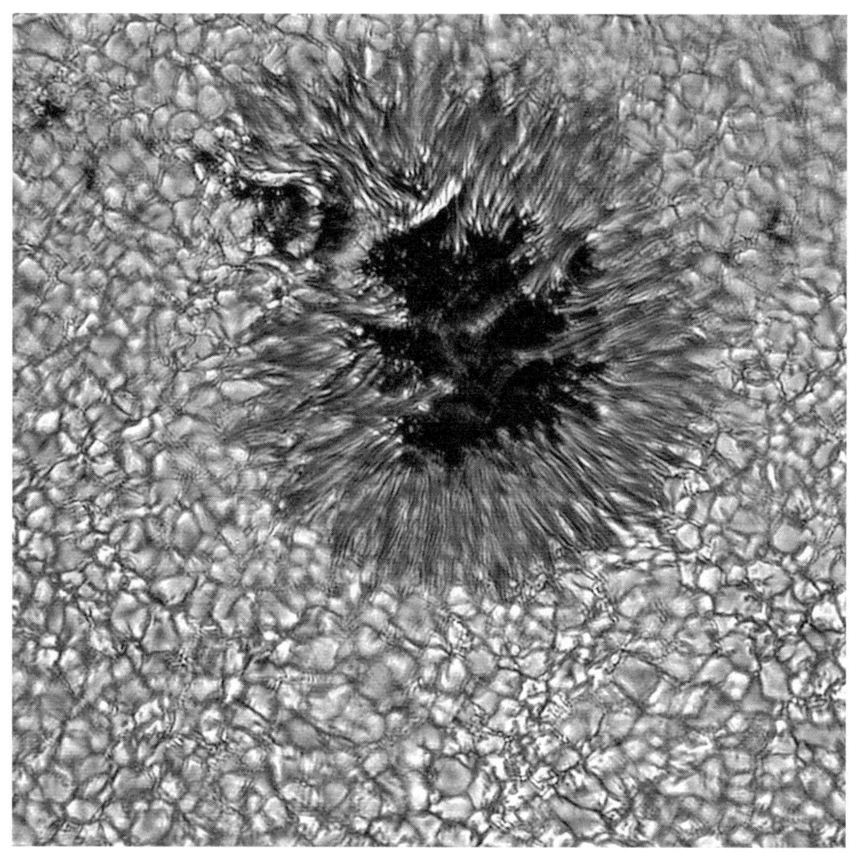

Figure 7.2. The cells and granulation seen in this image of the surface of the Sun reflect seething, turbulent motions of high temperature plasma. Sunspots occur in pairs and are slightly cooler areas where the photosphere is threaded by an organized magnetic field and high-energy particles and energy can escape. This one is about the size of the Earth (NASA/SOHO).

The Weather Out There

On a typical summer's day in 2011, the current conditions at Spaceweather.com read: "Solar Wind speed 515.9 kilometers per second, 0.3 protons per cubic centimeter." That's a speed of 1,125,000 mph and just one atom in the volume of a sugar cube, or a vacuum a billion billion times thinner than the air we breathe. The website also predicts "X-Ray Solar Flare: 24-hr max: B4." That's a puny flare, thousands of times weaker than the 1859 event.[9] In space weather rankings, B-class solar flares are completely "below the radar" and negligible, C-class X-ray solar flares are small and have very little impact on Earth, M-class flares cause momentary radio blackouts and can affect Earth's Arctic and Antarctic regions, and X-class flares are major events causing planet-wide blackouts and week-long radiation storms. NOAA, the National Oceanic and Atmospheric Administration (more commonly known as the National Weather Service), provides detailed space weather reports updated every ten minutes. But what, one might ask, is the relevance of solar flares for our protected oasis of life on Earth, 93 million miles from the Sun?

Like a lonely sentinel stationed roughly a million miles from Earth at the Lagrangian point L1, NASA's Solar and Heliospheric Observatory (SOHO) keeps a steady finger on the Sun's pulse. Its game is called helioseismology, or the study of the "hum" of the sound waves reverberating throughout the Sun. Scientists use these acoustic probes to better understand its interior structure and energy production as well as possibly predict disturbances at the surface, like sunspots. SOHO has demonstrated with great clarity that solar flares—the sudden explosions of matter, electromagnetic radiation, and high-energy particles from the Sun—can be unimaginably energetic cataclysms (plate 11). As University of Michigan Space Physics professor Mark Moldwin notes, "Flares release tremendous amounts of energy in a few minutes and can reach temperatures of 100 million K (much hotter than even the core of the Sun)."[10] Space telescopes like SOHO and the new Solar Dynamics Observatory (SDO) are giving scientists a front row seat in observing the powerful dynamo of our star.

The Sun has magnetic poles. Approximately every eleven years, the direction of the poles reverses—imagine an object trillions of times bigger than a bar magnet completely reorienting its magnetic field. Such large-scale restructuring unleashes incredible magnetic forces, which lead over a few years to an exponentially greater degree of solar activity and solar storms. The tremendous energetic output at the most volatile stage of the process is known as the solar maximum. John Weiley's IMAX film *Solarmax* (2000) has captivated audiences at planetaria and science museums with SOHO footage of the Sun's violent storms leading up to the solar maximum of 2000 and 2001. The film's amazing time-lapse footage demonstrates our Sun's variability and the explosive energy of coronal mass ejections, plasma and magnetic fields blown off from the Sun's corona that can cause severe magnetic storms on Earth.

The Earth's magnetosphere largely protects us from the solar wind,[11] the Sun's outpouring of highly charged particles. More than that, this protection has been essential for the survival of life as we know it. Mars has a much weaker magnetic field and long ago lost most of its atmosphere and all of its oceans in large part due to the solar wind. However, when they reach dangerous levels, storms in space can cause power grid blackouts leaving millions without electricity, and may produce a rapid buildup of powerful electrical charges on satellites that can fry their delicate instruments. It has only been in the last few decades that we've realized the impact of space weather.

Space weather researchers highlight "evidence of the influence of solar activity on the terrestrial climate."[12] Mark Moldwin points out that in the 350-year span between AD 900 and 1250, during an extended warming period produced by increased solar activity, the North Atlantic experienced much milder temperatures. This prompted Nordic people to establish communities in Greenland and assign the region a "name that seems peculiar now since it is covered by one of the world's largest ice sheets year round."[13] When solar activity subsequently decreased, Greenland froze over and settlers there either migrated or died. Later, during what is often called the Little Ice Age, lasting roughly between 1550 and 1750, historians report "winters were so cold . . . the principal rivers in mid-latitude Europe froze over."[14] This has been attributed to a solar minimum and apparently an extended period in which there were few sunspots.

The Earth naturally has gone through cycles of ice ages, but as Moldwin explains, the most recent, the Quaternary, extended from 2.5 million years ago to about 10,000 years ago, when large reaches of ice receded. The problem we're currently facing is global warming, where temperatures consistent for the last 10,000 years are now on the rise. Moldwin writes, "Model predictions indicate that Earth's climate will be drastically different than it is today in less than 100 years because of the burning of hydrocarbon fuel for transportation and energy. That is very quick compared to the normal timescale of climate change."[15] Added to this is the fact that solar storms can seriously damage the Earth's ozone layer, which protects all life on Earth from damaging ultraviolet rays and could, in turn, further contribute to warming Earth's atmosphere.

NASA created its "Living with a Star" program in 2001 to gather better data for understanding the effects of the Sun on the Earth.[16] If we ever move beyond our planet and engage in routine space exploration, we must leave the protective "bubble" provided by the Earth's magnetic field and atmosphere. Our spaceships and astronauts will have to deal with the Sun and its stormy weather directly. A small armada of spacecraft has been involved in this effort, but the most notable is a very sturdy device that's still going strong seventeen years after its launch.

SOHO and Its Eyrie

On December 2, 1995, the Solar and Heliospheric Observatory (SOHO) spacecraft was launched onboard an Atlas rocket from Cape Canaveral in Florida. It's the size of a minivan, or, with its solar panels extended, about the size a school bus, and it weighs roughly two tons. SOHO was conceived by the European Space Agency; fourteen countries and more than three hundred engineers were involved in its design and construction. NASA was responsible for the launch and ground operations. It's a testament to the power of collaboration that so many nationalities can work together to produce a state-of-the-art scientific experiment.

SOHO is like a Swiss army knife, reminiscent of Cassini in its size and complexity. Its available space is crammed with twelve instruments that can measure everything from magnetic fields to X-rays. Nine instruments are led by European scientists and three are led by scientists from the United States, but all of them have teams that are a patchwork quilt of nationalities. Unnoticed by the fractious world of politics and tribalism, science is one field of endeavor where national distinctions and borders are almost meaningless. Some of the instruments have intimidating names, like the "Comprehensive Suprathermal and Energetic Particle Analyzer" built by the University of Kiel in Germany, but generally they all measure the location, intensity, and spectrum of either high-energy X-ray and ultraviolet radiation or cosmic rays. Multinational harmony does not, however, extend to gender parity. Space astronomy is still male-dominated; all twelve instrument principle investigators and 80 percent of the science team members are men.[17]

SOHO moves around the Sun in step with the Earth, by slowly rotating around a point in space called the first Lagrangian point (L1), where the sum of the Sun and the Earth's gravity combine to keep the satellite locked onto the Sun-Earth line.[18] This position is about a million miles away from the Earth in the direction of the Sun, about four times the distance to the Moon. SOHO's eyrie gives it a ringside seat for watching solar activity. Other space missions are using or plan to use L1, but it's perfect for solar observations since a spacecraft in this orbit is never shadowed by the Earth or the Moon.

SOHO has been beaming data to the Earth from twelve scientific instruments at a gigabyte, or two CD's worth, per day. Analysis of this data has yielded some exciting insights into the Sun, including the first images of a star's convection zone, where energy is carried from the fusion core to regions near the surface, and the structure of sunspots just below the photosphere. SOHO has also provided the best measurements to date of the temperature, rotation, and gas flow patterns within the Sun, and it has revealed new types of solar activity, such as waves in the corona and tornadoes on the surface. As of late 2012, SOHO data had been used to discover over 2,350 comets.[19] The spacecraft had a nominal lifetime of two years. In 1997, it was extended for five years due to its great success. In 2002, it got another five-year extension, and in 2009 it got a third extension, until the end of 2012. SOHO is now well into its second solar cycle of observations. However, it was not always smooth sailing.

Spinning Out of Control

It's a space scientist or engineer's worst nightmare. Your mission is working flawlessly and then, due to oversight or unforeseen events, its starts tumbling out of control. Several hundred million dollars' worth of hardware turns a blind eye to its scientific mission. Deaf to telemetry and far beyond the reach of astronauts, it languishes uselessly in deep space.

On a June evening in 1998, mission controllers at NASA Goddard Space Center in Maryland took note as SOHO put itself into Emergency Sun Reacquisition mode. This mode is a safe "holding pattern" autonomously entered by the spacecraft any time it encounters anomalies. From this point, the ground controller sends a special sequence of commands to recover normal science operation, the first step of which involves pointing the spacecraft at the Sun so it knows where it is. Since SOHO had entered this mode five times since its December 1995 launch, the controllers weren't too worried. But this time was different. Over the next few hours, a series of mistakes seemed to doom the spacecraft, first causing loss of attitude control, followed by an interruption in crucial telemetry, then loss of power, and finally loss of all thermal control. It was a bleak day, and it appeared to be the death knell for the young mission. In the control room, the tension grew in waves.[20] At the first entry into Emergency Sun Reacquisition (ESR) Mode, engineers calmly issued the commands to point SOHO at the Sun by carefully orchestrating its three roll gyroscopes. Project scientist Bernhard Fleck said there was no undue concern. Even when a second ESR was triggered a few hours later, there was no panic; engineers had seen this before too. But as the roll thrusters were firing to point SOHO at the Sun, a third ESR was encountered. Out in deep space, a million miles from Maryland, SOHO was spinning faster and faster. Then the communication link went down, presumably disrupted by the spacecraft's wild motion. SOHO was unreachable.

The subsequent investigation showed that the spacecraft performed as designed; all the errors were human. In hindsight, Fleck thinks they gained a false sense of security after two and a half years of operations.[21] Like pilots who had done many takeoffs and landings and who had flown on sunny days and through thunderstorms, they thought they had seen it all. They didn't realize they were in a nosedive.

With the spacecraft seemingly lost, NASA and ESA quickly convened a review board to issue a diagnosis and post mortem.[22] The review board concluded that seat-of-the-pants decision making in the control room exacerbated the problem. Controllers erroneously removed the functionality of SOHO's normal safe mode and misdiagnosed the state of two of the three gyros. At any time during the mishap, if they had verified that Gyro A was not controlling the roll angle properly, they could have avoided the serious problem. The mundane and somewhat sheepish conclusion: when a very complex system is made and operated by human beings, sooner or later they're going to do something wrong.

Ground controllers kept trying to send messages to SOHO using the Deep Space Network, but they had heavy hearts. The spacecraft was in an uncontrolled spin, possibly at a rate that could cause structural damage. Engineers believed it was spinning with its solar panels edge-on to the Sun and so not generating any power. As a result, the batteries and the onboard fuel would have frozen into a state from which they could not be recovered. But there was a window of opportunity that admitted a sliver of hope. SOHO's steadily changing orbit with respect to the Sun was increasing the illumination on the solar panels for a few months and they would give the batteries a chance to recharge. Twelve hours a day, controllers "pinged" SOHO in the hopes of getting a response. Then, a month after contact with the spacecraft had been lost, a break. The huge 305-meter radio dish at Arecibo was able to bounce radar off SOHO and the refection was picked up by the Deep Space Network dish at Goldstone. SOHO was spinning at a modest rate of one revolution per minute. Communicating with the frozen spacecraft was not easy, but six weeks after contact had been lost a feeble signal was received. SOHO was alive.

Over the next two months, engineers clawed SOHO back from the brink. They thawed out its batteries and fuel lines and were gratified that none of the scientific instruments seemed worse for wear after having been subjected to temperature variations from +100°C to –120°C. Some of the instruments even improved their performance after experiencing this bracing range of temperatures. SOHO was back, but two of the three gyros were not responsive due to the series of mishaps. As luck would have it, the third gyro failed a few months later. Although the gyroscopes were not needed to gather science data, maintaining a Sun-oriented attitude used valuable fuel and lowered the margins of safety if anything else should go wrong. Undaunted, the engineers made lemonade with their lemons. They developed special software to enable the spacecraft to point using the control reaction wheels, the first time an ESA spacecraft had ever been operated without gyroscopes. For all the detailed planning, SOHO ended up living by its wits in a series of knife-edge decisions.

Harmonies of the Sphere

The Greek mathematician Pythagoras imagined a universe described by numbers. When he talked about the "harmonies of the spheres," he meant music that enlightened individuals might hear resulting from the translucent shells that carried the celestial objects. Kepler continued this line of thought and applied it to the elliptical orbits of the planets. These ideas sound archaic but they're not misguided. The universe contains many periodic and oscillatory phenomena, and the formalism to understand them involves studying the resulting harmonic frequencies.[23] Situations as different as planet, moon, and ring systems and their orbits and interactions, the spiral arms of the Milky Way, and the interactions between matter and radiation in the early universe are well described in terms of coupled frequencies and harmonics. We've seen beautiful examples of this in the complex gravitational dance of Saturn's rings and moons, discussed in the Cassini chapter.

A hundred years ago, nobody imagined that the Sun could be studied in terms of harmonics. Apart from the sunspot "blemishes," the surface seemed smooth and featureless. Solar properties vary smoothly through the region we see as the "surface." The edge marks the distance out from the center at which the density of gas reduces to the point where light no longer interacts with particles and travels freely. Inside this region, which is called the photosphere, light is trapped and so the Sun's interior is hidden from view. The first inkling that the interior was pulsating came when George Ellery Hale built his heliostat on Mount Wilson in the early twentieth century. High-magnification photographs showed a surface mottled with fine structure, and time sequences revealed that the Sun's surface was a seething sea on which the pairs of sunspots floated like large lily pads. The field of helioseismology matured as the century progressed, and solar scientists identified thousands of different oscillatory modes—the Sun "rings" like a bell. The Sun is also like an echo chamber (plate 12). Sound travels through the plasma and sets up standing waves, like the vibrations of the head of a drum or the air inside an organ or woodwind instrument.[24]

Just as seismologists can infer the internal structure of the Earth from the way sound waves and earthquake tremors pass through the planet, helioseismologists can study the Sun by seeing how interior sound waves manifest at the surface. This work has led to measurements of the density, temperature, and chemical abundance of the interior, as well as inference of the age of the Solar System and the constancy of the gravitational constant.[25] The workhorse instrument on SOHO is the Michelson Doppler Imager since it shows the oscillations of the entire Sun. This instrument discovered a layer about a third of the way to the Sun's center where the orderly interior, within which energy flows radially, transitions to the turbulent outer region, where energy moves in convective loops. This is the place where the solar magnetic field is created. Just as large-scale flows like the Gulf Stream and the jet stream are important for the Earth's climate, SOHO data have shown that such flows are important for solar weather.

SOHO's data is of such high quality that 3D maps of the Sun were derived for the first time. The maps answered questions that puzzled Galileo: how deep do sunspots extend and how can they survive for weeks at a time? The answers: they are fairly shallow but they are rooted in places where the plasma converges and strongly flows downward. SOHO scientists have managed the amazing trick of holographically reconstructing features on the far side of the Sun.[26] All these images are available daily on the web. In fact, a plethora of solar data is available online, since a small armada of satellites is monitoring our life-giving star all the time.

The Rowdy and Quiet Sun

People have become familiar with the fact that climate change not only involves global warming but also greater extremes in weather. Rare but damaging events like tornados and hurricanes and floods appear to be increasing.[27] Another type of weather that can impact human affairs is space weather. In addition to emitting a continuous stream of superhot hydrogen and helium called the solar wind, the Sun periodically ejects billions of tons of plasma in a cataclysm called a coronal mass ejection. Researchers think this happens when magnetic fields are stretched and then snap into a different arrangement, like rubber bands pulled to the breaking point.[28] The ejections are huge clouds of material, as much as 10 billion tons of hot gas, and when they head toward us, they can cause fierce magnetic storms in the upper atmosphere. This extreme "weather" in space can impact the functioning and the reliability of technology on the ground and in low Earth orbit. Strong electromagnetic fields resulting from space weather induce surging currents in electrical wires, disrupt power transmission lines, cause widespread blackouts, and affect the infrastructure that forms the backbone of the Internet. Power grids are particularly vulnerable to electrical overloads caused by space storms. Extremes of space weather also dislocate the radiation belts, and damage the satellites used for essential functions like telecommunications, weather forecasting, and GPS or global positioning systems.

Throughout history, people at high latitudes have experienced the Sun-Earth interaction through auroras, but the effect is not always benign. We've already seen that in the dog days of summer in 1859, telegraph wires in the United States and Europe spontaneously shorted out and caused numerous fires. Soon after, the northern lights (or Aurora Borealis) were observed as far south as Rome and Hawaii. This was a "perfect storm" in space, where several changes conspired to produce a mammoth solar flare. The 1859 solar storm was the strongest ever recorded and it occurred when electrical technology was in its infancy. A weaker storm in 1921 induced ground current strong enough to shut down the New York City subway system. Much weaker storms in 1989 and 1994 knocked out communication satellites and parts of the North American power grid.

Arthur C. Clarke's visionary proposal in 1945 in a letter to the editor of *Wireless World* to place communication and television satellites in geostationary orbit helped launch the Information Age. As of October 2009, the U.S. Satellite Database listed over 13,000 operating satellites in Earth orbit.[29] In 2001, it was estimated that the satellite fleet in Earth orbit was worth $100 billion.[30] During a single solar storm, billions of dollars of this orbiting hardware can be destabilized or destroyed. After the 1989 storm, Sten Odenwald reports that "satellites in polar orbits actually tumbled out of control for hours" as others attempted to "flip upside down."[31] John Freeman explains why solar storms can decimate satellites despite their sophisticated electronics: "Storms in space produce miniature lightning storms on the surface of the satellite" and "immerse the satellites in a 'cloud' of hot electrons" that mimic commands from ground operators but are in fact phantom commands.[32] If the cost of a fleet of satellites seems detached from our everyday existence, consider days or weeks without Internet access or GPS for vehicles, or communication and navigation systems for airline and ocean liner transport, cargo shipping, and road transport. Solar storms can even affect oil pipelines, which undergo accelerated corrosion from the bombardment of charged particles streaming from the Sun, eventually causing breaches in the pipeline.[33]

Space weather prediction remains an emerging science spearheaded by NASA, ESA, and the United Nations Office of Outer Space Affairs. Data from SOHO, and increasing concern over the impact of space weather, caused NASA to commission a new study in 2009. The resulting report provides clear economic data to quantify the risk to the near-Earth environment from episodes of intense solar activity. Extreme space weather is in a category with other natural hazards that are rare but have far-reaching consequences, like major earthquakes and tsunamis.[34] It's likely that more than once in the next twenty years there will be an "electrojet disturbance" that disrupts the national power grid. In the 1989 event, the loss of some portions of the grid put stress on others and led to a cascade affect. The end result was power outages affecting more than 130 million people and covering half the country.

SOHO cannot prevent these natural disasters, but it can give two or three days' notice of Earth-directed disturbances. And as we become more accurate in anticipating space storms, operators can place satellites in protective modes, shut down or limit power grids, redirect commercial flights, warn oceanic cruise and cargo ships, and place astronauts working on the International Space Station in the safest possible location on the station. Such steps will not only save lives but also protect the information systems that sustain our electronically fragile and networked global community.

Although SOHO's bread and butter is looking for the high-energy radiation that's a hallmark of solar activity, there was a recent concern that the Sun is becoming *too* inactive. Through 2008 and 2009 there were fewer and fewer sunspots. In part this was due to the natural solar cycle, but the calm was eerie. Entire months would go by without a single blemish on the Sun, and the solar minimum was as deep as it has been in a century. Bill Livingston and Matt Penn of the National Solar Observatory made the unsettling discovery that the magnetism of sunspots is declining by about 50 Gauss per year.[35] Since sunspots have no substance—they're just dim, cool markers of a concentration of magnetism—the worry is that the magnetic field would drop below 1500 Gauss, at which point no sunspots can form at all. Did decline like this presage a return to the Maunder Minimum of the late seventeenth century, when the European climate chilled markedly? Luckily, in 2010 normality asserted itself and the sunspot numbers started gradually rising.

The Sun is also fairly quiet in terms of the variation of its energy output. From peak to trough, the variation in total solar flux is only 0.1 percent and no long-term variation has been seen over thirty years. The extreme ultraviolet tail of the dog wags much more: 30 percent within a few weeks and a factor of 2 up to 100 over the whole solar cycle, depending on wavelength.[36] This has important implications for our understanding of the Earth's climate, since human-induced change must be distinguished from natural variations in solar output.

Perhaps one reason the Sun is so equable is the fact that it steadily snacks on comets. One of the surprises of the SOHO mission has been its prowess as a comet-hunter; the trawl is already well over two thousand. The Large Angle and Spectrometric Coronograph blocks the Sun's face with an occulting disk, creating an artificial eclipse (figure 7.3). Most of the comets are "sun-grazers" that fly like Icarus very close to the Sun and then grow tails as their icy cores are heated. Toni Scarmato, a high school teacher from Calabria in Italy, discovered the one-thousandth comet in 2006, and two-thirds of them have been found by scouring coronograph data just after it's taken and put on the web.[37] SOHO found its two-thousandth comet in late 2010 and it has discovered half of

Figure 7.3. Comets travel on high elliptical orbits and some are swallowed by the Sun on one of their inward trajectories. This image uses an occulting disk to block out the Sun's light, revealing a faint comet, the 2,000th to be detected by SOHO. It was discovered by an amateur astronomer who has found over a hundred using SOHO data (NASA/SOHO/Karl Battams).

the comets for which orbits have been measured since 1761. The dance of fire and ice has become a way that the public has embraced this protean space mission.

Aurorae, the Sun, and the Arts

The Aurora Borealis and Aurora Australis occur when electromagnetic particles from the solar wind collide with or excite atoms in the Earth's magnetosphere (figure 7.4). We now understand the

Figure 7.4. Aurora Australis as seen from the Space Shuttle, near the solar activity maximum that occurred in 1991. The aurora is caused by sheets of glowing gas, energized by high-energy solar radiation, and it extends up to an altitude of 300 miles. The light show is accompanied by more damaging effects to orbiting spacecraft and satellites (NASA/Earth Science and Image Analysis Laboratory).

explicit connection between auroras and magnetic storms, and recently have confirmed the "three-century old theory that auroras in the northern and southern hemispheres are nearly mirror images—conjugates—of each other."[38] Very few people in human history have inhabited the high (or low) latitudes where these brilliant light displays are usually visible. For thousands of years the ethereal, vertical curtains of light, draped from 60 to 200 miles above the Earth's polar regions, have inspired wonder, as well as poetic and musical responses. The Vikings first recorded the Aurora Borealis in AD 1250. About seven hundred years later, in 1897, Norwegian explorer Fridtjof Nansen wrote of the northern lights: "It was an endless phantasmagoria of sparkling color, surpassing anything that one can dream. Sometimes the spectacle reached such a climax that one's breath was taken away; one felt that now something extraordinary must happen—at the very least the sky must fall."[39]

The twentieth-century composer Edgard Varese worked with "found sound," incorporating sounds with a non-musical origin into his music, and with electronic instruments such as the Theremin. He once composed a piece based on having seen the Aurora Borealis. His wife Louise recalled in her memoir of the composer, "Nature in its most magnificent and terribly impersonal aspects moved him passionately." Titles for many of his compositions were drawn from astronomy or science. Of the aurora, Louise writes that Varese claimed he "not only saw but heard" the majestic luminescent curtains dancing across the night sky and later notated "the sounds that had accompanied the movements of the light."[40] Electromagnetic waves or natural radio waves can only be detected with a radio receiver.[41] If we take Varese at his word, perhaps he experienced some form of synesthesia, so that in fact he had heard the northern lights. In her memoir, Louise Varese speculated that the score of the aurora her husband mentioned was either *Les Cycles du Nord* or *Mehr Licht*, two of several compositions that were later lost or destroyed.

More recently, Terry Riley and the Kronos Quartet's chamber music composition *Sun Rings*, inspired by the Sun's dynamic magnetic field, was written to be performed against a backdrop of IMAX-sized images from SOHO and the TRACE (Transition Region and Coronal Explorer) satellite. Riley's score is accompanied by magnificent sunspots tracking across the face of the Sun as well as finely detailed footage of magnetic field lines breaking through, looping above, and re-submerging into the Sun's surface. Such contemporary compositions speak to our deep, primal entanglement with the Sun, which is no less important to our lives now than it was in the ancient past.

From prehistory to the modern era, humans intuitively understood the Sun as integral to their very existence. Paleolithic and Neolithic communities scattered across the globe knew that the Sun powerfully shaped their lives. Evidence of their sense of the Sun's significance is found at the Mnajdra temple on Malta, at Newgrange in Ireland, at Stonehenge on Salisbury Plain, and at many other megalithic structures. Such monuments speak of an ancient past in which peoples from disparate times and places were attentive and accurate observers of our star.

On the island of Malta, just off the coast of Sicily, are the remains of a complex of stone temples known as Mnajdra, dating to

5500 BC. One of these limestone edifices marks precise alignments with the Sun during fall and winter equinoxes.[42] Physicist Guilio Magli describes Mnajdra as a "stone calendar" marking spring and autumn equinox: "In the course of the seasons, one can follow the movement of the Sun, which rises on the horizon, observing day by day at which point the light strikes the altar inside the temple."[43] In walking through Mnajdra, sited as it is overlooking the Mediterranean Sea, it's easy to recognize that the location and the megalithic temple were sacred to its ancient architects. Built with cleverness but no metal tools, the effort involved in creating the edifice was prodigious, and a reminder of the investment ancient peoples made in giving homage to the Sun.

Stonehenge, the remarkable structure ancient Britons built on Salisbury Plain sometime between 3015 and 2400 BC, is so well aligned that "the general orientation of the axis of the monument [looks] . . . towards sunrise at the summer solstice in one direction, and towards sunset at the winter solstice in the other."[44] Caroline Alexander observes that, in coming upon Stonehenge, "the great-shouldered silhouette is so unmistakably prehistoric that the effect is momentarily of a time warp cracking onto a lost world." Alexander contends that the architraves atop the monoliths, "bound to their uprights by mortise-and-tenon joints taken straight from carpentry, [are] an eloquent indication of just how radically new this hybrid monument must have been. It is this newness, this assured awareness that nothing like it had existed before, this revelatory quality that is still palpable in its ruined stones." Though the stone temples at Göbekli Tepe in Turkey are far older and date to approximately 9600 BC, no prehistoric structure like Stonehenge exists anywhere else in the world. Alexander surmises that the Britons who constructed the monument "had discovered something hitherto unknown, hit upon some truth, turned a corner—there is no doubt that the purposefully placed stones are fraught with meaning."[45] Emphasizing the point that the architects deliberately aligned Stonehenge to mark the winter and summer solstices, Magli writes, "This is therefore the only information that the builders left us in writing. Granted it is written in stone, and with stone, and in the language of the sun and of the stones. But it is nevertheless written."[46]

Diodorus of Sicily, a Greek historian from the first century BC, supposedly commented on "a lost account set down three centuries earlier, which described 'a magnificent precinct sacred to Apollo and a notable spherical temple' on a large island in the far north, opposite what is now France."[47] Stonehenge, Diodorus suggested, was constructed to pay homage to the Sun. As with the relics of any prehistoric culture, multiple interpretations are possible. Archaeologist Mike Parker Pearson of the Stonehenge Riverside Project has recently reinterpreted Stonehenge as a burial monument. He contends that the people who raised the monoliths on the chalk downs apparently gathered each winter solstice to mark the setting Sun as the beginning of a new year and to remember their ancestors.

Astronomer Edward Krupp has investigated the sophisticated and sizeable burial chamber in Ireland known as Newgrange or Bru na Boinne: "Newgrange is a megalithic surprise," writes Krupp, who characterizes the structure as emerging from the landscape "like a highway tourist attraction."[48] During winter solstice at this stone monument, the first rays of the Sun illuminate a room deep in the structure to mark the beginning of the new year. Dating from about 3700 to 3200 BC, Newgrange was designed so that "two weeks either side of the winter solstice, the Sun, on rising, shown down the length of the entrance passage and illuminated the central chamber—as it still does."[49] The beam of sunlight, as it travels deep into the monument, is thinned and sculpted by the megaliths' calculated placement. To offer a vivid picture of the massive effort required to construct the edifice, Magli emphasizes that "5000 years ago, someone built a monument involving thousands of tons of earth and rock, covered it with quartz like a giant jewelry box, [and] carefully measured the direction of the sunrise at winter solstice to line up a corridor built with stones as heavy as many elephants together."[50] Not surprisingly, the legend associated with Newgrange recounts tales of the earliest known Irish gods, "The Lords of Light."[51]

The Chankillo complex in Peru predates the Inca by two thousand years, offering insight into the precursors to Sun worship among the Inca and their official Sun cult. In 2007, Ivan Ghezzi, Peru's national director of archaeology, and astronomer Clive

Ruggles reported discerning the layout of what is considered the oldest solar temple in the Americas. Ghezzi was first to surmise the purpose of thirteen towers on a ridge near Chankillo, a fourth century BC ceremonial complex in the northern coastal, desert region of Peru. Traditionally considered a fortress, Chankillo is now understood to be a very precise solar observatory built by the people who predated the Inca. Ghezzi realized that the prominent line of stone towers were markers that indicated the Sun's position throughout the months of the calendar year. He contacted Ruggles and together these researchers, using hand-held GPS devices, determined that the location of the towers as projected against the horizon "corresponds very closely to the range of movement of the rising and setting positions of the Sun over the year." In particular, winter and summer solstice alignments are clearly marked by the towers. Ghezzi and Ruggles have shown that the towers and the gaps between them offered "a means to track the progress of the Sun up and down the horizon to within an accuracy of two or three days."[52]

One import of the findings at Chankillo, write Ghezzi and Ruggles, is that "sunrise ceremonies, at a sanctuary on the Island of the Sun in Lake Titicaca, surrounding a crag regarded as the origin place of the Sun, almost certainly had pre-Incaic roots."[53] Either the Inca, or the peoples who predated them, named an island in Lake Titicaca as Isla del Sol, or Island of the Sun, in honor of the god who created the Sun, Moon, stars, and humankind. It is well known that at the high Andean city of Machu Picchu, constructed in the late 1400s, the Inca kept accurate records marking the winter solstice, the first day of the Inca year. But the Inca may have adopted their attentiveness to the Sun from peoples who far predated their great civilization.

Long before the Neolithic Britons were raising those remarkable trilithons on Salisbury Plain, the Egyptians had already developed writing and record keeping and were constructing the Great Sphinx at Giza, apparently in recognition of the god Horus or Horemakhet, believed to be a personification of the Sun on the horizon.[54] An even earlier instantiation of the Sun god Horus was Ra, the patron god of the ancient Egyptian city Heliopolis, located in the Nile delta. In approximately 2400 BC, Ra was combined with the god of Thebes to become Amun-Ra, the highest deity in the Egyptian pantheon. These are a few of the varied civilizations that recognized the importance of the Sun to their survival. In the Information Age, we're just as dependent on our star as ancient peoples—maybe even more so.

The Once and Future Sun

Talan Memmott's *Lexia to Perplexia* is an online fictional hypertext project that interleaves conventional writing with programming code for Html and Javascript to explore the ways human culture has been shaped by emerging digital communication technologies. Memmott writes: "The Earth's own active crust we are, building, building—up and out—antennae, towers to tele*."[55] Memmott evokes an interesting concept. Humans have produced an electronic crust, or an information and technology layer, over Earth's surface and extending into orbit. This layer of electronics is comprised of technologies ranging from radio and television relay stations perched on mountain tops to fiber-optic lines in homes and businesses, from cell towers dotting the high ground to transoceanic cables in the ocean depths. This data-rich envelope extends from backyard satellite dishes to powerful astronomical radio telescopes lined across the desert in Socorro, New Mexico, to billions of dollars in satellite hardware in low Earth orbit.

This infrastructure transmits information to cell phones, radios, televisions, computers, global positioning devices, emergency service centers, hospitals, and weather reporting stations, etc. Memmott writes, "I spread out—pan—send out signals, smoke and otherwise, waiting for Echo."[56] The point is that since the ancient past, humans have extended their communication capabilities over larger and larger distances, through technologies that today transmit information around the globe at the speed of light. But this diaphanous "skin" is sensitive to the conditions of the space environment just as our skin is sensitive to the Sun. Never before have we been so dependent upon an understanding of the inner workings of the Sun and its impact on the electronics that sustain our information-based culture. As we continue to expand and rely on this electronic, information layer encasing the Earth, we're increasingly impacted by the Sun's powerful magnetic reach.

Evolutionarily, we've adapted to living with our star. As John Freeman points out, humans, like other mammals, insects, and plants, have evolved "sense organs that can make use of the Sun's outward flood of electromagnetic radiation. It's not an accident that our eyes are sensitive to the same portion of the electromagnetic spectrum where solar radiation is most intense."[57] Similarly, our skin is well adapted to sunlight in a number of ways, one being that in about fifteen minutes of exposure our skin absorbs the daily recommended amount of vitamin D. Our lives are intimately bound up with the Sun, and not just because of our need for its light and warmth. The iron in our blood was forged inside massive stars over 4.5 billion years ago and then surfed the blast waves of supernovae into interstellar space. Stars like the Sun spewed heavier elements into space that eventually coalesced into our Sun and Solar System. Harlow Shapley popularized this concept in the early 1900s by claiming that we are made of "star stuff." The human body, as all life on Earth, is comprised of carbon, calcium, oxygen, and other heavy elements forged in the cores of stars that exploded long before our Sun was born. Given how much more there is to it than meets the eye, it's fitting that one of NASA's initial Braille books for the blind focused on the Sun.[58]

The Solar Dynamics Observatory, launched in early 2010, is carrying on SOHO's work with even greater accuracy in examining the Sun's interior and interpreting the sound waves traveling inside and across its surface. Part of NASA's "Living with a Star" program, SDO is tracking magnetic fields within the Sun in hopes of discovering the mechanism that drives the Sun's eleven-year cycle. SDO is sending data to Earth at a rate a thousand times faster than SOHO, equivalent to downloading 300,000 songs a day. The satellite is 50 percent heavier than SOHO and views the Sun in high definition, or nearly IMAX quality, taking a picture in eight different colors every ten seconds.[59] It's serving as a first alert against magnetic storms sweeping over our fragile home in space.

We've learned from SOHO and other missions that the rock-steady light from the Sun, varying by less than a percent from year to year or decade to decade, is not the whole story. In invisible forms of radiation, the Sun is epic and Byzantine in its behavior, and scientists have not fully understood this apparently simple, middle-aged and middle-weight star. Scattered through the Milky Way galaxy, there are an estimated hundred million habitable Earth-like worlds orbiting Sun-like stars, and each will have its own complex relationship with its parent star.[60] Our Sun and the space weather it produces will determine the future of our species as well as that of all life on Earth, even the planet itself. Having sustained our world for billions of years, the Sun is still a devoted protector and guardian. It reaches out across a hundred million miles to cradle, caress, stroke, and occasionally, scold us.

8 Hipparcos

MAPPING THE MILKY WAY

A 1939 BIOGRAPHY OF ALBERT EINSTEIN offers a poignant example of the perspective that helped shape the famous scientist's relativistic view of Earth in space: "The world is moving along rapidly in space: your office in the morning will not be where it was when you left it at the close of business. It will never be in the same place in space again!"[1] Indeed, the Sun plunges daily some 12.5 million miles through the empty wastes of space, never to return to its former location. The Earth orbits the Sun at roughly 67,100 miles per hour even as the Sun roars around the Milky Way galaxy at about 490,000 miles per hour. Meanwhile our galaxy is reeling toward the Virgo Cluster, the largest and nearest cluster of galaxies, roughly 54 million light-years distant, at a speed of about 864,000 miles per hour. But in relation to the background radiation of the universe, the Milky Way is racing through space at approximately 1.3 million miles per hour.[2] As astronomers grapple to understand our place in the universe, every second, our planet, Sun, and Solar System, as well as the Milky Way galaxy, whirl blindly into the depths of outer space.

It's some consolation, however, that our neighboring stars and galaxies are barreling into the unknown abyss along with us. The sameness of the night sky has been recorded for millennia and is of great value to humankind. Long before written records, the positions of stars were used as directional aids in traversing the unmapped and largely unpopulated expanse of Earth's surface, its deserts and wastelands, and in navigating uncharted seas. The same stars traced the seasons of Earth's passage on its perdurable orbit around the Sun. When humans as a species were first forming words, the rising, setting, and annual return of the stars and constellations in the night sky must have offered a sense of permanence in a savage world. Deeply embedded in our primal imagination were the stars as guides in navigating novel landscapes, locating seasonal fruits and vegetables, following migratory animals that provided food and pelts, and preparing for oncoming seasons.

Much older than our Sun, the Milky Way galaxy began to form approximately 13 billion years ago and is a barred-spiral comprising some 200–400 billion stars. Its spiral arms, strewn with massive clouds of gas and dust coalescing into newborn stars, sweep in magnificent arcs around the millions of stars comprising the galaxy's central bulge. Because of the scale of this whirlpool of dust and planetary systems, from our perspective neighboring stars appear to form a fixed pattern of constellations and they only change their relative positions over millennia (plate 13).

Imagine if our planet circled a star that in turn was orbiting within a globular cluster (among millions of stars clustered to form a soft-edged spherical structure), all the nearby star patterns would change over a lifetime, possibly never to recur! The night sky would be disorienting rather than a familiar point of reference. The seeming sameness of the night sky is a result of nearby stars hurtling along with our Sun as it circumnavigates the Milky Way every 226 million years, their high speeds and small relative motions reminiscent of racing cars on a circular track, but on a galactic scale. Of these stars, even the closest are so unimaginably far away, their actual motion through interstellar space is all but imperceptible. Hipparcos mission lead scientist Michael Perryman explains:

> The bright stars forming Ursa Major, for example, one of the largest and most prominent of the northern constellations, known variously as the Big Dipper or the Plough, look the same now as they did hundreds of years ago—Ptolemy listed it, Shakespeare and Tennyson wrote about it, and Van Gogh painted it. And they will look just the same to our children, and to theirs. But to earliest humanity, a

> hundred thousand years ago, and to those equally far in the future, the constellation would be unrecognisable, grossly distorted from its present shape.[3]

In the prehistoric past, the stars were beyond human investigation and perhaps even comprehension. Their extreme remoteness compared to our neighboring planets prevented us from initially realizing their true nature. Over time we recognized that stars are fundamentally like our Sun, replete with worlds we are only now surveying. As will become clear, the apparent sameness of the night sky over long periods of time has been an invaluable natural phenomenon for humankind coincident with our very survival.

Ancient Star Catalogs and Sky Maps

From time immemorial we have projected our stories, myths, and legends onto the night sky. Seeing patterns among the stars was a serious preoccupation from primordial time. Our shared narratives about the constellations, across cultures and millennia, served as a survival mechanism not to be underestimated or discounted as simple tales of mythological figures. Biocultural theorist Brian Boyd claims that human attention to pattern emerged as an evolutionary adaptation. Boyd and other scholars contend that arts relying on patterns such as storytelling, song, and cave painting emerged via natural selection and allowed individuals and clans to better collaborate and share information that enhanced survival.[4]

Recent analyses of prehistoric paintings and markings in caves in France have revealed a series of twenty-six symbols that may reflect humankind's earliest attempts at pictographic writing, traditionally thought to begin about 3,000 BC. Anthropologists Genevieve von Petzinger and April Nowell collated a database of cave signs from 146 sites in France dating from 35,000 to 10,000 years ago. "What emerged was startling: 26 signs, all drawn in the same style, appeared again and again at numerous sites."[5] The symbols range from straight lines, to circles, spirals, ovals, dots, Xs, wavy lines, and various hand symbols among others. Similar symbols have been found at Paleolithic sites around the world.

Certain symbols studied in France frequently appear in deliberate groupings, such as the occurrence of dots with a particular hand symbol, which may indicate the beginnings of a system of writing. As Nowell speculates, "We are perhaps seeing the first glimpses of a rudimentary language system."[6]

The celebrated Lascaux cave paintings not only incorporate these symbols, but may also include representations of the Pleiades and Hyades star clusters and an ice age panorama of the night sky. Clive Ruggles and Michel Cotte, who in 2010 headed the International Astronomical Union's Working Group on Astronomy and World Heritage, reported to UNESCO that some archaeoastronomers contend a series of dots above the aurochs in the Hall of the Bulls represents the Pleiades star cluster and that one of the auroch's eyes and adjacent dots may depict the star Aldebaran and the Hyades cluster.[7] The French government's website on Lascaux's cave art notes that in all of the renderings, the horses were painted first, then the aurochs, and then the stags. These animals apparently correspond to the seasons of spring, summer, and autumn, respectively, providing "a metaphoric evocation that, in this setting, links biological and cosmic time."[8] Even more fascinating is Chantal Jèques-Wolkiewiez's assertion that two panels of these cave images may depict the night sky as perceived by Magdalenian people from the top of Lascaux hill during a summer solstice roughly 17,000 years ago. Besides comparing the cave paintings to computer models of the night sky in the last ice age, she also found that light during sunset at the summer solstice stills enters the cave to illuminate some of the paintings.[9]

While it may be very difficult to determine whether the dots in the Lascaux paintings are indeed asterisms, the markings ice age peoples at Lascaux left next to their remarkable paintings enticingly suggest they were meaningful forms of communication. Extrapolating from such biocultural and archaeological research, it seems likely that our attention to patterns seen in constellations in the night sky extends back to our earliest days.

The names of stars and designation of constellations as we know them in Western culture are so ancient that their origins remain elusive.[10] Long before the earliest written records, humankind told narratives about the stars and clustered them into constellations, which served as mnemonics for travel and navigation and as a repository of knowledge about the seasons, but also of legends and myth. Only in the last 3,600 years do we find unequivocal evidence of tracking the stars. Chinese rulers employed court astronomers to record information regarding stellar motions, transients, and astronomical events such as supernovae. But even older than ancient Chinese records is the Nebra sky disc. Discovered near the town of Nebra in Germany, the Nebra sky disc is believed by archaeoastronomers to be a Bronze Age durable sky map dating back to 1600 BC.

The disc was uncovered in 1999 by treasure hunters who hit upon an ancient burial site in a circular earthwork enclosure at the top of Mettelberg Hill. The disc, 30 centimeters or 11.8 inches in diameter, was found with two bronze swords among other items. The sky disc is made of bronze with gold overlays of the Sun, the Moon in phase, and multiple stars. Gold bands on its sides indicate the east and west horizons and mark an angle of 82.5 degrees. At Nebra, sunset at the winter and summer solstices is visible on the horizon 82.5 degrees apart. As the angular separation of those setting points varies at differing latitudes, some archaeoastronomers are convinced that the disc was constructed in the Nebra region and is the oldest extant sky map in the world. A cluster of seven gold dots on the disc are thought to be the earliest known representation of the Pleiades star cluster, used in the ancient past for identifying seasons of planting and harvest (figure 8.1). Investigation of its metal composition traces the disc to a Bronze Age mine in the Alps. The site where the disc was found is only 15 miles from Goseck, Germany, the location of a Neolithic ceremonial woodhenge dating to 7,000 years ago that researchers say clearly marks the position of sunrise on the horizon on the summer and winter solstices. Archaeologist Harald Meller, who posed as a buyer and worked with Swiss police in a sting operation to capture underground traders attempting to sell the sky disc, points out that it predates "the beginning of Greek astronomy by a thousand years."[11]

The ancient Greek writers Homer and Hesiod knew the names of recognizable stars and star clusters like the Pleiades and Hyades, and of constellations such as Ursa Major, the Bear. *The Iliad* and *The Odyssey*, attributed to the poet known as Homer, remain the

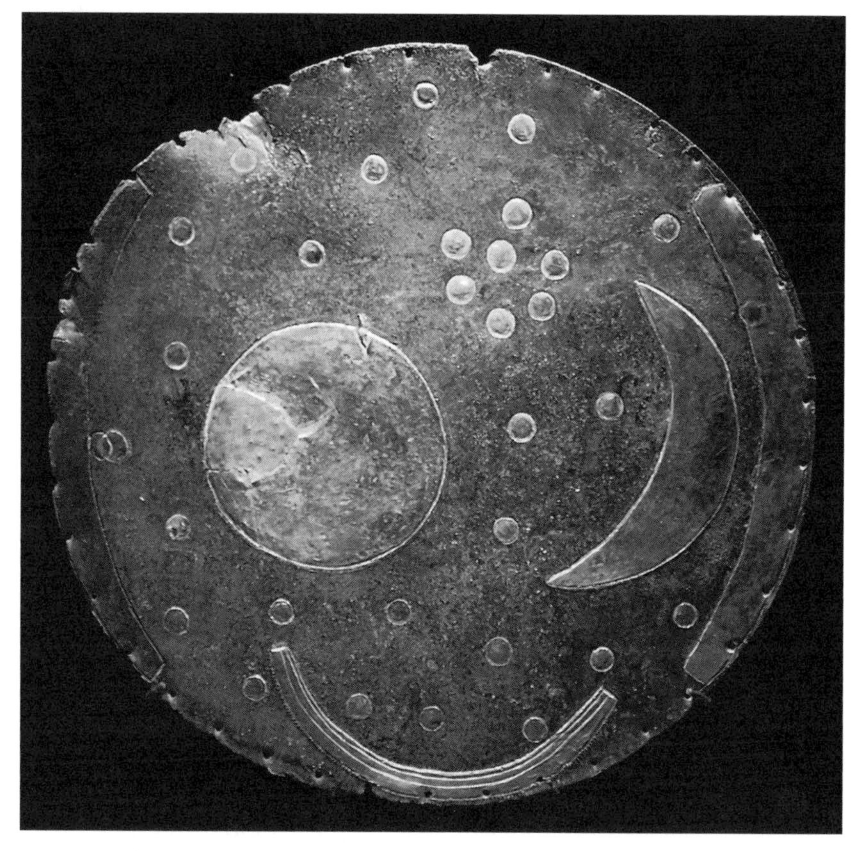

Figure 8.1. The Nebra Sky Disc dates to 1600 BC and is considered the oldest extant map of the Pleiades. Star-mapping was practiced by many ancient cultures, some of which left artifacts with realistic representations. The Pleiades is a nearby, young star cluster still embedded in the gauzy, glowing gas from which it formed, shown as a clump of points just above and to the right of the center of the disc (Wikimedia Commons/Dbachmann).

oldest extant Greek texts we have. Homer's tales were rooted in an oral tradition that extends back centuries prior to their being recorded. In *The Iliad*, traditionally dated to approximately 700 BC, some of the constellations we know today appear on the shield that Hephaistos forged for Achilles:

> He made the Earth upon it, and the sky, and the sea's water,
> and the tireless Sun, and the Moon waxing into her fullness,
> and on it all the constellations that festoon the heavens,
> The Pleiades and the Hyades and the strength of Orion
> and the Bear.[12]

James Evans also notes Homer's description of Odysseus orienting his ships by keeping the constellation Ursa Major, which turns about the celestial north pole, left of his vessels as he sails East.[13]

One significant reason early agrarian societies marked the stars in the night sky was to develop an agricultural calendar, initially important for planting and harvesting. Evans points out that a few generations after Homer, Hesiod wrote *Works and Days*, the opening lines of which directly associate the rising and setting of star groups with agriculture:

> When the Pleiades, daughters of Atlas are rising,
> Begin the harvest, the plowing when they set.[14]

Evans explains that winter wheat was the only wheat planted in Greek antiquity, so that when the Pleiades were setting in the west in late fall it was time to plow the ground and plant the wheat.

Nick Kanas points out that ancient Greek astronomers supplemented their knowledge of the night sky with what they could glean from the Egyptians and Babylonians: "From the Egyptians, they learned about the length of the year, its break-up into a 12-month calendar, the division of day and night into 12 hours each. . . . From the Mesopotamians, they learned a sophisticated system of constellations[,] especially involving the zodiac along the ecliptic."[15] Babylonian temple scribes conducted serious astronomical observation and carefully preserved their records on cuneiform tablets. Many of these astronomical records have been recovered, some of which date to the seventh century BC. A few of the most well known of these tablets are titled MUL.APIN, meaning "Plow Star," the title of which apparently refers to the stars of the Triangulum constellation and the star Gamma Andromedae, not Ursa Major, also known as the Plough, Big Dipper, or the Bear. Evans observes that the MUL.APIN tablets, copies of much older texts, begin with a list of dozens of stars and provide a star calendar indicating the rising and setting of stars at particular times of the year. Also included in the tablets, which are more accurate than Hesiod's agricultural calendar, are observations of various constellations as well as the planets Mercury, Venus, Mars, Saturn and Jupiter, the planet associated with the primary god of Babylon.

Ancient Greek astronomers thought of the stars as fixed or unmovable. However, the Greek astronomer Hipparchus, after whom ESA's Hipparcos mission is named, was an accurate observer who "suspected that one of the stars may have moved, and . . . wished to bequeath to his successors data against which any future suspected movements might be tested."[16] Hipparchus was interested in what today is called astrometry, or the science of measuring the position and motions of stars and other astronomical objects. He produced a star catalog now lost or destroyed. According to Floor van Leeuwen: "The oldest catalog of stellar positions we know of is the compilation made around 129 BC by Hipparchus, a catalog that is still being investigated. Its only surviving copy appears to be a map of the sky on a late Roman statue, and is known as the Farnese Atlas."[17]

For thousands of years, all we've known of Hipparchus's star guide were descriptions by Ptolemy. But astronomer Bradley Schaefer asserts that, indeed, the Farnese Atlas (figure 8.2), a statue of the Greek figure Atlas kneeling while holding on his shoulders a globe of constellations, represents the stars and constellations known to the ancient Greeks. He contends that the statue "is the oldest surviving depiction of the set of the original Western constellations, and as such can be a valuable resource for studying their early development."[18] Schaefer realized after a detailed study of the globe that the constellations depicted match the night sky in the era and from the location where Hipparchus lived in 129 BC. As evidence in favor of this possibility, Schaefer writes: "First, the constellation symbols and relations are identical with those of Hipparchus and are greatly different from all other known ancient sources. Second, the date of the original observations is 125 ± 55 BC, a range that includes the date of Hipparchus's star catalogue (c. 129 BC) but excludes the dates of other known plausible sources." Schaefer concludes that "the ultimate source of the position information [of the constellations on the globe] used by the original Greek sculptor was Hipparchus's data."[19]

Hipparchus was the first to identify the Earth's precession, produced by the gravity of the Sun and Moon on the Earth's equatorial bulge. The "precession of the equinoxes" refers to the gravitationally induced, gradual shift in the Earth's axis of rotation, so

Figure 8.2. The Farnese Atlas is a marble statue from Roman times made from a Greek original, standing seven feet tall. Bradley Schaefer has argued that the star catalog of Hipparcus served as inspiration for the detailed constellations represented on the globe. This interpretation, and dates attached to the statue, remain controversial (Wikimedia Commons/Gabriel Seah).

that the equinoxes occur earlier each sidereal year over the course of 25,765 years, when this cycle of precession begins again. Precession is a changing view of the stars caused by a subtle variation in the Earth's orbital orientation relative to the Sun; it's not related to the kind of stellar movement that ESA's Hipparcos mission has charted.[20] As Michael Perryman explains, the Hipparcos mission is based on the concept of parallax: "The key to measuring stellar distances is actually based on the classical surveying technique of triangulation. It simply makes use of the fact, known since the time of Copernicus, that the Earth moves around the Sun, taking one year to complete its orbit. This yearly motion provides slightly different views of space as we speed around the Sun." We experience the same effect in observing an object by first closing one eye and then the other. Perryman points out that "this stereo vision gives us depth perception and allows us to estimate distances, at least to nearby objects. . . . Astronomers use the same stereo technique, but with views of the celestial sky separated by hundreds of millions of kilometers as the Earth moves around the Sun. In this way, Nature has generously and serendipitously granted us the possibility of measuring distances stretching across the vast expanse of our Galaxy."[21]

Astronomy's Human Genome Project

Michael Perryman has dubbed the Hipparcos mission astronomy's equivalent of the Human Genome Project.[22] Perryman explains that as astronomers more accurately map the location, velocity, and vector of stars in our galaxy we can understand the age and morphology of the Milky Way, how our galaxy has evolved in the past, and what the future holds for our Solar System and the galaxy. For instance, the Hipparcos mission has contributed to our better understanding of the galaxy's current structure. We know our galaxy is not a perfect spiral, but is instead a barred spiral that's warped so that the limbs at one end curve up and at the other bend down (figure 8.3). Another major contribution of Hipparcos, for astronomers and popular audiences, is that the mission improved the estimates of distances to stars harboring exoplanets. In this way, it has crystallized our sense of the growing number of distant worlds in space. We've seen in the earlier chapters on the Solar System that planets and moons are potential abodes for life. As the Human Genome is a project to map the underlying structure of terrestrial life, so Hipparcos is a tool to help astronomers map plausible sites for extraterrestrial life. The search for life beyond the Earth is a foundational scientific pursuit, and it has attracted attention from some unlikely quarters.

The Vatican has maintained an observatory over the centuries in order to officially determine dates of the calendar year; the Gregorian calendar has been used in the Western world since 1582. However, astronomers of the Vatican Observatory more recently

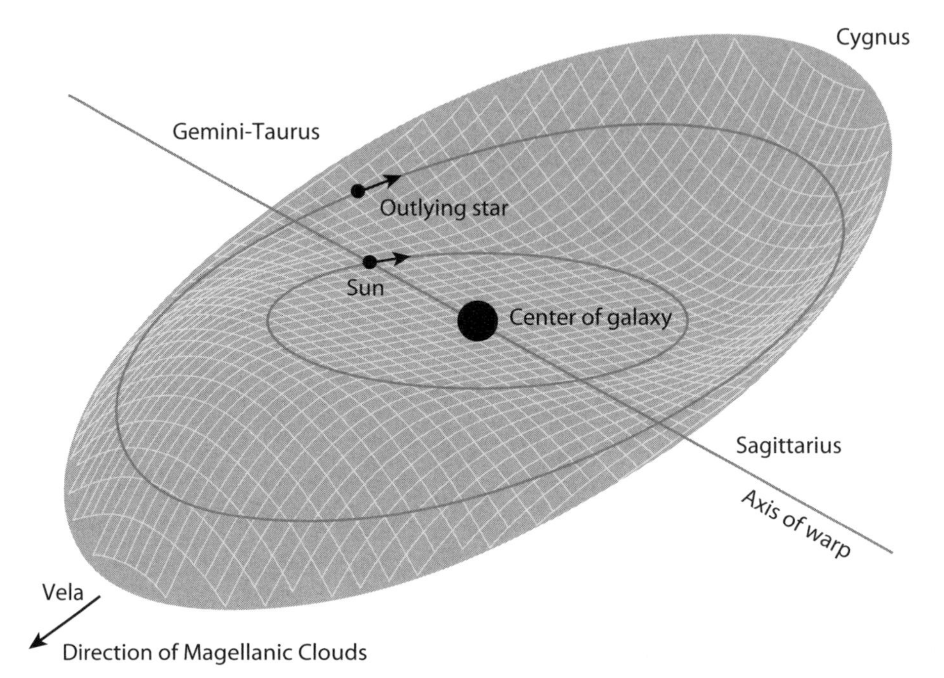

Figure 8.3. Hipparcos measured the positions for hundreds of thousands of stars and so was able to map out the disk of the Milky Way over 500 light-years. This was enough to detect a subtle warp in the disk, exaggerated in this schematic view. The shape of the disk is like a brimmed hat with the brim turned down on one side (ESA/Hipparcos).

have been focusing on other concerns. In November 2009, Pope Benedict XVI called leading astronomers, astrobiologists, and cosmologists to Vatican City to spend a week presenting recent findings regarding exoplanets orbiting nearby stars and to discuss the possibilities of intelligent life in those star systems.

Of the Vatican's interest in exobiology, science reporter Marc Kaufman noted: "Just as the Copernican revolution forced us to understand that Earth is not the center of the universe, the logic of astrobiologists points in a similarly unsettling direction: to the likelihood that we are not alone, and perhaps that we are not even the most advanced creatures in the universe. This . . . may conflict with the stories we tell about who and what we are."[23] During the five-day meeting scientists addressed subjects such as the origins of life, extremophiles and their habitats, the likelihood of such life thriving on moons in the outer solar system, and whether life's biosignatures could be detected on exoplanets.

As yet, exoplanets are mostly gas giants with little chance of life on them, but as the detection limit has reached Earth mass with NASA's Kepler satellite, research spurs scientists, philosophers, and theologians alike to contemplate the implications for our place in the universe. "The questions of life's origins and of whether life exists elsewhere in the universe . . . deserve serious consideration," explained José Gabriel Funes, a Jesuit priest who is also the director of the Vatican Observatory. Co-author Chris Impey, who presented a paper at the meeting and co-edited the written proceedings,[24] comments: "Both science and religion posit life as a special outcome of a vast and mostly inhospitable universe. There is a rich middle ground for dialog between the practitioners of astrobiology and those who seek to understand the meaning of our existence in a biological universe."[25] Reporter David Ariel, who also covered the meeting, aptly noted, "The Church of Rome's views have shifted radically since Italian philosopher Giordano Bruno was burned at the stake as a heretic in 1600 for speculating, among other ideas, that other worlds could be inhabited."[26]

For the moment, most of the vast inventory of stars remains out of reach. But several hundred relatively nearby stars are known to have planets, and Hipparcos has been an essential tool in measuring their distances. These new and potentially habitable worlds range from a dozen to a few hundred light-years away. Spanning the entire galaxy, one estimate is of 8 billion terrestrial habitable worlds around Sun-like stars, each of which has the potential to host life.[27] This number is the same order of magnitude of the number of base pairs derived from the Human Genome Project, making literal the analogy of a vast mapping project to parse life in the Milky Way.

Unsung Heroes of Astronomy

To judge the scientific contributions of Hipparcos, we start by recognizing that measuring the positions of stars is both fundamental and unglamorous. It's fundamental because it's the key to measuring the physical properties of celestial objects. Positions are the keys to the trigonometric determination of distance, and distance is needed to calculate the size, mass, and intrinsic brightness of any planet, star, or galaxy.[28] Without distances we're stuck with the appearance of stars in the sky, and a star that's far away and luminous can appear to be the same brightness as a star that's nearby and dim. That ambiguity is fatal to any reliable understanding of the denizens of the night sky. It's unglamorous because measuring a position is the simplest and most obvious way to characterize a star: there's no image, just two angles to identify a unique spot on the sky, with no units. Needless to say, the people who do such prosaic work don't always get their due.

It was not always that way. On the spinning Earth, the measurement of star positions is critical for keeping time and navigating. Early cultures noticed and tracked star positions as if their lives depended on it, which they did! In the third century BC, Timocharis and Aristillus produced the first star catalog in the Western world while working for the Great Library at Alexandria. About a century later, Hipparchus extended their work, generating a catalog with 850 star positions. He also divided the stars into intervals of logarithmic brightness that form the basis for a system that astronomers still use. This was a natural way to classify brightness since the eye has a nonlinear or logarithmic response to light. Ptolemy increased the catalog to 1,022 stars.[29] These star catalogs are among the most impressive intellectual achievements of antiquity; later generations of admiring astronomers called Ptolemy's stellar compendium Almagest, which means "greatest" in Arabic.[30]

As in many other aspects of astronomy, the torch for mapping stars was then taken up by Arabs for a millennium. Around AD 964, the Persian Abd al-Rahman al-Sufi wrote his *Book of Fixed Stars*, which depicted the constellations in glorious, natural color. Al-Sufi was the first to catalog the Large Magellanic Cloud and the Andromeda Nebula, two distant star systems whose true nature would not be fully understood until the 1930s. The pinnacle of pre-telescopic observations was reached by Tycho Brahe in the sixteenth century. Through relentless attention to detail and the control of systematic errors, he improved on the positional errors of earlier catalogs by a factor of fifty. His reputation didn't suffer from doing these mundane measurements; Brahe was celebrated in his lifetime and is considered the greatest observer before Galileo.

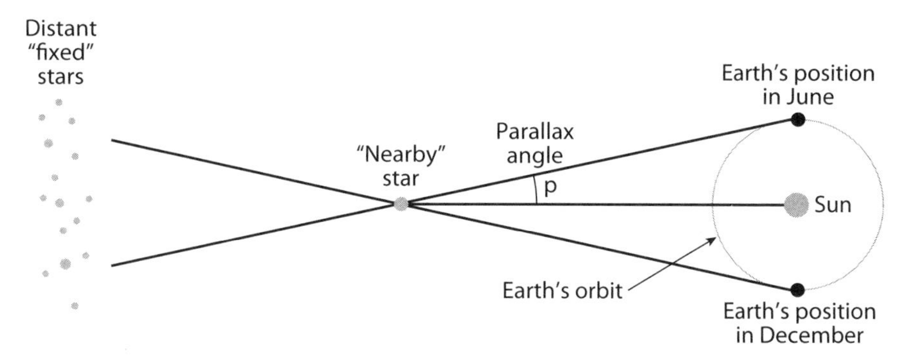

Figure 8.4. Distances to stars could not be measured for over two hundred years after the invention of the telescope. The technique that succeeded for the first time in 1803 and was used by Hipparcos to detect the small angular shift in perspective of a nearby star relative to more distant stars caused by the Earth moving around the Sun (Chris Impey/University of Arizona).

The big prize in astrometry was its use to measure the distance to a star. Stars are so far away that the apparent seasonal shift of a nearby star with respect to more distant stars—the effect called parallax—was not observable for the first two centuries of use of the telescope. Friedrich Bessel won the race to detect parallax by showing that 61 Cygni, one of the closest stars, was nearly 10 light-years away, or a staggering 60 trillion miles.[31] Bessel didn't have a university education, but his meticulous calculations elevated him to fame as one of the most noted scientists and mathematicians of the nineteenth century. The parallax shift is extremely subtle, and far more difficult to detect than the large-scale migration of constellations through the night sky as the Earth spins on its axis and orbits the Sun (figure 8.4). Almost all stars have parallax shifts over the course of a year of about one second of arc or less, and most stars visible to the naked eye have parallax shifts smaller than 0.1 second of arc. For comparison, each of the letters on an eye chart that defines 20/20 vision spans an angle of five minutes of arc, a 3,000 times larger angle.

Thereafter astrometry lapsed into the status of a worthy but dull aspect of astronomy. In part, this was because it was so challenging to measure parallax; in the half century after Bessel's measurement, new star distances were only added at the rate of about one per year. Through the twentieth century, photographic plates made it easier to capture and measure star positions, and Herschel's project to map the Milky Way galaxy was carried out by researchers in Europe and the United States. But the air blurs out the light of all stars to about 1 arc second in diameter (1/3600 of a degree), larger than the size of the angle that had to be measured to detect parallax. Refraction and telescope flexure also complicate a parallax measurement. Astronomers were bumping up against the limitations of the atmosphere and the only solution was to go into the pristine environment of space.

Hipparcos Scans the Skies

By the 1980s, astronomers had convinced their funding agencies that a space observatory to measure star positions in the vacuum of space would be a good investment. The High Precision Parallax Collection Satellite (Hipparcos) went through a series of design studies with the European Space Agency and was launched in 1989 on an Ariane 4 rocket from French Guiana.[32] Hipparcos is the only facility in this book not supported or operated by NASA, but its importance transcends its country of origin, and U.S. astronomers have used it extensively for their research. National boundaries melt away in the night sky and international collaboration is the lingua franca of astronomy. Indeed, the stars belong to no one and yet to everyone.

The telescope that transformed the precision with which astronomers can map the sky was only 29 centimeters in diameter, not much larger than a dinner plate. Many amateur astronomers use bigger glass for the mirrors of their handmade telescopes. Its mission lasted for just three and a half years, from August 1989 to March 1993, yet the data are still generating scientific results and publications twenty years later.[33] Hipparcos was one of the last space missions before the advent of CCD detectors. The satellite swept its gaze across two widely separated patches of sky and the starlight fell on a set of alternating transparent and opaque bands and then onto an old-fashioned photomultiplier tube. The primary goal of the mission was to measure the positions of 100,000 stars with an accuracy of 0.002 arc seconds. How small is this angle?

Five hundred times smaller than the typical angle by which a star image is blurred out by the Earth's atmosphere, or equal to the angle made by lines to the two opposite sides of a penny in New York as seen from the apex of the triangle in Paris.

Imagine a great city ringed by a fence. It's nighttime and you're outside the tall fence looking in. As you walk around the fence, the lights of the city will appear to flicker on and off as they pass behind the slats of the fence and then reappear in the gaps. Now imagine a somewhat different situation: you're inside the city and wearing a hood. The hole for each eye is covered with extremely thin vertical slats, like a miniature fence. As you turn, the lights from the streets and buildings brighten and fade as they pass in between and behind the slats. Hipparcos worked in this way, scanning a great circle on the sky every two hours, with its two imaging fields, or eyes, seeing a particular star 20 minutes apart. The precision of the measurement came about because the angle between adjacent slats was only one arc second, and then combining a hundred or more observations of the same star gave a much smaller angular error. In addition to using the data to measure positions, astronomers used the repeated observations to search for variability in the light of hundreds of thousands of stars as Hipparcos pivoted to scan the entire sky.

Hipparcos by the Numbers

For a direct sense of what Hipparcos learned, ESA offers a sky map on its website that can easily fit in the palm of your hand.[34] The "Hipparcos Star Globe" represents the brightest stars and the major constellations measured by the satellite as charts that can be printed out on two sheets of paper and then assembled into a sky sphere. In fact, for ease of construction, the sky is projected onto an icosahedron, a polyhedron with 20 triangle-shaped faces. Instructions for constructing this astronomical origami are also provided on the ESA website. This simple sky chart conveys no more than the "bony skeleton" of the night sky's stars; Hipparcos mapped the anatomy in exquisite detail. Its main instrument charted the positions of 118,218 stars with the highest precision.[35]

In addition, a beam-splitter was used with a secondary detector to map out the sky with slightly lower precision—the resulting Tycho Catalog lists 1,058,332 stars.[36] Years after the satellite ceased operation, astronomers produced the definitive Tycho-2 Catalog, containing a prodigious 2,539,913 stars.[37] That number is 99 percent of the stars down to 11th magnitude, which is a level 100,000 times fainter than the brightest star, Sirius. With this exquisite level of detail, Hipparcos has mapped our location in the "city of stars" called the Milky Way galaxy (plate 14).

Space missions produce data of such complexity and abundance that it's often years before all the results are known. Hipparcos is no exception. The Tycho-2 Catalog was published in 2000, seven years after the satellite returned its last data. As recently as 2007, Dutch astronomer Floor Van Leeuwen re-analyzed the Hipparcos data. He diagnosed many small effects that had been overlooked in the original analysis, such as tiny jogs in the spacecraft's orientation due to micrometeorite impacts and subtle changes in the image geometry each time the satellite went into Earth's shadow and then emerged into sunlight. He also took advantage of great gains in the power of computers to improve the calculation of positions. With a million stars, the number of angles between any star and all of the others is a million squared, or a trillion. To pin down the errors in positions, those trillion angles must be calculated many times, a procedure that took six months at the end of the mission but only a week when Van Leeuwen did his work using much faster processors. His analysis has shrunk the errors by a factor of three from the initial goal of 0.002 arc seconds, and a factor of ten for the brightest stars.[38] This minuscule angle would be formed by drawing lines from the top and bottom of Lincoln's eye on a penny in New York and having them come to an apex in Paris.

The trick of Hipparcos is to measure positions across the entire sky rather than picking off stars one by one. Thus it gains from the power of large numbers. Imagine you had to cover the floor of a large room with irregular but similar-sized tiles. You could lay them out tile by tile with a good chance of keeping the separation of adjacent tiles uniform, but as you covered a larger region it would become very difficult to control the uniformity. Whether you worked from the center out or the edges in or from side to side, it's likely you'd either have tiles left over or leave a gap. The optimum solution would be to be aware of the distance between any two tiles and regulate it over the whole area, thereby filling it uniformly. If the tiles are now stars on the sky, that's what scientists did with the Hipparcos data. They made an optimal solution for all stars simultaneously. In practice it was a calculation that taxed the best computers at that time.

Behind the numbers are the people who work to make a mission successful, often devoting their entire scientific careers to the task. Michael Perryman was born in the dreary industrial town of Luton, just north of London, and he was interested in math and numbers from an early age. His math teacher at school advised him to study a subject with better employment potential. He ignored the advice and studied theoretical physics at Cambridge University. For his PhD he stayed at Cambridge but switched to radio astronomy, joining a group that was still buzzing with excitement from the discovery of pulsars and the award of the Nobel Prize in Physics to Martin Ryle and Tony Hewish in 1974. At this point he has spent thirty years of his career on the unglamorous but very important work of mapping star positions, exceeding the amount of time the illustrious Tycho Brahe spent on his observations.[39] Appropriately, in 2011 he was awarded the prestigious Tycho Brahe Prize by the European Astronomical Society.

Perryman was just twenty-six when he was selected to be the project scientist for the Hipparcos mission, a great honor and responsibility for someone so young.[40] Soon he found himself responsible for the coordination of two hundred scientists and for all the headaches that go along with a complex multinational project. The biggest challenge came soon after launch when a motor on the Ariane launch rocket malfunctioned and the satellite didn't reach its desired geostationary orbit. The unplanned orbit exposed Hipparcos to high levels of radiation twice a day and it was thought the satellite might not last more than a few months. The team adjusted and made the best of the situation, but for over two years Perryman lived under a "sword of Damocles" as gyros were knocked out by the radiation. In the end, the mission exceeded its design goal of both lifetime and science. Away from the project, Perryman enjoyed hiking and caving, choosing to escape underground from his upward-looking day job.

Stepping Out into the Universe

It sounds implausible that a small telescope with a single mode of observing could touch every area of astrophysics from planets to cosmology, but that's the legacy of Hipparcos. By measuring the positions of more than 100,000 stars two hundred times more accurately than ever before, Hipparcos redefined and recalibrated the basic ingredients of stars, and that's fundamental because stellar properties lie at the base of a pyramid of methods used to establish the distance to nearby galaxies and on into the universe.[41] Without a tether in nearby stars, the entire edifice of the cosmological interpretation of redshift, where the shift of galaxy radiation to longer wavelength is used to ascribe a distance according to cosmic expansion, would be suspect.

In the Solar System, distances to the planets are known with very high precision. We can bounce radar off Mercury, Venus, and Mars, and measure the time of its round-trip journey to derive distance. Kepler's laws provide the relationship between orbit period and distance that lets us calculate the distance to the outer planets. Beyond the Solar System, even the nearest star is trillions of miles away. The most direct method of measuring distance uses the slightly different perspectives on a nearby star between when the Earth is six months apart in its orbit of the Sun. As we've seen, this is stellar parallax. A skinny triangle in space is created with the base equal to the Earth's orbital diameter and the long sides hundreds of thousands of times longer. Distances measured by trigonometry involve no assumptions (except that three-dimensional space is flat and Euclidean) so they form the secure base for all other distance determinations in the universe.[42]

As astronomers step out through the universe, no single method of measuring distance works on all scales. So they have to step out with a series of overlapping indicators, each of which has a range of applicability. Like a series of ladders climbing high into the sky, where a wobbly lower ladder causes the whole edifice to sway, a tighter local distance scale ensures more accurate distances to galaxies. Before Hipparcos, astronomers had measured the distances to several dozen stars with just 1 percent precision. Hipparcos increased that haul by more than an order of magnitude to more than four hundred stars. At the precision level of 5 percent, Hipparcos increased the number of stars with reliable triangulated distance from one hundred to over seven thousand. Good distance determinations are now widely available out to nearly five hundred light-years from the Sun. That's a small patch of the Milky Way, which is 100,000 light-years across, but it's a patch large enough to include all stellar types and almost all known exoplanets. The precision of the distances maps into a similar precision for the derived parameters such as size, luminosity, and mass.

Hipparcos measured the distance of the bright star Polaris as 432 light-years and tightened the error from thirty to seven light-years. Polaris is important for the distance scale because it's the closest and brightest Cepheid variable star, whose properties define a distance indicator that can be used from our neighborhood in the Milky Way out to galaxies tens of millions of light-years away. Unfortunately, Polaris turns out to be anomalous and a recent study revised the Hipparcos distance down by a third to 323 light-years.[43] Another bright star, Deneb, had an estimated distance of 3,200 light-years that was almost completely indeterminate; its distance firmed up to 1,400 light-years with an error of 230 light-years. Hipparcos also measured distances to a few open star clusters like the Pleiades and the Hyades, which allows bridges to be built to more remote regions of space.[44] First, the trend line of main sequence stars—a relationship between luminosity and temperature when the energy source is hydrogen fusion—can be fit for all stars within the cluster. The offset between the main sequence best fits for two clusters gives the relative distance because the brightness of the stars goes down with the square of the increasing distance. Second, a cluster will contain rare variable stars like Cepheids and RR Lyraes that can be seen to very large distances (Polaris is a nearby Cepheid variable). Cepheids have a useful linear relationship between their luminosity and the period of their variability, so knowing the period and the relative brightness of two Cepheids gives the relative distance between them.[45]

Nearby variable stars with parallax measurements can be used to reach out to analogous variables in galaxies as far away as the Virgo Cluster, 54 million light-years away. Once in the realm of galaxies, global properties of galaxies such as their rotation rates and sizes are used to calculate relative distances. By now, the ladder is getting rickety and errors are 10 percent or more. Thereafter, the expansion of the universe imprints a redshift on all galaxies, and distances can be derived in the context of the big bang model, which relates distance to recession velocity for all galaxies. Exploding stars called supernovae are also used to estimate distances in the remote universe, billions of light-years from the Earth, but all these methods are rooted in the work of the modest Hipparcos telescope.

Hipparcos Touches All of Astronomy

Astrometry may be the "Cinderella" of modern astronomy, but astronomers in all fields are continually reminded that everything starts with mapping brightness and position. Hipparcos leveraged historical measurements by providing the most accurate reference frame.[46] With the invention of the photographic plate in the mid-nineteenth century, comparing photographs of star positions from different eras could in principle reveal star motions, but it's usually unclear which set of plates has the largest errors. With Hipparcos as the rock-steady "gold standard," astronomers gleaned new insights from century-old data.

The hundred or more observations that the satellite made of each star allowed it to detect variability. Over 12,000 variable stars were found in the database, about 10 percent of all the stars studied, two-thirds of which were previously unknown.[47] The observation of the variables was coordinated with a network of amateur astronomers, who filled in with data when a star was temporarily out of the satellite's viewing zone. Hipparcos was also able to resolve or distinguish over 24,000 double and multiple star systems.[48] Binary stars were actually a headache for the science team, because they mimicked problems in the photometry and a faint companion could throw off the position of the brighter star in a pair if the two images were not well separated.

A sampling of projects will give a sense of the dizzying range of investigations enabled by the Hipparcos data. Galactic archaeology is a good example. Hipparcos data showed that some of the stars in the neighborhood of the Sun are part of a disk that's ten times thicker than the disk where most of the Milky Way's star formation takes place.[49] Differences in the heavy element abundance in the two components are consistent with a model where the Milky Way was assembled from smaller galaxies over billions of years.[50] Ten percent of the stars in the spherical halo as well as some in the "thick" disk seem to come from a single "invader" galaxy that was disrupted soon after the Milky Way formed. Also, the fine view of stellar motions provided by Hipparcos allows astronomers to turn back the clock and trace the Sun's passage around the galaxy and in and out of the galactic disk over the past 500 million years. During that time the Sun has passed through spiral arms four times, each corresponding to an extended cold spell in the climate history of the Earth. It is speculated that exposure to high cosmic ray flux in the spiral arms leads to more cloud cover and longer Ice Ages.[51]

Hipparcos data were used to show that the dim companions in some stellar systems are brown dwarfs. These elusive objects are gas balls less than 8 percent of the mass of the Sun; too cool to shine by nuclear fusion, they emit a feeble infrared glow and slowly contract as they leak their energy into space. In 1991, a star being observed by Hipparcos dimmed slightly on five occasions due to the shadow of a giant planet passing in front of it. This was four years before Mayor and Quleoz stunned the world with their discovery of the first planet beyond the Solar System. But nobody was looking for such signals in the Hipparcos data, so the eclipses remained undetected until 1999.[52] Since then additional exoplanets have been dug out of the database.

Hipparcos also produced a beautiful confirmation of general relativity (we earlier described a test of relativity by Cassini). Einstein's theory states that mass bends light, and it was first confirmed in 1919 by observations of the deflection of starlight as it grazed the limb of the Sun while observed during an eclipse. General relativistic bending is 1.7 arc seconds at the limb of the Sun, and it declines with the projected distance from the Sun's gravity but is still a detectable 0.004 arc seconds at right angles to the sight line toward the Sun.[53] This subtle measurement shows that the paradigm of curved space applies everywhere. The curvature is so slight that it doesn't negate the use of Euclidean triangles to measure distances. To test general relativity, while firming up measurements of the size and expansion rate of the universe, is quite an achievement for a small and often-overlooked space mission.

Passing the Baton to Gaia

In a sense, not much has changed since ancient humans first looked toward the skies. Astronomers today are equally compelled to interpret the patterns in the stars within our galaxy and beyond. In late 2013, ESA plans to launch its next astrometry mission with the Gaia spacecraft (Global Astrometric Interferometer for Astrophysics), which will expand the work of Hipparcos.[54] The Gaia mission, however, will be far more extensive and accurate in its survey of stars, taking in data on a thousand million stars in the Milky Way, charting the position, distance, brightness, and movement of each one (figure 8.5). It will achieve this by using much larger mirrors than Hipparcos and much more sensitive CCD detectors. Its goal is a precision of 20 micro-arcseconds for stars at magnitude 15, or 10,000 times fainter than the eye can see, and 200 micro-arcseconds for stars at magnitude 20, or a million times fainter than the eye can see. To extend an earlier analogy, the smaller of these tiny angles would be formed by drawing lines from the top and bottom of Lincoln's eye on a penny in New York and having them come to an apex not in Paris but on the surface of the Moon!

It is anticipated that Gaia will measure and characterize several thousand exoplanets, detecting them by the subtle motion they induce in their parent star. Among these distant worlds, some that are Earth-like must surely exist. However, Gaia's primary mission will be to produce an extraordinarily precise three-dimensional

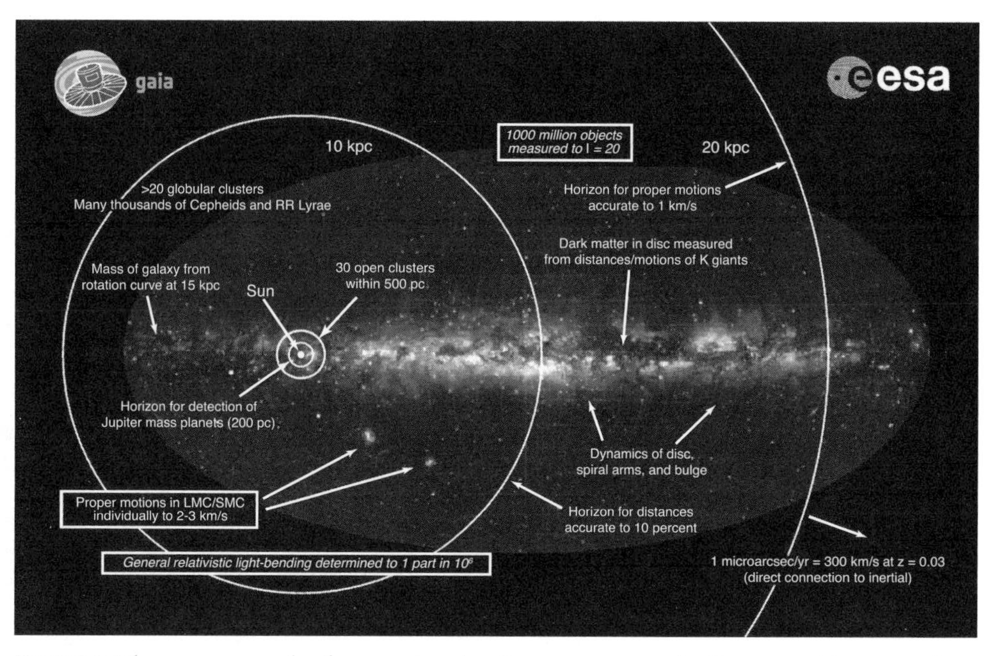

Figure 8.5. The protean work of measuring the distances to stars facilitates a wide range of astronomy projects. Where Hipparcos reached several million stars, ESA's Gaia mission will reach and measure nearly a billion, enabling everything from a local test of general relativity to a better understanding of the archaeology of the Milky Way (ESA/Gaia).

map of the Milky Way galaxy and sharpen the crispness of our 3D view.[55] As explained on the Gaia website, "In the process, Gaia will also map the motions of stars, which encode their origin and subsequent evolution. Through comprehensive photometric classification, Gaia will provide the detailed physical properties of each of the billion stars observed." Such information will include data on luminosity, temperature, gravity, and elemental composition, which can be used to unravel information on the origin, structure, and evolution of our galaxy. "Gaia's expected scientific harvest is of almost inconceivable extent and implication. . . . Amongst other results relevant to fundamental physics, Gaia will follow the bending of star light by the Sun, over the entire celestial sphere, and therefore directly observe the structure of space-time." As a means of comparison with its predecessor, Gaia planners note: "The 16 volumes of Hipparcos would instead be 160,000 volumes and instead of filling one normal bookshelf, that bookshelf would have to stretch the equivalent distance of Paris to Amsterdam."[56]

Perhaps most important, Gaia will dramatically alter our sense of humankind's place in the Milky Way and in the universe. Far exceeding the *Apollo 8* image of "Earthrise from the Moon," or the *Apollo 17* image of the "Whole Earth," the Gaia mission has the potential to entirely overhaul our thinking about our grain of sand, the Earth, in the chasm of interstellar space. Astronomers have already discussed the extent to which the mission will powerfully impact nonspecialist audiences. In the coming decade, our children will have the opportunity to see in exquisite detail where we are in relation to our nearest star neighbors as educators plan virtual astronomy courses for educational institutions and planetaria. Based on new and rich data from Gaia, virtual flights into a 3D version of a larger portion of the Milky Way galaxy are just one of the expected educational outcomes.[57] With Gaia, and the virtual galaxy its data will help build, the scientific and public understanding regarding our place in the Milky Way is poised to be radically reconfigured.

Roughly half the world's population live in urban areas. Light pollution inherent to cities means that 50 percent of humankind has to a large extent lost access to, and knowledge built upon, the night sky. However, scientific missions like Hipparcos and Gaia are restoring our relationship to the stars and generating remarkable new knowledge regarding Earth's place in our galaxy. Hipparcos has already rewritten the narratives we will tell our children about the past and present morphology of the Milky Way and the exoplanets we are currently finding orbiting nearby stars. What Gaia promises to contribute will rewrite the textbooks again. The story of our galaxy that we hand on to future generations will disclose the number and locations of many other worlds like ours and will inevitably be vital to the human narrative of future ages.

9 Spitzer

UNVEILING THE COOL COSMOS

Space is mostly empty, but a thin gruel of gas and dust that occupies regions between stars dims and reddens light.[1] Thousand-trillion-mile wide clouds containing gas and microscopic dust grains absorb and attenuate visible light and reradiate it at infrared wavelengths. NASA's Spitzer Space Telescope has the remarkable ability to see through interstellar dust and has allowed us to look into the vast clouds in which stars are born, like those of the Orion Nebula, our nearest star-forming region. Spitzer can also peer into the dark, dust strewn plane of our Milky Way galaxy that previously had been nearly impossible to penetrate. Anything that radiates heat, such as living bodies, and any cool object in space, such as planets or moons or even tiny silicate (rocky) and carbon (sooty) grains 1/10,000 to 1/100 of a millimeter across, emits infrared radiation. Only an infrared telescope can image objects that glow in light waves too long for the human eye to see. The Spitzer Space Telescope can detect such light billions of light-years from Earth and has revealed this cool and invisible universe with unprecedented clarity.

With more than 850 exoplanets known as of early 2013, and another 2,700 candidate planets identified by the Kepler telescope, astronomers estimate that there are at least 50 billion exoplanets in the Milky Way galaxy. Scientists calculate that 500 million of those planets orbit in the habitable zone, the distance from their star that could allow for life.[2] NASA's Spitzer Space Telescope is helping to detect and characterize these extrasolar worlds. In December 2010, Spitzer discovered the first carbon-rich exoplanet, named WASP-12b, the geology of which may be comprised largely of diamond and graphite.[3] Whereas rocks on Earth generally consist of silicon and oxygen in the form of quartz and feldspar, Spitzer's observations suggest that WASP-12b, about 1,200 light-years from here, has nothing like terrestrial geology. Astronomer Marc Kuchner of NASA Goddard Space Flight Center, who has helped theorize carbon-rich planets, explains that increased carbon in a planet's composition can entirely alter its geology: "If something like this had happened on Earth, your expensive engagement ring would be made of glass, which would be rare, and the mountains would all be made of diamonds."[4] Spitzer has been instrumental in analyzing the geological makeup of exoplanets, and what astronomers are finding exceeds their wildest expectations.

At the other extreme from dim worlds in the nearby universe, Spitzer has discovered massive galaxies billions of light-years away that are forging stars at a prodigious rate. An infrared-bright, star-forming galaxy might be making thousands of stars each year compared to a couple for the Milky Way. These "starburst" galaxies are infrared beacons from the early construction phase of the universe, during which many large galaxies were being assembled from smaller galaxy "pieces" for the first time.[5] Deep within the dust-obscured hearts of distant galaxies, new worlds were being forged at a fantastic rate. A single, modest-sized telescope in space has seen directly into the center of these galaxies and provided insights on the full range of their creation stories.

Working with the Palette of Light

We inhabit a physical universe filled with electromagnetic radiation spanning a factor of trillions in wavelength. The human eye sees only a sliver of this radiation, designated as visible light. It's as if there was a full piano keyboard in front of us with eighty-eight keys and we were restricted to making music with two adjacent keys or notes. Space telescopes like Spitzer and the Chandra X-ray Observatory, covered in the next chapter, have opened up the palette of sensation and given us an octave or more with which to view the universe. Infrared astronomy, in particular, has been instrumental in revealing how stars and their planetary systems form. Spitzer can see through the vast clouds of dust, within which myriad stars are being born, and can penetrate the granular torus or dense disk that typically surround newborn stars. This capability has literally opened people's eyes to the worlds that lie beyond the red end of the rainbow.

Isaac Newton first identified the colors of the visible spectrum by observing sunlight passed through a glass prism. William Herschel later investigated temperatures associated with each color of visible light. Having in 1781 discovered Uranus, the first new planet since antiquity, Herschel was already the greatest astronomer of his age when he began experimenting with light. He noticed that heat passed through various colored filters used to observe the Sun and carried out experiments to understand this phenomenon. Using a thermometer with a soot-blackened bulb to better absorb heat, Herschel measured the temperature of each color band of visible light and noticed that the temperature increased from the violet to the red end of the visible spectrum.[6] Surprised to record the highest temperatures beyond the edge of the red band, Herschel realized that electromagnetic radiation existed in wavelengths we can't see. He had detected infrared light. Such early discoveries regarding the electromagnetic spectrum paved the way for remarkable new possibilities in studying the universe.

Light is conventionally subdivided by wavelength into radio, microwave, infrared, visible, ultraviolet, X-ray, and gamma ray, with radio designating cool, long wavelengths and gamma rays representing the highest energy, short wavelengths. These subdivisions are merely convenient demarcations for differences along a continuum of electromagnetic radiation. While most forms of radiation can't reach the ground from sources in space (figure 9.1), all forms of electromagnetic radiation, from radio to gamma rays, travel at the same speed as light in a vacuum, approximately 300,000 kilometers per second or 186,000 miles per second. We're intimately familiar with most of these bands of radiation. Radio waves transmit music and news on your favorite radio station. Microwaves make possible communication technologies including cell phones and, of course, in microwave ovens. Ultraviolet radiation can quickly give you a suntan or sunburn and is the means by

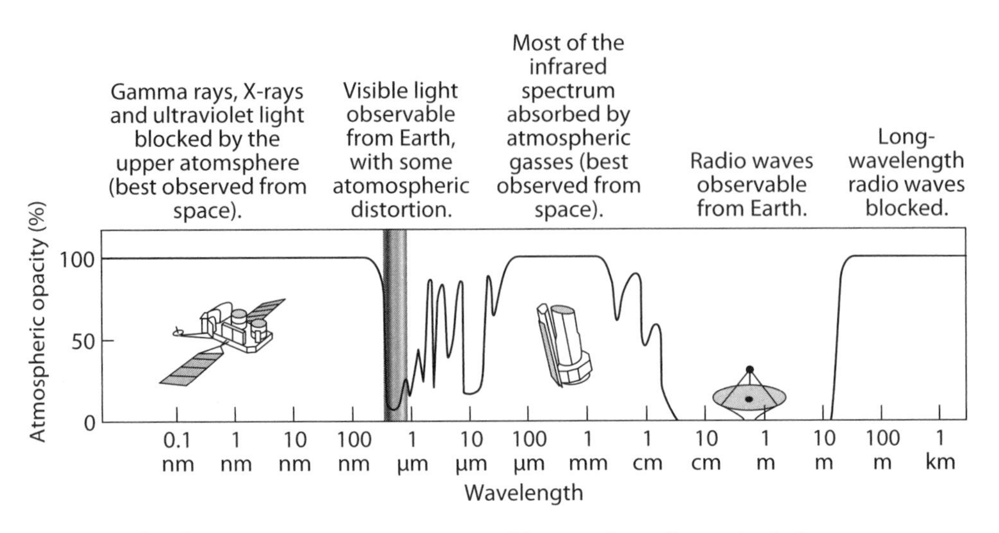

Figure 9.1. The electromagnetic spectrum spans fifteen orders of magnitude from meter-length radio waves to gamma rays the size of an atomic nucleus. Only radio waves, microwaves, and a small slice of optical and near infrared radiation can reach the ground from space. Spitzer measures long infrared waves from its vantage point in Earth orbit (NASA/Wikimeida Commons/Mysid).

which our bodies produce Vitamin D. We became familiar with UV light from the 1970s black lights. X-rays have been invaluable in medical and dental practice as well as in fluoroscopy, which offers real-time views through living tissue, while high-energy lasers are used in brain and eye surgery and other delicate medical procedures. Radiation at infrared wavelengths also has powerful therapeutic effects. Low-level, "cold laser" therapy contributes to the healing of wounds as noted in a report cited by the National Center for Biotechnology Information.[7] A near infrared light-emitting diode developed by NASA has been used to heal and reduce pain for chemotherapy patients who subsequently suffer from mouth lesions.[8] In the past fifty years, we've harnessed this invisible radiation to vastly improve our lives.

The Emergence of Infrared Astronomy

The Earth and its atmosphere emit infrared radiation in all directions,[9] swamping any signals from space. It's analogous to trying to see stars in the noontime sky. For a faint, cool source to stand out against a warmer backdrop, an infrared telescope must lower thermal backgrounds. Which is to say, the longer the wavelength of light being observed, the colder a telescope and its detector need to be to see an infrared source from space. By the 1960s, infrared astronomy took off with the development of a new type of solid-state detector that had greater sensitivity than old-style radiometers.[10] Another obstacle to detecting infrared radiation from cosmic sources is the fact that those waves are absorbed by water vapor in the air we breathe. Astronomers initially chose high mountaintops for their observations, to be above as much of the infrared-absorbing atmosphere as possible. For example, some of the best infrared observations are made 4,200 meters high atop the dormant volcano Mauna Kea in Hawaii, and new telescopes are being constructed in the arid Atacama Desert in Chile, including a major new millimeter array on the Chajnantor plateau at an altitude of 5,000 meters. An extreme version of this strategy was the Kuiper Airborne Observatory, which operated from 1974 to 1995. This 36-inch telescope in a converted Air Force C141 jet made observations for a few hours at a time at an altitude of 45,000 feet. The successor to this facility is a 2.5-meter telescope in a converted Boeing 747-SP, the Stratospheric Observatory for Infrared Astronomy.[11] High-altitude balloons and rockets carrying small telescopes have been used to reach even higher altitudes of over 100,000 feet.

It proved essential to cool infrared telescope detectors to liquid nitrogen or even liquid helium temperatures so that radiation from the detector wouldn't wash out signals from the sky. On ground-based telescopes observing the near infrared, two to six times the wavelength of visible light, liquid nitrogen is used to cool the detector to 77 Kelvin or –177°C. Though seemingly exotic, liquid nitrogen is basically liquefied air and costs no more than milk. For observing at the mid-infrared, 20–200 times the wavelength of visible light, liquid helium is used to cool the detector to 4.2 Kelvin or –269°C, a hair's breadth from absolute cold. But the Earth environment emits so much radiation at these wavelengths that astronomical signals are swamped by background radiation and deep mid-infrared observations can only be made from space.

Space infrared astronomy was kicked into gear with the very successful NASA Infrared Astronomical Satellite (IRAS) mission, which in 1983 surveyed the entire sky in the infrared for the first time. In less than a year, IRAS detected about 350,000 infrared sources. Around the same time, ESA launched the Infrared Space Observatory, which had higher sensitivity and much broader wavelength coverage.[12] In 1997, an infrared camera was added to the instrument suite of the Hubble Space Telescope. This camera has been very successful but it has a small field of view and shares time with four other instruments, so infrared astronomers had always planned for their own facility.

Spitzer, from its earliest inception, was especially designed for infrared astronomy and is sensitive enough to detect infrared signatures of stars and galaxies billions of light-years away. The space telescope has been instrumental in unveiling small, dim objects like dwarf stars and exoplanets and can even determine the temperature of their slender atmospheres. Originally proposed in the late 1970s as NASA's Space Infrared Telescope Facility, the Spitzer Space Telescope suffered from uncertainty, a delay after the loss of the space shuttle *Challenger*, near-cancellation, congressional limbo, budget cuts, and "descoping." Nevertheless, in 2003 the telescope was finally launched, after being renamed subsequent to a public opinion poll conducted by NASA. The last of NASA's four Great Observatories, the $800 million telescope was named after Lyman Spitzer, an early advocate of the importance of orbital telescopes.[13] After launch, the spacecraft took about 40 days to cool to its operating temperature of 5 Kelvin. Once cooled, it took just an ounce of liquid helium per day to maintain its detectors at their operating temperature. A solar panel facing the Sun serves to gather power and protect the telescope from radiation. On the anti-solar side, a series of concentric shells and a black shield radiate heat into space. Spitzer's most precious resource was its liquid helium, used to cool the telescope and its instruments to the phenomenal temperature of just 1.2 Kelvin, within an iota of the temperature where atoms and molecules have no motion. This extra cold relative to the liquid helium operating temperature comes from venting the helium into the vacuum of space, much as your skin would cool if a liquid evaporated from it. Background radiation is reduced by a factor of a million relative to a similar-sized telescope on the Earth's surface. The telescope's cryostat held

350 liters at launch, as much as the gas tank of a large minivan. Helium evaporates and is boiled away by tiny heat inputs from the instruments and from the telescope structure, so the act of observing steadily depletes the cryogen.

The goals of cooling the telescope and preserving the liquid helium as long as possible dictated the telescope's orbit and design. Careful design minimized these heat inputs, which are typically measured in milliwatts. As a reference, your fingertip radiates 25 milliwatts. The Spitzer facility is 4.5 meters tall and 2.1 meters in diameter and weighs about as much as a minivan. Previous space observatories had used either a low Earth orbit to be serviceable by the Space Shuttle (such as the Hubble Space Telescope) or a high Earth orbit with a period of one to two days (like the Chandra X-Ray Observatory). Spitzer is in an unusual Earth-trailing solar orbit. The big benefit is in escaping from the Earth's heat and being situated in a good thermal environment where the telescope can cool in the vacuum of space. The orbit also avoids the Earth's radiation belts, and there's excellent efficiency of observing because the Earth and Moon rarely intervene. But one disadvantage is that Spitzer's radio signals are getting weaker as it moves away from Earth. Spitzer slips away from Earth at about 10 million miles per year and is now farther away than the Sun. Only the huge 70-meter dishes of the Deep Space Network are sensitive enough to gather its precious data. Mission planners estimate that they will lose touch with Spitzer in early 2014.

Penetrating the Murk of Star Formation

As we peer through the plane of the Milky Way, the center of our galaxy is completely hidden due to interstellar dust.[14] The galactic center is 28,000 light-years away and only one in a trillion optical photons can escape—it's like trying to look through a closed door. Unless this cold and grainy dust is accounted for, intrinsic stellar properties can be badly misjudged, a point first noted by Robert Trumpler in 1930. Star formation takes place in chaotic, dense, swirling regions of gas and dust where the concentration of dust is so high the action is mostly hidden from view. There's a dramatic difference, for instance, between the optical and infrared views of the Orion nebula. In the optical view the bright nebula contains a dark region that appears to be devoid of stars, while in the infrared view that penetrates the dust a dense star cluster shows up in that dark region.[15] The youngest stars are invisible even to the most powerful optical telescope. However, with Spitzer's ability to penetrate dust, the broad sweep of star formation in the Orion Nebula (M42) and the Omega Nebula (M17) and throughout the Milky Way has been revealed (plate 15).

Spitzer's detectors are so sensitive that the telescope has lifted the veil on some of the earliest star formation in the universe. Black holes that inhabit the centers of all galaxies are generally starved of fuel and are quiet. But it was not always this way. When the universe was 7 percent of its present age, nearly 13 billion years ago, it was seven times hotter and 350 times denser than it is now. There was a furious phase of star formation and galactic construction in the first 3 billion years after the big bang, during which many of the largest galaxies were put together. Spitzer has two unique advantages for studying the distant universe. The first is associated with the redshift caused by expanding space. By the time the light from a galaxy 13 billion light-years away reaches us, its light has been stretched and mostly shifted to the infrared. The second advantage is that light from star formation and black hole growth shrouded in dust in the early universe is absorbed and reradiated at longer wavelengths where Spitzer works best. As a result, Spitzer has provided a more reliable measure of the star formation history of the entire universe.[16] The most distant, infrared-luminous galaxies have a rate of star formation thousands of times larger than the Milky Way. Such a rate was sustained by the ready availability of gas in the denser, primordial universe. Gas was consumed so quickly that the fireworks didn't last long.

One provocative and tantalizing study claimed to use Spitzer to reveal the first objects produced by gravity in the universe, at the end of the cosmic "dark ages." Alexander Kashlinsky and his group studied the diffuse infrared background in ultra-deep Spitzer images.[17] They first carefully removed light from foreground stars and nearby galaxies, leaving only the most ancient diffuse radiation. They then analyzed the fluctuations of the faint infrared emis-

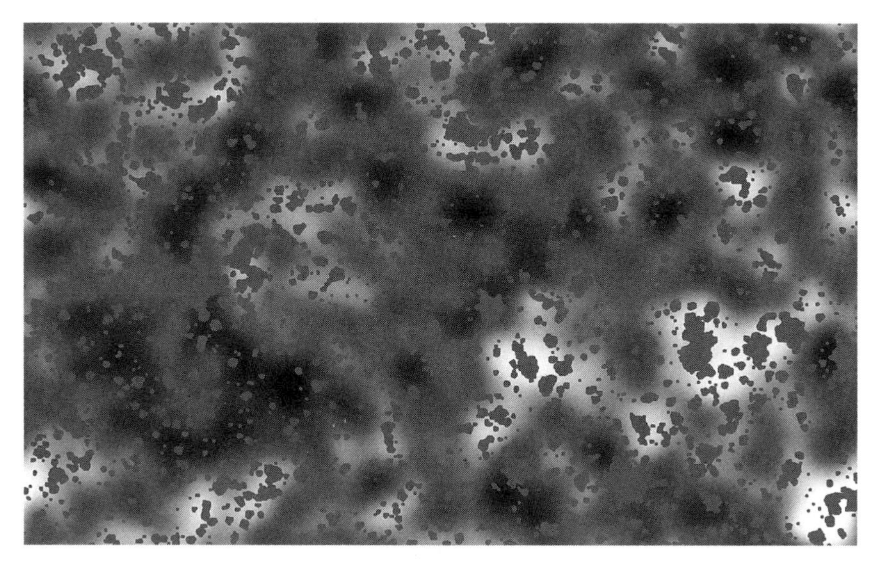

Figure 9.2. Early phases of the evolution of the universe are probed in this very deep image from the Spitzer Space Telescope. After blanking out stars and galaxies (uniform grey areas), there is a faint varying glow of infrared radiation (pale to dark shading) that represents structures in the universe that formed 13 billion years ago (NASA/SSC/Alexander Kashlinsky).

sion that remained and matched them to simulations of clustering in the early universe (figure 9.2). "Imagine trying to see fireworks at night from across a crowded city," explained Kashlinsky. "If you could turn off the city lights, you might get a glimpse of the fireworks. We have shut down the lights of the city to see the outlines of its first fireworks."[18] It's unclear whether the faint infrared glow in Spitzer's ultra-deep images actually comes from a first generation of huge stars hundreds of times the mass of the Sun, or from the first wave of supermassive black holes. Either way, this truly ancient radiation carries information about the universe when it was a fraction of its present age.

Closer to home, Spitzer conducted a sky survey of our own galaxy dubbed GLIMPSE (Galactic Legacy Infrared Mid-Plane Survey Extraordinaire) by making images of swaths spanning 65 degrees on either side of the galactic center and one degree on either side of the galactic plane.[19] A hundred times the sensitivity and ten times the resolution of any previous survey, the GLIMPSE mosaic represents 110,000 different pointings of Spitzer and 840,000 individual images, with the postage stamps painstakingly stitched together to make a seamless map. Spitzer's GLIMPSE photos are as scientifically meaningful as they are aesthetically stunning. But since Spitzer takes data at invisible wavelengths, there's no direct correspondence between its images and the colors of the rainbow. The filters Spitzer used for the GLIMPSE mosaic are sensitive to thermal emission from dust at temperatures from 30 K to 1600 K.

In the maps, blue represents radiation typically coming from older stars. Green colors trace stellar embryos by the emission of complex molecules called polycyclic aromatic hydrocarbons. On Earth, these molecules are found in car exhaust and barbeque grills, or anywhere carbon is incompletely burned. Red colors depict radiation from the youngest stars, still in their cool placental dust, as well as diffuse emission from graphite grains similar to tiny pieces of pencil lead. While no simple coding scheme can convey all the subtleties of the astrophysics of star formation, this three-color scheme is useful for researchers and armchair connoisseurs alike. Polycyclic aromatic hydrocarbons are too thinly dispersed for anyone to be able to savor their sooty aroma, but they're excited by ultraviolet radiation so they map out newborn, massive stars.[20] New stars reside in bubbles or on the edge of green ridges carved out by stellar winds. Many of these stars die as supernovae, which further excite the diffuse gas. The young stars appear as yellow or red dots, some in tight knots where they formed together; others are more randomly sprinkled throughout the disk of the galaxy. Older stars register as blue in the combined image and they're so abundant that they merge into a blue-white haze.

The GLIMPSE survey has contributed to new understanding of the morphology of our Milky Way galaxy. Teams working on the GLIMPSE images cataloged over 100 million stars, and used them to trace out the Milky Way's spiral arms and stellar bar. Spitzer project scientist Michael Werner reports that the survey "strongly confirmed previous evidence that the Milky Way is a barred spiral galaxy."[21] Prior to this survey and that of Hipparcos, it was believed that our galaxy had four spiral arms: Norma, Perseus, Sagittarius, and Scutum-Centaurus. Adding its data to the Hipparcos survey, GLIMPSE confirms that the Milky Way has only two major arms, Scutum-Centaurus and Perseus that trail from the ends of the stellar bar at the galactic core. Additionally, as a result of GLIMPSE, astronomers have determined a new obscuration relationship for the galaxy, allowing more accurate estimates of distance to stars even in the presence of dust. Most fundamentally, they've measured the "pulse" of the galaxy by counting how fast new stars are forming. Our galaxy generates, on average, one star like the Sun per year, using mostly recycled gas from the death of older stars. That doesn't sound like much for a system with several hundred billion stars, but as the Milky Way settles into the steady rhythm of middle age its wild times of forming thousands of stars a year are long over.

Given the volume of GLIMPSE data, the public has been recruited via the website Zooniverse.org to assist astronomers in analyzing Spitzer's views of our galaxy. Launched in 2007, Zooniverse developed the Milky Way Project, by which citizen scientists scour Spitzer's GLIMPSE images to identify gaseous nebula and newly forming stars.[22] Zooniverse offers a variety of mission data sets to volunteers who assist in identifying new solar system formation, track ancient weather patterns, report solar storms, and hunt for massive explosions of stars called supernovae. For the Milky Way Project, volunteers use simple online tools to circle huge bubbles of gas and dust in which stars are currently forming. The GLIMPSE survey imaged thousands of these globe-like bubbles where high-energy radiation from newborn stars collides with surrounding dust clouds. As the Milky Way Project site explains: "Right now our best understanding is that these are regions around young massive stars that are so bright that their light has caused a shock wave to affect the cloud around them and blown a bubble which we can see in infrared light."[23] Identifying massive bubbles in Spitzer images assists scientists in determining the mechanisms occurring in star birth.

Of course, it was always understood that the day would arrive when the last drop of Spitzer's 350 liters of liquid helium would evaporate away. This occurred on May 15, 2009. Without liquid helium, the telescope's long wavelength instruments have lost all their sensitivity. However, the two shortest wavelength channels of one of the cameras can operate at their design sensitivity, so Spitzer was approved for a "warm mission" phase where it is continuing to do important science. Even in its "diminished" state, Spitzer is more sensitive than any ground-based telescope in the infrared. Also, "warm" is a relative term; touching any part of the spacecraft would give you a very serious cold-burn. At the balmier operating temperature of –242°C, Spitzer is continuing work with two channels of one instrument as part of its "warm mission."[24]

Given that its near-infrared channels are able to do science from comets to cosmology, more than 10,000 hours of programs were approved in the first two years of the "warm mission." Exoplanets are "hot" so some of the largest programs were to characterize exoplanets where our line of sight is aligned with the orbital plane of the exoplanet. As the planet passes behind its parent star, the combined infrared emission from the star plus the exoplanet dips, and measurements made at different wavelengths can give the size and temperature of the exoplanet. By the end of the mission, the number of exoplanets characterized will have increased from a dozen to 50 or 60 and we'll have a much better understanding of the extraordinary diversity of other worlds in space. To set the stage for this exciting Spitzer science, we take a detour into the role of light and darkness in the evolution of life. We can only imagine the flora and fauna that might possibly exist on planets orbiting other stars, but life's adaption to light here on Earth might offer some scant clues.

Stars, Life, and Sight

Exoplanets have quickly become the focus in the search for life beyond Earth and the quest to know whether we are alone in the unfathomable depths of space. Finding a large number of habitable exoplanets would suggest that life abounds in the universe. For now, astronomers are scouring stars in the Milky Way most likely to host habitable worlds. Stars range in size from small, cool dwarfs no bigger than Jupiter to colossal stars topping out at about 200 solar masses and 10 million times brighter than our Sun. Blue giant stars are very hot, bright suns, such as those in the Pleiades Cluster or in Orion's Belt, that emit mostly UV light. About half of all stars are binaries, in which one star orbits another. Binary

Figure 9.3. An artistic rendering of exoplanet Kepler-16b that orbits a double star system and experiences dual sunsets as on Tatooine in George Lucas's *Star Wars*. The exoplanet has roughly the mass of Saturn and orbits the twin stars about twice a year. Astronomers were surprised that planets could form around a binary star (NASA/JPL-Caltech/T.Pyle).

star systems can host planets with stable orbits around both stars, called circumbinary planets (figure 9.3). The first of such exotic exoplanets, reported in 2011, was dubbed the Tatooine planet for its dual sunsets as depicted in the film *Star Wars*.[25] But in some binary systems gas interactions between the stars produce intense X-ray radiation deadly to life on their attendant planets. The most common stars in the Milky Way galaxy are M dwarfs. With less than half the mass of our Sun, M dwarfs emit most of their light at infrared wavelengths. Brown dwarfs, too small to generate nuclear fusion in their core, emit light solely in infrared. Among the many types of stars, it seems we couldn't have wished for a better star to orbit, nor one more suited for life, than our Sun.

Any type of star can host planets, but the light stars pump out determines whether, and what kind of, life might survive on their satellite worlds. Astronomers Ray Wolstencroft and John Raven contend that life works most efficiently with blue light, which they argue triggered photosynthesis on Earth.[26] If that's true, one might assume that planets orbiting hot, blue stars would be the most likely to host organisms. However, large blue stars emit dangerous ultraviolet radiation and expend their fuel in a mere 10–20 million years. By contrast, our Sun has existed for 4.5 billion years, and astronomers expect the Sun will not expend its fuel for another 4 billion. It's our Sun's long life span, in part, that allowed time for photosynthesizing organisms to emerge. As it happens, our Sun is the type of main sequence star that emits its greatest amount of radiation in the range of visible light. At some point 3 billion years ago, living organisms began to harness this light for processes in photosynthesis.[27] Cyanobacteria or its progenitor began to use sunlight and carbon dioxide to generate food in the form of sugars, and in turn release a tiny gulp of oxygen. That seemingly simple development slowly, but radically, began to alter the chemistry of Earth's atmosphere from one conducive to anaerobic life to an atmosphere rich in oxygen. But oxygen levels sufficient to sustain complex life as we know it didn't evolve until approximately 550 million years ago. As Leslie Mullen reports, "If the evolution of photosynthesis is the same everywhere, then the lifetime of blue stars is just not long enough" for life to emerge on surrounding planets.[28]

Though no one knows for sure exactly how life emerged, a group of scientists at the Santa Fe Institute posit that the reductive Krebs cycle may have brought together chains of carbon molecules to produce the first organisms.[29] The Krebs cycle, or citric acid cycle, describes how organisms produce energy by breaking down carbohydrates, fats, and proteins. In what has been called the "metabolism first" theory, the reductive Krebs cycle is proposed as the mechanism through which the first organisms emerged. "In some primitive single-celled organisms, [the Krebs cycle] operates in reverse—it takes in energy and puts small molecules together into big ones. Scientists term the reactions in this mode as 'the reductive Krebs cycle.'"[30] Eric Smith, a professor at the Santa Fe Institute, has argued that geochemical processes on the early Earth might have combined smaller molecules into larger ones because it was energetically favorable to do so, similar to the chemical reactions in the reductive Krebs cycle. In other words, life might be *built into*, or be an expected outcome of, terrestrial geology.

Plants stretch their stems and leaves toward the Sun, and even the earliest organisms would have used a single photon of light for fuel and to obtain valuable information about the immediate environment. "The very pigments that led to photosynthesis probably also allowed those earliest creatures to react to light," writes Ben Bova, who notes that even "single-celled organisms have light receptors, sensitive spots of pigment that absorb light." Michael Sobel's book *Light*, Michael Gross's *Light and Life*, and Bova's *The Story of Light* explore the myriad ways life on Earth is enabled by, and is adapted to, the light of our Sun. Photoreceptor cells in bacteria, plants, and animals demonstrate the selective pressure of visible light, as such cells evolved due to the survival advantage they provided organisms on a planet with a star that emitted its greatest radiation in visible light. "The very earliest creatures certainly did not have a sense of vision," notes Bova. "All they could do was react to light the way a scrap of paper reacts to a puff of wind—move either toward the light or away from it."[31]

By the Precambrian period, soft-bodied marine animals emerged that had at least a rudimentary sense of light and darkness. Under such conditions, finding prey and avoiding being eaten were difficult prospects. But life on Earth would suddenly, on geological timescales, change. Neurobiologist Michael Land and zoologist Dan-Eric Nilsson note that "something remarkable seems to have happened in the interface between the Precambrian and the Cambrian eras. Within less than five million years, a rich fauna of macroscopic animals evolved, and many of them had large eyes."[32] Land and Nilsson posit that a simple strip of photoreceptive cells in organisms could, in time, become curved to better sense the orientation of the incoming light. If that curved space became filled with a liquid, the light could be focused on a lens to create image-forming vision. Nilsson and Land conservatively estimate that the evolution of image-forming eyes could occur in a half million years, and perhaps as little as 300,000–400,000 years.

Land and Nilsson, as well as zoologist Andrew Parker, have researched how eyes may have emerged in response to the light produced by our Sun. The first sophisticated eyes, dating to 530 million years ago, appeared at the beginning of the Cambrian period and they're evident in the extraordinarily well-preserved fossils of the Canadian Burgess Shale. Parker refers to the Cambrian as the "big bang" in evolution: "It paved the way for the emergence of the vast diversity of life found today, whether in Australia's Great Barrier Reef or Brazil's tropical rainforests. It involved a burst of creativity, like nothing before or since, in which the blueprints for the external parts of today's animals were mapped out. Animals with teeth and tentacles and claws and jaws suddenly appeared."[33] Parker argues that visible light acted as a strong selection pressure for the development of eyes in most species. In what he calls the "Light Switch" theory, Parker contends that sunlight has been the primary factor in determining animal evolution, behavior, and morphology. With the emergence of more complex eyes, Cambrian species could colonize niches within the water column of ancient oceans that in turn impacted their morphology. Trilobites, for instance, thrived in Cambrian seas (there were some 4,000 species) and apparently were the first animals to develop compound eyes. The trilobite could clearly see its environment, navigate, and seek prey. As a result, they and many other species developed exoskeletons to protect themselves. Coincident with the evolution of eyes that could focus, Parker argues, the fossil record reveals the emergence of a radical diversity of all phyla of animals. Land and Nilsson agree: "If eyes had not evolved, life on earth would have come out very differently. More than any other organs, eyes have shaped the evolution of animals and ecosystems since the Cambrian explosion."[34]

The Light of Our Sun

Human eyes, and all vertebrate eyes, are actually protrusions of the brain. Michael Sobel explains, "Embryological studies show that the eyes begin to develop very early as two small buds on the neural tube that eventually becomes the brain."[35] We often think of ourselves as the pinnacle of evolutionary adaptation, but the human eye is adapted to detect a very narrow slice of the overall electromagnetic spectrum. This is not at all accidental. Had our Sun been a brown dwarf, emitting most of its light as infrared, our eyes would likely be adapted to infrared wavelengths. Some animals and insects, in fact, can see in wavelengths altogether invisible to us. Pit vipers have a sensory organ located near their eyes that allows them to detect infrared or thermal images of prey. Butterflies are thought to have a wide range of vision and can identify potential mating partners by ultraviolet markings on their wings. Bees, many types of fish, and some birds also see in ultraviolet light. Scientists have recently discovered that flowers and plants, when viewed in ultraviolet light, often display patterns invisible to us and color quite distinct from the color we see. By imaging flowers in ultraviolet light, scientists realized that the patterns revealed in ultraviolet apparently serve as "landing strips" or markers useful in guiding insects to pollinating portions of the flower.[36]

At least forty to sixty different types of eyes, with up to ten distinct means of forming images, have independently evolved, from simple constructions like a patch of skin acting as a photoreceptor, to the compound eyes of flies and spiders, to the sophisticated eyes of the hawk or squid. "The result is an enormous range of eye types using pin-holes, lenses, mirrors, and scanning devices in various combinations to acquire information about the surrounding world," write Land and Nilsson. "Not all eyes are paired and placed on the head: there are chitons with eyes spread over their dorsal shell, tube worms with eyes on their feeding tentacles and clams with eyes on the mantle edge."[37] Various cave-dwelling animals, because of a lack of light, have evolved without eyes. Some lizards, frogs, and fish, though they have complex eyes, also sport at the top or back of the head a third eye, or parietal eye, that at the least can indicate the presence or absence of light caused by the passing of a predator's shadow. Even such a rudimentary eye offers valuable survival information.

At the other end of the spectrum of eye development are box jellyfish that have twenty-four eyes of four different types, two of which are extremely similar to human eyes. Existing for 600–700 million years, jellyfish have survived five mass extinctions but they're often incorrectly characterized as not having a brain. Researchers recently found that neurons in jellies occur in neuronal centers distributed throughout their bodies and that they purposefully navigate within their environment. Box jellyfish use the acute vision of their pseudo-human eyes to see under and above water in navigating to and from underwater mangroves in order to feed, as a story in the *New York Times* reports: "Not only are the eyes equipped with a cornea, lens and retina, as human eyes are, but they are also suspended on stalks with heavy crystals on one end, a gyroscopelike arrangement that ensures the eyes are focused unerringly skyward. . . . Every morning they must return to the roots or risk starvation. They rise toward the surface and their upturned eyes scan the sky, until at last they spy the mangrove canopy."[38] Beyond even these complex eyes are those of the cephalopods, like the octopus or squid, believed to be the most sophisticated eyes on the planet. Cephalopods focus their eyes with fine movements of the lens, unlike vertebrates, and they can automatically keep their pupils horizontal and even sense the polarization of light.

Seeing in the Dark

The largest and most densely populated habitat on Earth is the ocean realm. As Claire Nouvian observes, "The deep sea, which has been immersed in total darkness since the dawn of time . . . forms the planet's largest habitat." Zoologists and oceanographers are far from documenting the welter of sea life. "Current estimates about the number of species yet to be discovered vary between 10 and 30 million," writes Nouvian. "By comparison, the number of known species populating the planet today, whether terrestrial, aerial, or marine, is estimated at about 1.4 million."[39] Down-welling light, whether sunlight or moonlight, in the upper reaches of the ocean's water column can reveal an animal's location or hideout. To avoid predators, marine animals largely forage for food at night, under cover of darkness. The result is that the ocean's water column is the site of the largest daily animal migration known, with marine animals migrating vertically as much as tens of meters to a kilometer. Harbor Branch Oceanographic Institute scientist Marsh Youngbluth explains:

> Each evening and morning all the oceans and even the lakes of the world are a theater to mass movements involving billions and billions of creatures swimming from deep waters to the surface and then back again to a colder, darker world. . . . Sixty years ago, when active sonar was first available, captains of fishing vessels thought that the bottom was rising under the boat. This phenomenon, called the "vertical migration," is the largest synchronized animal movement on Earth. . . . Like nomads in deserts, they move to and from the surface waters in search of an oasis: the first 100m of the sea, where the sunlight still penetrates, the "photic" zone, abounding with food. . . . The signal that triggers and resets this ritual is the daily waning and waxing of sunlight, so travel frequently starts at dusk and ends at dawn.[40]

Although the deep sea is one place sunlight cannot reach, light and vision are no less critical there than for landlocked creatures. The Spookfish, for instance, lives at 1,000 meters' depth in cold, dark Pacific Ocean waters and has reflective telescopes for eyes.[41] Spookfish weren't exactly the first astronomers, but unlike any other vertebrates, their eyes include crystal mirrors that focus light on the retina. And some marine animals have evolved the equivalent of wearing Polaroid sunglasses—they can see polarized light, which assists in their discerning jellyfish and other nearly transparent species moving through the water column.

Deep below the coral reefs, ocean animals produce their own light to see and lure prey, signal distress or otherwise communicate, attract a mate, repel and stun predators, and counter-shade themselves.[42] Nouvian contends that bioluminescence is the most common mode of communication on the planet. Approximately 90 percent of all ocean animals produce or deploy bioluminescence in some capacity. Also referred to as cold body radiation or "cold light," nearly 98 percent of the energy used in generating this living light affords only a minimal release of heat, presumably to protect the luminescent organism from being detected by predators who sense infrared.

Bioluminescence has evolved independently some forty to fifty times as a result of its clear survival advantages. While fireflies provide the most familiar example, a handful of land-based organisms luminesce, including fly larvae as well as some earthworms, snails, mushrooms, and centipedes. In the ocean realm, bioluminescence is the rule rather than the exception. Ranking among the biolumi-naries are bacteria, snails, krill, squid, eel, jellies, sponges, corals, clams, sea worms, Green Lanternsharks, Megamouth Sharks, and an array of fish. Inhabitants of mid-water habitats luminesce at varying intensities to countershade themselves and blend in with down-welling light so informative in the ecology of the water column. Lanternfishes, of which there are at least two hundred species, can illuminate photophores along the entire length of their bodies to provide countershade. "Photophores are often highly evolved structures, consisting not only of a light-generating cell, but also a reflector, lens, and color filter," explains Scripps Institute oceanographer Tony Koslow. "Most photophores emit light in the blue end of the spectrum to match the wavelength of down-welling light."[43]

Viperfish and deep-sea anglerfishes have attached to their bodies a kind of a fishing rod with a lighted lure that attracts sizable prey directly into their toothy mouths. Marine biologist Edith Widder notes other deployments of bioluminescence in the ocean depths:

> There are many fishes, shrimps, and squids that use headlights to search for prey and to signal to mates. . . . And as with some cars, some headlights can be rolled down and out of sight when they are not in use—a handy way of hiding that reflective surface and allowing the fish to better blend into the darkness. Most headlights in the ocean are blue, which is the color that travels furthest through seawater and the only color that most deep-sea animals can see. But there are some very interesting exceptions like dragonfish with red headlights that are invisible to most other animals but that the dragonfish can see and use like a sniper scope to sneak up on unsuspecting, unseeing prey.[44]

Deep-sea prawns and some squid emit bioluminescent fluids to startle predators and evade being eaten. Koslow reports, "Some jellyfishes jettison bioluminescing tentacles before making their escape, much as a lizard may let go its still-writhing tail to occupy its predator while effecting its escape."[45] Other jellyfish can spray their predators with a luminescent substance that makes the

Plate 1. More than a century after Percival Lowell mapped Mars, the Hubble Space Telescope turned its best camera onto the red planet, producing this true color image in March 1997, just after the last day of Martian spring in the northern hemisphere. Mars was near its closest approach to Earth, 60 million kilometers away (NASA/David Crisp and the WFPC2 Camera Team).

Plate 2. Mars as seen by the Viking 2 lander. Images like this dashed the more fevered speculation of Mars as a living planet stemming from Percival Lowell and subsequent science fiction; it was revealed as a frigid and arid desert with a tenuous atmosphere, where water cannot exist for more than a moment on the surface before evaporating (NASA/Mary Dale-Bannister, Washington University in St. Louis).

Plate 3. Artist's conception of one of the Mars Exploration Rovers on Mars. Spirit and Opportunity each far exceeded their life expectancy and have performed at a very high level in the unforgiving conditions on the surface of Mars. The rovers were identical as pieces of hardware, yet had quite different experiences and adventures. Opportunity has exceeded its design lifetime by a factor of more than 35 (NASA/Jet Propulsion Laboratory).

Plate 4. Victoria Crater at *Meridiani Planum* on Mars, which is about 730 meters in diameter. Opportunity spent over two years exploring the rim of the crater, occasionally venturing inside, and by mid-2011 had traveled over 20 miles across the Martian terrain. Opportunity had a lucky bounce on landing and extricated itself from the kind of sand dune that disabled Spirit (NASA/Jet Propulsion Laboratory).

Plate 5. The four Galilean moons of Jupiter, visited by the Voyager spacecraft in 1979. In decreasing size order, the moons are Ganymede, Callisto, Io, and Europa; Jupiter is not shown at the same scale. Ganymede is 5300 km in diameter. The spacecraft made many discoveries in the Jovian moon system, including volcanism on Io and a subsurface ocean on Europa (NASA Planetary Photojournal).

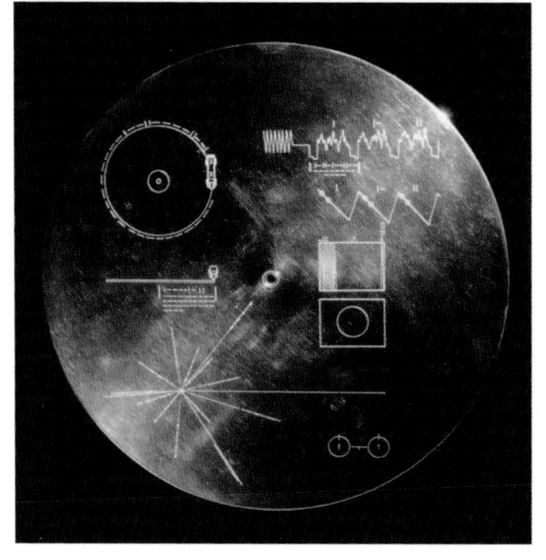

Plate 6. Attached to the body of each Voyager spacecraft is a gold-plated, copper phonograph record. The record contains musical selections, images, and audio greetings in many world languages, along with instructions on how to retrieve the information. The analog technology will be very durable in the far reaches of space (NASA/Jet Propulsion Laboratory).

Plate 7. An artist's impression of a close-up view of Saturn's rings. The rings are thought to be made of material unable to form a moon because of Saturn's tidal forces, or to be debris from a moon that broke up due to tidal forces. The particles are made of ice and rock and range in size from less than a millimeter to tens of meters across (NASA/Marshall Image Exchange).

Plate 8. A panoramic view of Saturn's moon Titan, from the Huygens lander during its descent to the surface in late 2005. In this fish-eye view from an altitude of three miles, a dark, sandy basin is surrounded by pale colored hills and a surface laced with stream beds and shallow bodies of liquid composed of methane and ethane (NASA/ESA/Descent Imager Team).

Plate 9. Stardust collected samples of a comet and interstellar dust samples using a particle collector with cells containing aerogel, which is an amorphous, silica-based material that is strong yet exceptionally light. Particles entered this solid foam at high speed and were decelerated and trapped. The spacecraft returned its samples to Earth in 2006 (NASA/Jet Propulsion Laboratory).

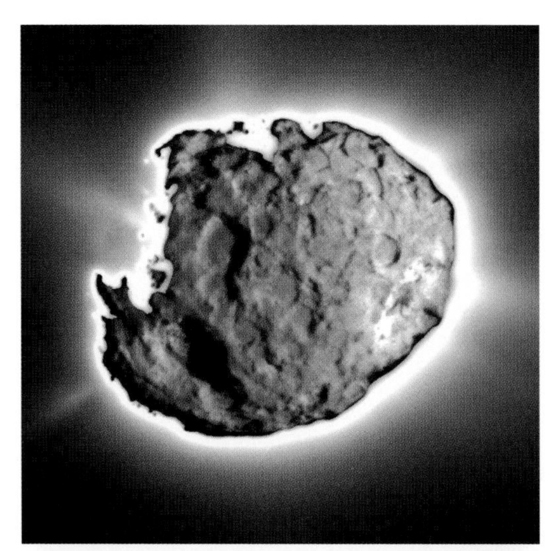

Plate 10. This composite image was taken of comet Wild 2 by Stardust during its close approach in early 2004. The comet is about 3 miles in diameter. Its surface is intensely active, with jets of gas and dust spewing millions of miles into space. This image is a hybrid: a short exposure captures the jet while a long exposure captures the surface features (NASA/Jet Propulsion Laboratory).

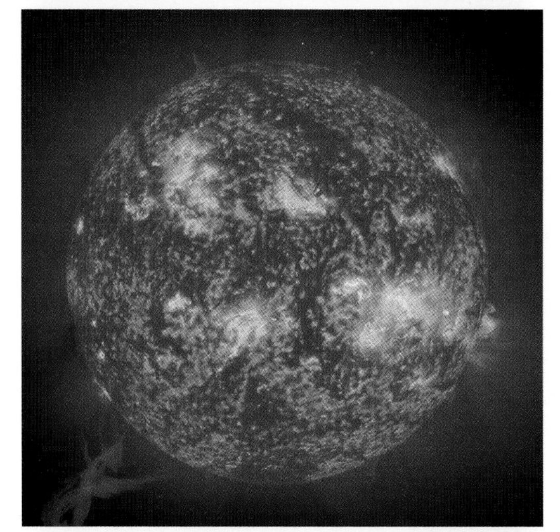

Plate 11. SOHO took this image of the Sun in January 2000. The relatively placid optical appearance belies the intense activity seen in this ultraviolet image, where a huge, twisting prominence has escaped the Sun's surface. When these events are pointed at the Earth, telecommunications and power grids can be affected (NASA/SOHO).

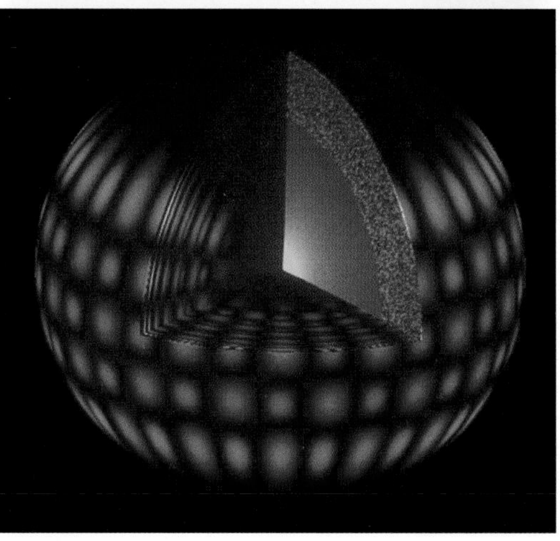

Plate 12. This computer representation shows one of millions of modes of sound wave oscillations of the Sun, where receding regions are colored red and approaching regions are colored blue. The Sun "rings" like a bell, with many complex harmonics, and study of the surface motions can be used to diagnose the interior regions (NSO/AURA/NSF).

Plate 13. People have used the sky as a map, a clock, a calendar, and as a cultural and spiritual backdrop since antiquity. This celestial map was produced in the seventeenth century by the Dutch cartographer Frederick de Wit. Constellations and star patterns are unchanging from generation to generation, so the stars were seen as being eternal (Wikimedia Commons/ Frederick de Wit).

Plate 14. We live in a city of stars, seen here in a full-sky panorama of the Milky Way photographed from Death Valley in California. The ragged band of light represents a view through the disk of the spiral galaxy we inhabit, and Hipparcos has mapped out the nearby regions of the disk and parts of the extended halo (U.S. National Park Service/Dan Duriscoe).

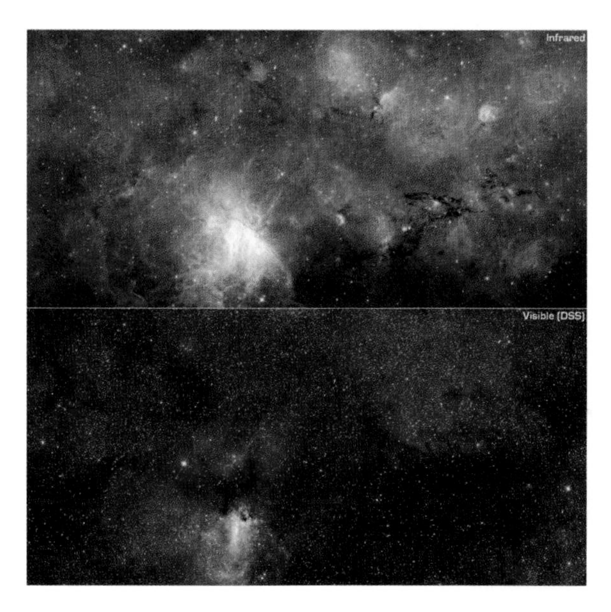

Plate 15. This Spitzer Space Telescope image shows star formation around the Omega Nebula, M17. This Messier object is a nebulosity around an open cluster of three dozen hot, young stars, about 5,000 light-years away. Spitzer records information at invisibly long wavelengths, and the difference between the view in the infrared (*top*) and the optical (*bottom*) can be dramatic. Colors in the infrared view represent different temperature regimes, with red coolest and blue hottest (NASA/JPL-Caltech/M. Povich).

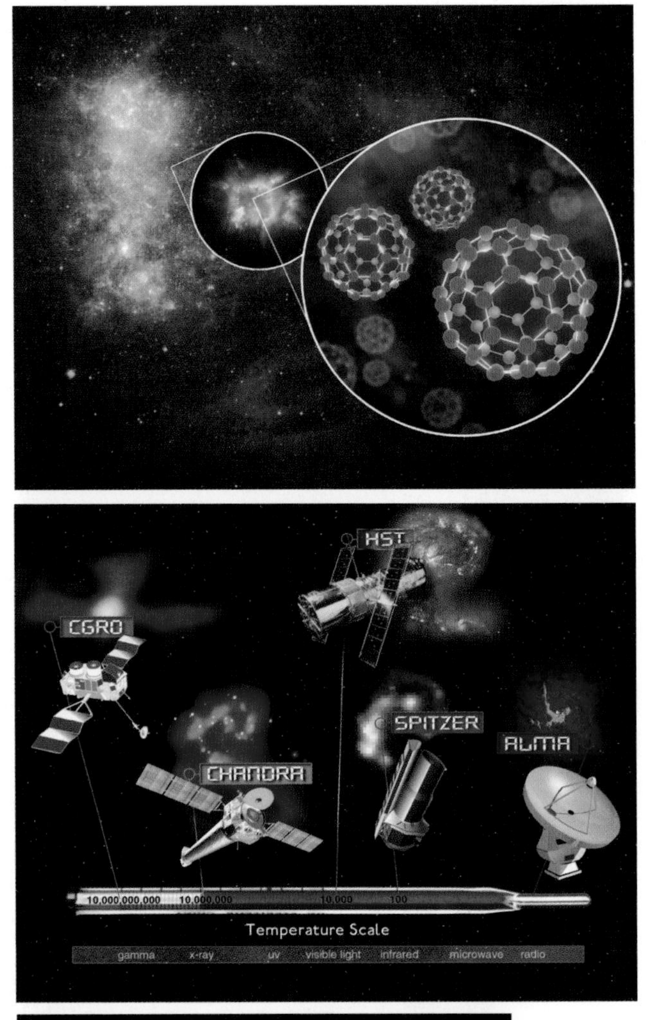

Plate 16. The Spitzer Space Telescope detected molecules of buckminsterfullerene, or "buckyballs," in a nearby galaxy, the Small Magellanic Cloud. The first zoom shows the type of planetary nebula where the molecules were found, and the second shows the molecule structure, where sixty carbon atoms are arranged like a tiny soccer ball (NASA/SSC/Kris Sellgren).

Plate 17. NASA's Great Observatories are multi-billion-dollar missions with complex instrument suites, designed to answer fundamental questions in all areas of astrophysics. Including the ground-based Atacama Large Millimeter Array (ALMA), they can diagnose the universe at temperatures ranging from tens to tens of billions of Kelvin (NASA/CXC/M. Weiss).

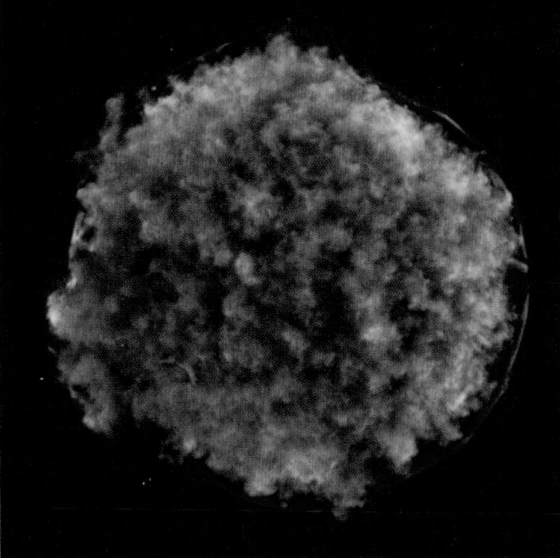

Plate 18. The dying star that produced this great bubble of hot glowing gas was first noted by Tycho Brahe in 1572. A white dwarf detonated as a supernova when mass falling in from a companion triggered its collapse; the shock wave from the subsequent explosion led to the blue arc. The surrounding material is iron-rich and highly excited iron atoms create spectral lines detectable at X-ray wavelengths (NASA/CXC/Chinese Academy of Sciences/F. Lu).

Plate 19. Two images of a pillar of star birth, three light-years high, in the Carina nebula, about 7,500 light-years away. Images taken through different filters select different wavelength ranges, which are combined into "true color" composites, where the colors convey astrophysical information in either visible light or infrared waves. Images like this have turned nebulae, galaxies, and clusters into "places" that resonate in the popular imagination (NASA/ESA/STScI/M. Livio).

Plate 20. This image of the towering gas columns and bright knots of young stars seen in the Eagle Nebula (M16) was probably the first Hubble Space Telescope image to achieve widespread public recognition. It was part of the inspiration for the Hubble Heritage project, which showcases a different high-impact color image on the web each month. New worlds are being born at the tips of these fingers of hot gas (NASA/ESA/STScI/J. Hester/P. Scowen).

Plate 21. This exquisitely accurate map of the microwave sky, a projection of the celestial sphere onto a plane, shows the universe when it was a tiny fraction of its present age. The temperature variation between red and blue "speckles" is about a hundredth of a percent. The tiny variations, on angular scales of about a degree, represent the seeds for galaxy formation. It took a hundred million years or so for gravity to form the first galaxies (NASA/WMAP Science Team).

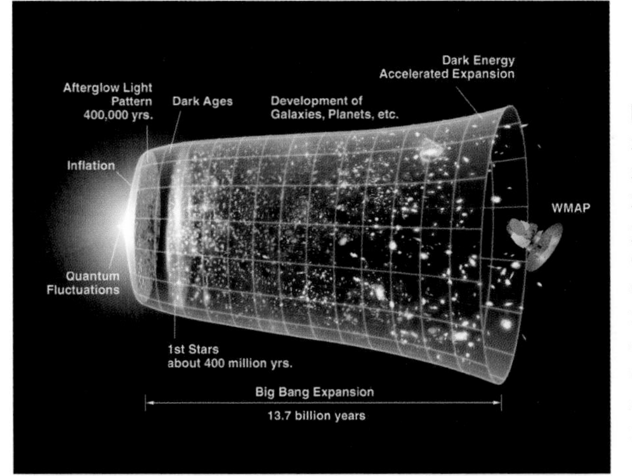

Plate 22. WMAP has played a major role in pushing the big bang model to the limit. The current model of the expanding universe posits an early epoch of inflation or exponential expansion, and subsequent expansion governed in turn by dark matter, causing deceleration, and more recently, dark energy, which is causing acceleration. WMAP has ushered in an era of "precision" cosmology (NASA/WMAP Science Team).

Plate 23. The Mars Science Laboratory, named Curiosity, will be exploring Mars for at least two years, starting with its landing in August 2012. The rover is the size of an SUV, compared to the Mars Exploration Rovers, which are the size of a golf cart, and the earlier Pathfinder, which is the size of a go-kart. Curiosity will study the past and present habitability of Mars by a detailed geochemical analysis of its rocks and atmosphere (NASA/JPL-Caltech).

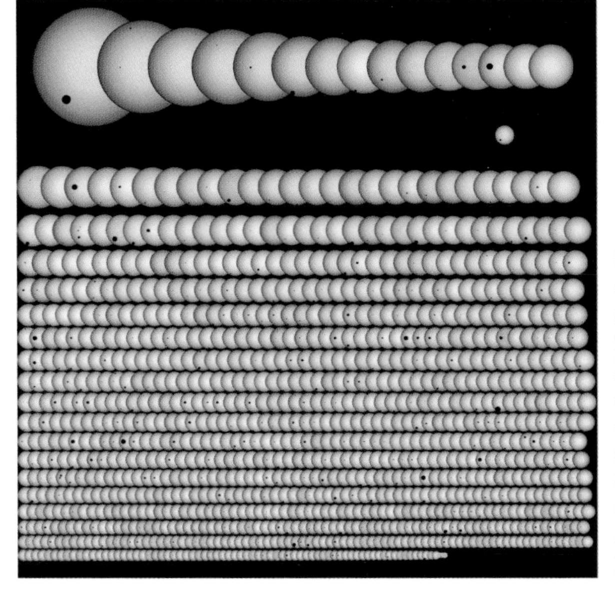

Plate 24. This montage of 1,235 exoplanet candidates from Kepler shows the planets projected against their parent stars, giving an idea of how they are detected by the slight dimming of the star's light. By the end of its mission, Kepler will have collected enough data to be sensitive to Earth-like planets in Earth-like orbits of their stars, many of which are expected to be habitable (NASA/Kepler Science Team).

predator in turn visible to fish looking for a quick meal. Medical researchers are only beginning to realize what we might learn from deep-sea bioluminaries. Off Puget Sound in the Pacific Northwest live the *Aequorea victoria* jellyfish, from which Green Fluorescent Protein (GFP) was first derived and used to generate other fluorescing marker proteins crucial in cancer and brain research and invaluable to cell biology and genetic engineering.[46] What zoologists discover about the variety of species deploying bioluminescence in the deep ocean can help astrobiologists anticipate the life-forms that might illuminate the icy oceans of Europa, or alien seas on exoplanets orbiting other stars. To that end, Spitzer and other telescopes nightly scour the skies in search of other worlds.

Forming New Solar Systems

The cold and dusty nebulae in which stars form are also places where planets form, and Spitzer has helped to tell this story. Newborn stars are embedded in a circumstellar disk of dust and grit coalescing around the star as a result of angular momentum conservation in the collapsing debris cloud. Just as an ice skater spins faster as they bring their arms to their chest, a diffuse cloud with slight rotation shrinks to a rapidly spinning compact disk.[47] Protoplanetary disks, like those in the Orion Nebula, are heated by a central star, reradiating all that energy in the infrared. Gas giant planets like Jupiter or Saturn form within the disk in as little as a few million years, the blink of a cosmic eye. The disk then dissipates and leaves behind a sparse gruel of debris made of dust particles that have been recycled though collisions of larger chunks of rock and evaporation (figure 9.4). Spitzer is perfectly suited to studying debris disks because the cool material easily outshines the star in infrared radiation. By comparing disks at various stages of their development, the evolutionary history of how solar systems are born can be pieced together.[48]

Spitzer's data indicates that the extremely dense, early phase of disk evolution lasts a couple of million years.[49] The process begins as microscopic particles aggregate into larger particles the size of dust grains. Rocky planets like Earth and Mars form by accre-

Figure 9.4. An artist's concept of a young star in its dense disk of gas and dust. Light cannot escape from the interior of the disk, but infrared radiation emerges unaltered and reveals the process of planet formation. Within the disk, theory predicts that matter can accrete from dust grains into several thousand Moon-mass objects in just a few million years (NASA/JPL-Caltech).

tion, as grit grows into boulders, then mountains, then planets—a process that takes 10 million years. Spectra show that the dust is mixed with gaseous organic molecules, including water (steam), carbon monoxide, carbon dioxide, and methane. Soot or pure carbon grains are an important ingredient of planets, and carbon-based life-forms. So it was with great excitement that astronomers announced the discovery, a few years ago, of Buckminster fullerenes, or buckyballs, in deep space. Buckyballs are spherical molecules made of sixty carbon atoms arranged in hexagons and pentagons like the panels of a soccer ball; the name alludes to Buckminster Fuller, who designed geodesic domes that look like these carbon molecules. Observations made with Spitzer have shown that buckyballs, or buckminsterfullerenes, are common in hydrogen-rich regions of space such as gaseous nebula where stars are born.[50] Buckyballs are surprisingly robust and may be linked to the emergence of life by forming stable cages for concentrating other molecules and so accelerating interaction rates and chemical complexity (plate 16).

Spitzer is well-suited to diagnose the chemistry of planet formation because molecules have most of their spectral transitions at infrared wavelengths. The spectra of dozens of planet-forming disks have revealed interesting anomalies. Hydrogen cyanide (HCN), which is a major repository of interstellar nitrogen, is rarer around stars less than half the Sun's mass than around solar-type stars. This implies that planets around low-mass stars are nitrogen-poor. Since nitrogen is a key biological ingredient, this implies that life might be rare, or fundamentally different, on planets of low-mass stars. Spitzer has also shown that rocky silicate material in the space between stars is amorphous in form, while silicates in the debris disks associated with planet formation have the sharper spectral features associated with crystalline silicates.[51] In the lab, amorphous silicates can only be converted into the crystalline forms by annealing at a temperature of 1000 K, which allows the molecules to gently reorient themselves. In proto-planetary disks, annealing requires recurrent flaring of the star during the first million years of disk evolution. In these cool regions that can only be observed with an infrared telescope, we can observe the early steps along the path to planets and life.

Diagnosing Distant Worlds

The discovery of planets orbiting stars other than the Sun was one of the most dramatic events of twentieth-century science. Until 1995, our Solar System was unique. Since then an accelerating pace of discovery has confirmed more than 850 exoplanets, with 2,700 more good candidates.[52] Many of these are Jupiter-mass planets, orbiting Sun-like stars within one hundred light-years.[53] But exoplanet detection limits are steadily approaching Earth-mass, and systems with as many as seven planets, have been discovered. The majority of these exoplanets were discovered by an indirect method, where the orbiting planet induces a "reflex motion" or a wobble in the parent star, and the wobble is detected by a periodic Doppler shift in the spectrum of the star. Some of the early discoveries were "hot Jupiters," or large planets on close orbits around their parent stars. But the Doppler detection method reveals nothing more than a planet's mass and orbital distance.

Spitzer, by contrast, can observe exoplanets as they pass in front of or behind their star. This is an eclipse or a transit. Only the small fraction of systems where the orbital plane is lined up with our sight line can show eclipses. The detectors can gather infrared light from large exoplanets that orbit close to their star and that are heated to at least 1000 Kelvin, if the star is within two hundred light-years of our Sun. As a planet traverses the face of its star it blocks some of the starlight. There's a temporary dip in the infrared signal of the combined system due to the planet blocking a portion of the stellar disk. This eclipse allows the size of the planet to be measured. Then when the planet passes behind the star (this is called a secondary eclipse), the infrared signal again dips because the planet's contribution is missing from the system's heat signal. The size of the drop is the amount of infrared emission coming from the planet. Spitzer's extended orbit allows continuous observations over an entire exoplanet orbit, and the telescope's ability to measure small changes of just 0.1 percent in brightness makes these observations possible (figure 9.5).

NASA's Kepler spacecraft, launched in March 2009, is also looking for transits caused by exoplanets by staring at a region in the Cygnus constellation. Kepler scientists want to determine whether Earth-like planets, small rocky worlds with long period orbits, are common. The exoplanet Kepler 22b, which is 2.4 times larger than Earth, was the first planet found by the telescope to orbit in the habitable zone of its sun-like star. As with the Milky Way Project that uses data from Spitzer's GLIMPSE survey, Zooniverse's Planet Hunters Project invites volunteers to assist NASA in digging through Kepler's data. Kepler sends data on more than 150,000 stars to Earth at regular intervals, and volunteers for Planet Hunters survey the stars' light-curves to identify possible transits. "Planet Hunters is an online experiment that taps into the power of human pattern recognition," organizers explain on the

Figure 9.5. A distant exoplanet is partially eclipsed by its parent star in this artist's impression. Spitzer uses observations like this to "taste" the atmosphere of an exoplanet by seeing which atoms or molecules are absorbed by the atmosphere. This particular planet is about 1000°F and the atmosphere has substantial amounts of carbon monoxide but surprisingly little methane (NASA/SSC/Joseph Harrington).

website. So far, two likely planets have been detected by citizen scientists in the Kepler data.[54] As has been the history of science from its earliest days, nonspecialists are making serious contributions to Kepler's scientific outcomes.

While Kepler can predict statistically the ubiquity of planets in our galaxy, it takes the sensitive instruments of an infrared telescope to characterize their composition. Spitzer's ability to diagnose the atmospheric composition of remote exoplanets was a big surprise. Observations have been made on more than two dozen exoplanets, with dozens more expected during the ongoing "warm" mission. Spitzer project scientist Michael Werner explains the telescope's contribution to this effort: "Because hot Jupiters are rich in gas, different wavelengths arise from different levels in the atmosphere of different chemical constituents. Spitzer data have allowed the determination of planetary temperatures and of constraints on chemical composition (including the identification of water vapor), atmospheric structure and atmospheric dynamics."[55] Though Spitzer's primary mission was to detect objects hidden by interstellar dust, characterizing exoplanets may become its greatest contribution.

In 2007, scientists first measured weather on a planet beyond the Solar System.[56] The planet HD 189733b races around its parent star in just over two days. Temperatures across the planet range from 970 K to 1220 K, which means winds of 6,000 mph must rage to keep the temperature variation that modest. By comparison, the Earth's jet stream sails along at only 200 mph. The exoplanet Upsilon Andromeda b is even more extreme; day to night temperature variations are 1400°C, that's 2550°F, and a hundred times larger than is typical on Earth![57] In the young, dynamic field of exoplanets, there have been many surprises, and theorists' expectations are often confounded. One hot Jupiter-like planet was found to have carbon monoxide but almost no methane. This violates models of hot gas giants, where most of the carbon should be in the form of methane. Another two planets are shrouded by dry, dusty clouds unlike anything seen in our Solar System. These planets show little of the expected water or steam, meaning that, if present, water might be hidden under the clouds in the form of a scalding ocean.

On Planets and Dwarfs

Somewhere in the twilight between stars and planets lie objects called brown dwarfs. Below about 8 percent of the Sun's mass, physical conditions will never allow the fusion of hydrogen into helium, as happens in the Sun. There may be a flickering of energy from the fusion of hydrogen into deuterium, but lower mass objects don't have a sustainable source of energy. When they're young, brown dwarfs are easy to observe in the infrared because they generate a lot of heat during their gravitational collapse. As they age, they get cooler and fainter. For example, a puny star 10 percent of the mass of the Sun would have a temperature of 3000 K and a luminosity 1/10,000 that of the Sun. By contrast, a brown dwarf 5 percent of the mass of the Sun would be three times cooler and a further 100 times dimmer. It would take a million of these feeble objects to equal the light of the Sun. Some astronomers consider Jupiter a failed star or a brown dwarf. Models indicate that brown dwarfs likely host moons like those of Jupiter and Saturn, worlds replete with weather systems, geysers, volcanoes, mountain ranges, and oceans—even if, as on Europa, the oceans are frozen over.

Set on a moon orbiting a gas giant planet in the nearby Alpha Centauri star system, James Cameron's film *Avatar* (2009) mesmerized audiences with its rendering of Pandora orbiting a Jupiter-like world whose swirled and rippled clouds resemble those of Jupiter and its Great Red Spot. Cameron's Pandora (one of Saturn's moons also happens to be named Pandora) abounds in colorful flora and fauna evocative of Earth's ocean environments, like the giant Christmas Tree Worms that suddenly retract when touched, the seeds of the sacred tree that float with the pulses of a jellyfish, and the photophores that dot the plants, animals, insects, and human-like inhabitants of Cameron's adventure tale.[58] That the Na'vi have blue skin seems no accident or random artistic choice. In ocean waters, blue light is least absorbed and can penetrate into the depths so that most deep-sea animals see solely in blue light. With the announcement in October 2012 of an exoplanet with a similar mass to Earth detected in the Alpha Centauri triple star system, Cameron's fictional depiction seems even more plausible.[59]

In part, what captivated film audiences was the bioluminescence with which Cameron painted his Pandoran forests. Having explored and documented deep ocean fauna in the film *Aliens of the Deep* (2005), Cameron is familiar with myriad bioluminescent sea life and readily projects similar plants and animals onto his imaginary world. Draped with waterfalls alight with bioluminescence, Pandora's forests teem with glowing fireflies and whirling fan lizards. The forest floor is carpeted with illuminating moss, while its streambeds are lit with the equivalent of sea anemones. In effect,

Cameron anticipates how life might be adapted on an exomoon of a gas giant planet or a brown dwarf star. On such worlds, we might expect to find luminous biota highlighted with photophores as Cameron predicts or species with eyes better adapted to night or low-light conditions. Even as felines have dark-adapted vision superior to humans, the Na'vi have feline features and navigate the night-time forest with far better facility than the character Jake. Cameron's bioluminescent world may have been inspired by Jules Verne's *20,000 Leagues Under the Sea*, in which the characters, strolling on the ocean floor, encounter bioluminescent jellyfish and note their "phosphorescent glimmers." Actually, nearly all jellies in the deep sea are luminescent. Verne's voyagers likewise chance upon corals, the tips of which glow. In stark contrast to the nineteenth-century perception that the ocean floor was devoid of life, Verne imagined the seafloor abounding in bioluminescence. An engraved illustration for *20,000 Leagues* titled "On the ocean floor" depicts a kelp forest with large corals, crustaceans, and a flotilla of giant jellyfish whose bodies and tentacles radiate light (figure 9.6). Cameron's vision of bioluminescent organisms flourishing on a nearby exomoon is similarly prescient.

Back on Earth, the first brown dwarf was discovered in 1995. Spitzer has contributed to this research by detecting some of the coolest and faintest examples known, including eighteen in one small region of sky sifted from among a million sources detected.[60] Despite their extreme faintness, Spitzer can detect the coolest brown dwarfs out to a distance of one hundred light-years. The outer layers of these substellar objects are cool enough that they're rich in molecules. The composition of the gas, and so the appearance of narrow lines that act as chemical "fingerprints" in the spectrum, changes as the brown dwarf evolves. Most of the eight hundred cataloged brown dwarfs have atmospheres with temperatures in the range 1200°C to 2000°C. The next category, with temperatures from 1200°C down to 250°C, has strong methane absorption in their atmospheres. There's overlap (and often confusion) between planets and brown dwarfs because some giant exoplanets are larger and hotter than some brown dwarfs. The very coolest category of the brown dwarf, and the end point of their evolution, has recently been discovered. NASA's Wide Field Infrared Survey

Figure 9.6. An original illustration from Jules Verne's *20,000 Leagues Under the Sea* depicting bioluminescence. A more modern media representation is in James Cameron's 2009 motion picture *Avatar*, where bioluminescent creatures inhabit the exomoon Pandora, where the action takes place. On Earth, thousands of species of sea creatures of all sizes employ bioluminescence (Jules Verne's *20,000 Leagues Under the Sea*).

Explorer (WISE), which was launched in 2009, has finished a scan of the entire sky at infrared wavelengths. In 2011, astronomers reported six of the coolest stars ever found, one of which has a surface at no more than room temperature.[61] The WISE data have the potential to reveal brown dwarfs closer than Proxima Centauri, the nearest star to our Sun.

Another promising mission in the search for dwarf stars and transiting exoplanets is headed by Harvard astronomer David Charbonneau and was designed largely by Philip Nutzman, an astronomer at the University of California, Santa Cruz. The MEarth Project is focusing eight small robotic telescopes on 2,000 M dwarf stars.[62] The project targets particularly M dwarfs as they are smaller than the Sun and transiting planets would block out a greater portion of the star's light, making them easier to detect. Studying nearby transiting exoplanets could afford astronomers a better sense of whether Earth-like planets are common and additionally allow astronomers to discriminate the chemistry of their atmospheres. Within six months of launching their project, the team detected their first transiting planet, a super-Earth named GJ 1214b in orbit around a star 13 parsecs from Earth.[63] In 2010, the journal *Nature* reported that the atmosphere of GJ 1214b was found to be comprised either of water vapor, or of thick clouds or haze as on Saturn's moon Titan.[64] Infrared investigations are planned to determine which of these options exist on the planet. With resources like Zooniverse.org, in the coming decade the public will likely contribute to the search for biomarkers such as ozone or water vapor in the atmospheres of these other worlds.

Awash as it is in newborn stars and exoplanets, the universe may be teaming with life. Perhaps in some far future humans will have physical, robotic, or other means of virtual presence on a nearby exoplanet or one of its moons that, similar to Cameron's Pandora, teems with bioluminescent life. Astronomer Carole Haswell cautions, "If any of this is to happen, however, we need to use our collective ingenuity to understand and repair the effects that our industrial activity and our burgeoning population are having on our own planet."[65] What we learn from exoplanets, Haswell suggests, might be invaluable in understanding Earth's climate evolution and in ensuring our own survival and that of our companion species. In the meantime, NASA and the astronomical community welcome the public's contribution to one of humankind's greatest adventures—locating possible habitable planets and, in time, the signatures of life in some dark and overlooked corner of the vast and silent wastes of interstellar space.

10 Chandra

EXPLORING THE VIOLENT COSMOS

When we dream of other worlds, our dreams may be vivid and real and colorful, but they're subject to the limitations of our senses. To visualize something, even if in our mind's eye, we use the visual sense. For most of the history of astronomy we learned about the universe exclusively through visible light. Tens of thousands of years of naked-eye observations were followed by the first night time use of the telescope by Galileo in 1610, followed by a steady march of successively larger telescopes, culminating with the 8–11 meter behemoths of the present day, with dreams of even larger ones to come.[1]

The historical focus on visible light was natural, since the most powerful human sense evolved to match the peak energy output of our life-giving star. The cool outer envelopes of stars emit most of their radiation in the visible range, and galaxies are just assemblages of stars. The night sky is visually rich. About 6,000 stars are visible to the naked eye at a dark site, but they are just an infinitesimal sliver of the stellar plentitude of the universe. We live in a system of 400 billion stars, and the observable universe contains roughly 100 billion stellar systems, for a staggering total of 10^{23} stars.[2] Also, the spectral transitions that are the fingerprints of different types of atoms fall mostly in the visible wavelength range. Cosmic chemistry is practiced with telescopes that gather light and disperse it into its constituent wavelengths.

Nevertheless, just over two hundred years ago experiments showed there are wavelengths longer than the reddest waves of light and bluer than the bluest waves of light. A hundred years later, Wilhelm Röntgen performed the first systematic experiments to understand the nature of mysterious radiation that was generated by high-energy particles moving in the vacuum of a sealed glass tube. In 1895, he announced the discovery of "X rays." The details of the discovery are poorly documented because Röntgen had his lab notebooks burned when he died. X-rays electrified the scientific community and within a year their potential for medical imaging was realized. It was a lot longer before anyone discovered that they could be used for astronomy. Röntgen received the first Nobel Prize ever awarded for Physics in 1901.[3]

Just as X-rays provided a window into the human body, they now provide a window in space onto worlds barely imagined. X-ray telescopes have revealed the nature of black holes and neutron stars for the first time, they've discovered sizzling hot plasma in the centers of galaxy clusters and in the diffuse space between galaxies, and they've let us make great strides in probing extraordinary black holes that lurk in the centers of all galaxies, ranging from a few million up to a few billion times the mass of the Sun. Their high energy relative to visible light means they're often created by events of unimaginable violence, phenomena unknown to us fifty years ago (figure 10.1).

X-rays in Popular Culture

News of Röntgen's discovery of X-rays rippled around the world. "When Röntgen discovered X-rays in November of 1895 by placing his hand between a Crookes Tube and a florescent screen, the medical applications of the technology were immediately apparent."[4] By January 1896, Röntgen had issued a report regarding his experiments to media in Europe and the United States. The public was immediately fascinated by the possibility of seeing through seemingly solid objects or inside the human body, and the topic made headlines in newspapers and weeklies. Apparently, X-ray

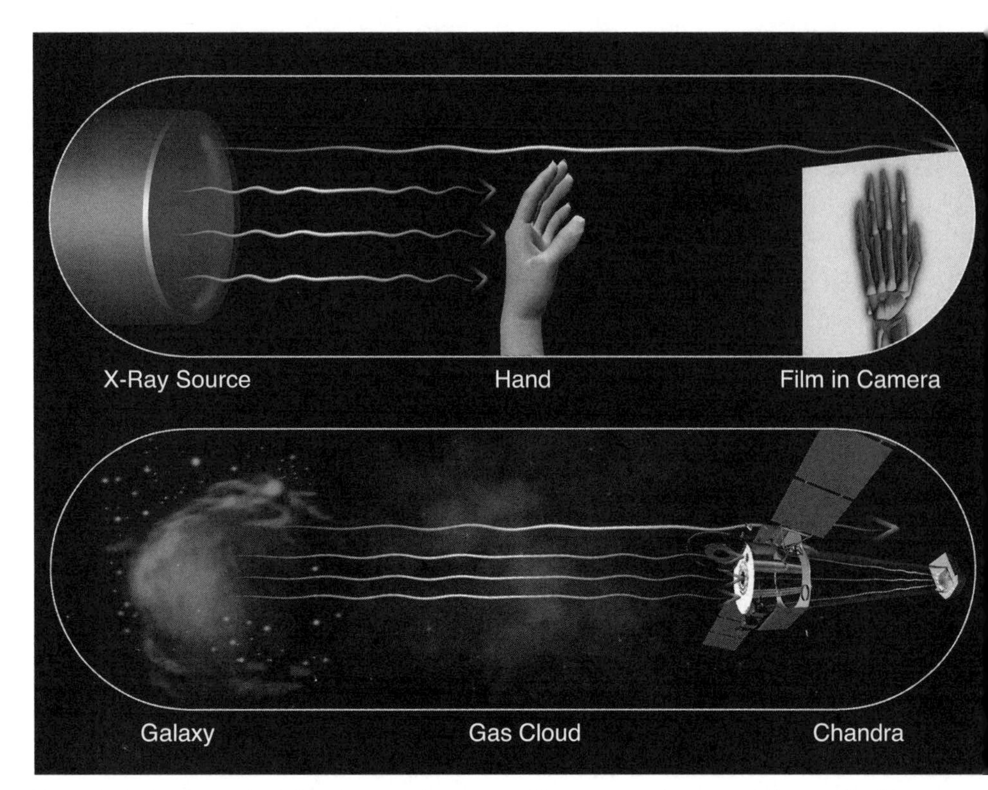

Figure 10.1. At the end of the nineteenth century, X-rays were harnessed for medical imaging, where dense material like bones cast a shadow in X-rays. In the astronomical realm, a very hot gas or plasma emits X-rays, as do regions near compact objects like black holes. The X-rays may also be absorbed by intervening material, and they can only be detected by a satellite above the Earth's atmosphere (NASA/CXC/M. Weiss).

"mania" swept through the public sphere but at a level far surpassing a simple cultural fad, explains Nancy Knight, who notes that "X-rays appeared in advertising, songs, and cartoons."[5] John Lienhard similarly reports, "Seldom has anything taken hold of the public imagination so powerfully and completely. . . . Right away, magazine cartoons celebrated the idea. A typical one showed a man using an X-ray viewer to see through a lady's hat at the theatre."[6]

Edwin Gerson has investigated the speed with which X-rays were incorporated into the branding of products in the United States, and he explores why so many marketers used the term X-ray to appeal to consumers. Being a medical doctor, Gerson began collecting household products dating back to the 1890s, none of which have anything whatsoever to do with X-rays. One could purchase Sniteman's X-Ray Liniment for horses and cattle, and Patt's X-Ray Liniment for people. X-Ray Golf Balls, ostensibly to improve the accuracy of one's game, were sold by John Wanamaker of New York. There was the D-cell X-Ray flashlight battery, X-Ray Stove Polish, X-Ray Cream Furniture Polish manufactured in New York, and X-Ray Soap made in Port Huron, Michigan. X-Ray Blue Double Edge razor blades were touted as "the finest blade known to science," while drinking fountains for chicken houses were marketed by the X-Ray Incubator Company in Wayne, Nebraska. Housewives could purchase an X-ray Coffee Grinder or the X-Ray lemon squeezer, and the X-Ray Raisin Seeder was available as early as November 1896. A company from Baltimore, Maryland, sold boxes of X-Ray headache tablets, with 8 pills for ten cents promising relief in just fifteen minutes.[7] Clearly, advertising by means of non sequitur is not just a phenomenon of the modern age.

The popular fascination with X-rays continued for decades. By 1924, an X-ray shoe fitter was deployed by shoe vendors to image both feet inside a new pair of shoes. David Lapp reports that in the United States "there were approximately 10,000 of these fluoroscopes in use, being made by companies such as Adrian X-Ray Shoe Fitter," and that by the early 1950s most shoe stores were using the devices.[8] As Gerson points out, "The public was simply astonished with X rays, and advertisers played off this spellbound attention by adding the name to almost any type of product. . . . Not only did the image of the X ray convey a sense of cutting-edge technology, it also functioned as a metaphor for 'powerful unseen truth and strength.'"[9]

Researchers and medical personnel working with X-rays were frequently exposed to dangerous or fatal levels of radiation, including Elizabeth Fleischmann, who at age 28 read a news article on how to build an X-ray device and by 1900 became the best medical radiologist in the United States. But in ten years' time, Fleischmann had been so badly overexposed to radiation by imaging her patients and developing the films that she did not survive.[10] At the time, the effect of exposure to radiation was poorly understood and this ignorance played out in dangerous exposure for both medical practitioners and those working with the new technology for purposes of entertainment. Nancy Knight, for instance, indicates that "papers reported daily" on Thomas Edison's efforts to X-ray a person's brain in order to figure out how the brain functions.[11]

Allen Grove reports that some speculated at the time that the new technology might afford a kind of X-ray vision that might make it possible to capture on film the images of ghosts, apparently a popular notion in England. He also comments that within months of Röntgen's announcement, theaters were headlining performances inspired by the popularity of X-rays.[12] Piano sheet music for American composer Harry L. Tyler's "X Ray Waltzes" from 1896 depicted on its cover a man holding his hand under an X-ray tube. Beneath the man's hand appears a skeletal X-ray image. In fact, in 2010, Tyler's waltzes were dusted off and featured in a BBC broadcast titled "Images that Changed the World." And, any discussion of X-rays in popular culture must recall that in the 1930s, when writer Jerry Siegel and artist Joe Shuster invented the prototypical superhero, it was inevitable that Superman would be gifted with X-ray vision. Though Superman didn't have such ability at the outset, in later adventures he is able to see in multiple wavelengths of the electromagnetic spectrum, including radio waves, infrared, ultraviolet, and eventually X-rays.[13]

Röntgen's discovery had a deep shaping effect on the arts. This was because X-rays were an entrée into a hidden world, the "world within the world" represented by the interior structure of atoms and molecules. In 1912, less than twenty years after the initial discovery, physicist Max von Laue showed that X-rays were a form of electromagnetic radiation because they diffracted when interacting with matter. A year later, the father and son team of William Henry and William Lawrence Bragg described mathematically how X-rays scattered within a crystal, winning the Nobel Prize in Physics for their work in 1915. This opened the door to X-ray crystallography, a powerful tool for measuring the spacing and arrangement of atoms within a solid. The impact of this discovery was particularly strong in the visual arts, which were undergoing their own revolution with the onset of Fauvism, Cubism, and Futurism. The X-ray was liberating because to artists it "proved that the external is not valid, it's just a false layer."[14]

Art historian Linda Dalrymple Henderson contends that X-rays suggested the possibility of stepping into the fourth dimension. She writes, "From the 1880's to the 1920's, popular fascination with an invisible, higher dimension of space—of which our familiar world might be only a section or shadow—is readily apparent in the vast number of articles and the books . . . published on the topic." Henderson has demonstrated that modern artists readily embraced X-rays, and their ability to expose realities and perspectives not perceptible to the human eye, as suggestive of four-dimensional space.[15] Cubist paintings, with their simultaneous depiction of all sides of a seemingly transparent object, and their suggestion of interior planes, explains Henderson, "are testaments to the new paradigm of reality ushered in by the discovery of X-rays and interest in the fourth dimension. Such paintings are new kinds of 'windows'—in this case, into the complex, invisible reality or higher dimensional world as imagined by the artist."[16] The inspiration was sometimes very direct. According to historian Arthur Miller, "Picasso's Standing Female Nude" (1910) was inspired by the power of X-rays to glimpse beyond the visible: what you see is not what you get. In this case, the inspiration was X-ray photographs taken to diagnose the illness of Picasso's mistress, Fernande Olivier. Superposed on a background of planes, her body lies open to reveal pelvic hip bones made up of geometrical shapes: forms reduced to geometry—the aesthetic of Cubism, inspired by modern science."[17] From the arts, to medicine, and particularly through astronomy and space exploration, X-rays have profoundly altered human culture.

A Gentle Astrophysical Giant

In 1923, Wilhelm Röntgen died as one of the most celebrated physicists of his time. A modest man, he had declined to take out a patent on his discovery, wanting X-rays to be used for the benefit of humankind. He also refused to name them after himself and donated the money he won from his Nobel Prize to his university.

That same year Subrahmanyan Chandrasekhar started high school in a small town in southern India. Chandra, as he was universally known, would grow into another great but modest scientific figure. Chandra means Moon or luminous in Sanskrit. He was one of nine children and although his family had modest means, they valued education; therefore he was home schooled since local public schools were inferior and his parents couldn't afford a private school. His family hoped Chandra would follow his father into government service but he was inspired by science and his mother supported his goal. In addition, he had a notable role model in his uncle, C. U. Raman, who went on to win the Nobel Prize in Physics in 1930 for the discovery of resonant scattering of light by molecules.[18]

Chandra's family moved to Madras and he started at the university there, but was offered a scholarship to study at the University of Cambridge in England, which he accepted. He would never live in India again. At Cambridge he studied, and held his own, with some of the great scholars of the day. As a young man he fell into a controversy with Arthur Eddington, at the time the foremost stellar theorist in the world, which upset him greatly. Eddington refused to accept Chandra's calculations on how a star might continue to collapse when it had no nuclear reactions to keep it puffed up. Chandra was a gentle man, unfailingly polite and courteous. In part due to this conflict, he chose to accept a position at the University of Chicago, where he spent the bulk of his career.

His name is associated with the Chandrasekhar limit, the theoretical upper bound on the mass of a white dwarf star, or about 1.4 times the Sun's mass. Above this limit, the force of gravity overcomes all resistance due to inter-particle forces, and a stellar corpse will collapse into an extraordinary dark object, either a neutron star, or if there's enough mass, a black hole. In the 1970s, he spent several years developing the detailed theory of black holes. Chandra was amazingly productive and diverse in his scientific interests. He wrote more than four hundred research papers, and ten books, each on a different topic in astrophysics. For nineteen years, he was the editor-in-chief of *The Astrophysical Journal*, guiding it from a modest local publication to the international flagship journal in astronomy. He mentored more than fifty graduate students while at the University of Chicago, many of whom went on to be the leaders in their fields. In 1983, he was honored with the Nobel Prize in Physics for his work on the theory of stellar structure and evolution.

Through his long career, Chandra watched the infant field of X-ray astronomy mature. The telescopes flown in sounding rockets in the early 1960s were no bigger than Galileo's spyglass. Just before its 1999 launch, NASA's Advanced X-Ray Astrophysical Facility was renamed the Chandra X-ray Observatory (CXO) in honor of this giant of twentieth-century astrophysics.[19] In a span of less than four decades, Chandra improved on the sensitivity of the early sounding rockets by a factor of 100 million.[20] The same sensitivity gain for optical telescopes, from Galileo's best device to the Hubble Space Telescope, took four hundred years—ten times longer!

The Chandra X-ray Observatory

Until the Chandra X-ray Observatory was launched, the best X-ray telescopes were only as capable as Galileo's best optical telescope, with limited collecting area and very poor angular resolution. With Chandra, X-ray astronomers gained several orders of magnitude of sensitivity, and the ability to make images as sharp as a medium-size optical telescope. Chandra was the third of NASA's four "Great Observatories." The others are the Hubble Space Telescope, launched in 1990 and still doing frontier science, the Compton Gamma Ray Observatory, launched in 1991 and deorbited in 2000 after a successful mission, and the Spitzer Space Telescope, launched in 2003 and currently in the final "warm" phase of its mission since its liquid helium coolant ran out in 2009 (plate 17).

It took a while to open the X-ray window on the universe because the Earth's atmosphere is completely opaque to X-rays. In the 1920s, scientists first proposed using versions of Robert Goddard's rocket to explore the upper atmosphere and peer into space. However, this idea wasn't realized until 1948, when a re-purposed V2 rocket was used to detect X-rays from the Sun.[21] The next few decades saw the development of imaging capabilities for X-rays and new detector technologies, and X-ray astronomy tracked the maturation of the space program. The first X-ray source beyond the Solar System, Scorpius X-1, in the constellation of Scorpius, was detected by physicist Riccardo Giacconi in 1962.[22] This intense source of high-energy radiation is a neutron star, the end result of the evolution of a massive star where gravity crushes the remnant to a state as dense as nuclear matter. The intense X-rays result from gas being drawn onto the neutron star from a companion and being heated violently enough to emit high-energy radiation.

Giacconi is another "giant" in astrophysics—as leading scientist for X-ray observatories from Uhuru in the 1970s to Chandra in the 1990s, first director of the Space Telescope Science Institute, and winner of the Nobel Prize in Physics in 2002. This last and ultimate accolade fittingly came almost exactly a century after the award of the first Nobel Prize in Physics for the discovery of X-rays to Wilhelm Röntgen. Born in Genoa, Italy, his early life was disrupted by the Second World War; as a high school student, he had to leave Milan during allied bombing raids. He returned to complete his degree and started his life as a scientist in the lab, working on nuclear reactions in cloud chambers. With a Fulbright Fellowship he moved to the United States and forged his career there. He had his hand in all the pivotal discoveries of X-ray astronomy: the identification of the first X-ray sources, characterization of black holes and close binary systems, the high-energy emission that emerges from the heart of some galaxies, and the nature of the diffuse X-rays that seem to come from all directions in the sky.[23]

The growth in the number of celestial X-ray sources gives a sense of how each new mission has advanced the capabilities: 160 sources in the final catalog of the UHURU satellite in 1974, 840 in the HEAO A-1 catalog in 1984, nearly 8,000 from the combined Einstein and EXOSAT catalogs of 1990, and about 220,000 from the ROSAT catalog of 2000. Chandra has had a wider-field but lower sensitivity counterpart in the European XMM-Newton mission, also launched in 1999. These two X-ray satellites have detected a total of over a million X-ray sources.

Chandra was launched by the Space Shuttle *Columbia* into a highly elliptical orbit that takes it a third of the way to the Moon. At five tons, it was the most massive payload launched up to that time by the Space Shuttle. The elongated orbit gives it lots of "hang time" in the perfect vacuum of deep space and lets science be done for 55 hours of the 64-hour orbit. This comes at the expense of the spacecraft being unserviceable by the Shuttle when it was still flying, so the facility has to work perfectly. In fact, the only technical problem was soon after launch when the imaging camera suffered radiation hits during passage through the Van Allen radiation belts; it is now stowed as the spacecraft passes through those regions. The spacecraft had a nominal five-year mission at time of launch but it's producing good science well into its second decade and is expected to last at least fifteen years.[24]

One of two different imaging instruments can be the target of incoming X-rays at any given time. The High Resolution Camera uses a vacuum and a strong electric field to convert each X-ray into an electron and then amplify each one into a cloud of electrons. The camera can make measurements as quickly as 100,000 times per second, allowing it to detect flares or monitor rapid variations. Chandra's workhorse instrument is the Advanced Camera for Imaging Spectroscopy. With 10 CCDs, it has one hundred times better imaging capability than any previous X-ray instrument. Either of these cameras can have one of two gratings inserted in front of it, to enable high- and low-resolution spectroscopy. Spectroscopy at X-ray wavelengths is a bit different from optical spectroscopy. The spectral lines seen by Chandra are usually very high excitation lines of heavy elements like neon and iron, coming from gas that's kept highly agitated by high-energy radiation or violent atomic collisions.[25]

There are three major differences between optical and X-ray detection of sources in the sky. The first is the way the radiation is gathered. X-rays falling directly on silvered glass have such high energy that they penetrate the surface and are absorbed, like tiny bullets. X-ray telescopes use a shallow angle of incidence, so that the photons bounce off the mirror like a stone skimming off water. Chandra uses a set of four concentric mirrors, six feet long and very slightly tapered so they almost look like nested cylinders. This method of gathering radiation makes it difficult to achieve a large collecting area. The second difference is the much higher energy of X-rays. Chandra measures photons in a range of energy from 0.1 to 10 keV, or 100 to 10,000 electron volts, which is a standard unit of measure for photons. On the electron volt energy scale, two numbers that bracket this range are 13.6 eV, the modest energy required to liberate an electron from a hydrogen atom and 511 keV, the rest mass energy of an electron. For reference, photons of visible light have wavelengths 10,000 times longer and energies 10,000 times lower.

With each photon packing such a punch, a typical astronomical source emits far fewer X-ray photons than visible photons, so each is very valuable. The goal in X-ray astronomy is to detect every photon individually. Very few photons are required to detect a source (this is helped by the fact that the "background rate" is low; X-rays are not created by miscellaneous or competing sources, so the X-ray sky is sparsely populated). In some of the deepest observations made by Chandra, two or three photons collected over a two-week period is enough evidence to declare that a source has been detected. There are papers in the research literature with more authors than photons!

Chandra unlocks the violent universe because celestial X-rays have high energy and can only be produced by extreme physical processes.[26] The Sun and all normal stars are very weak X-ray sources because their cool outer regions produce thermal radiation peaking at visible wavelengths. It would take a gas at hundreds of thousands or millions of degrees to emit copious X-rays; diffuse gas with this very high temperature is distributed between galaxies. Another way X-rays can be made is when particles are accelerated to extremely high energies; they release the energy in a smooth spectrum that extends to X-rays and even gamma rays.[27] Despite the million plus X-ray sources that have been cataloged, there are thousands of times more optical sources, so the X-ray sky is relatively quiet. But many of those X-ray sources are extremely interesting because they're situations where matter has been subject to extreme violence.

Heart of Darkness

Black holes are black. That seems like a self-evident statement. Nothing can escape the event horizon of a black hole; the event horizon isn't a physical barrier or a boundary but an information membrane, defining the region from which no particle or radiation can escape. Black holes are the ultimate expressions of general relativity, where mass curves space so much that a region is pinched off from view.[28]

Black holes are the final states of massive stars. Every star is in a life-long battle between the forces of light and darkness. The light comes from fusion reactions creating pressure that pushes outward, while the dark is the implacable force of gravity pulling inward. In the end gravity always wins. When a star twenty or more times the Sun's mass exhausts its nuclear fuel, the core collapses into a state so dense that nothing can escape, not even light. An isolated black hole would be black and undetectable. However, more than half of all stars are in binary or multiple systems, and that's also true of the most massive stars that collapse catastrophically when their nuclear fuel is exhausted, forming a black hole. The rotation of the star is amplified in its newly compact state, so the black hole spins very fast. As material from a companion star is pulled onto the black hole, it forms a disk of gas, like water swirling into the drain of a bathtub.[29] The disk is very hot, tens of thousands of degrees, and it glows in ultraviolet radiation and X-rays. Some hot plasma is accelerated along the pole of the spinning black hole, where it emits X-rays and gamma rays. So while a black hole is black, gas from a companion can be heated into pyrotechnic activity when it's falling into the black hole (figure 10.2). The accretion process is well enough understood theoretically that X-ray signatures can be used to identify black holes. Some of the radiation comes from no more than 100 kilometers from the event horizon. The Chandra Observatory has played a vital role in this work.[30]

Chandra has the sensitivity to detect stellar black holes hundreds of light-years away. Only about twenty binary systems have well-enough measured masses to be sure the dark companion is a black hole, but X-ray observations can be used to identify black

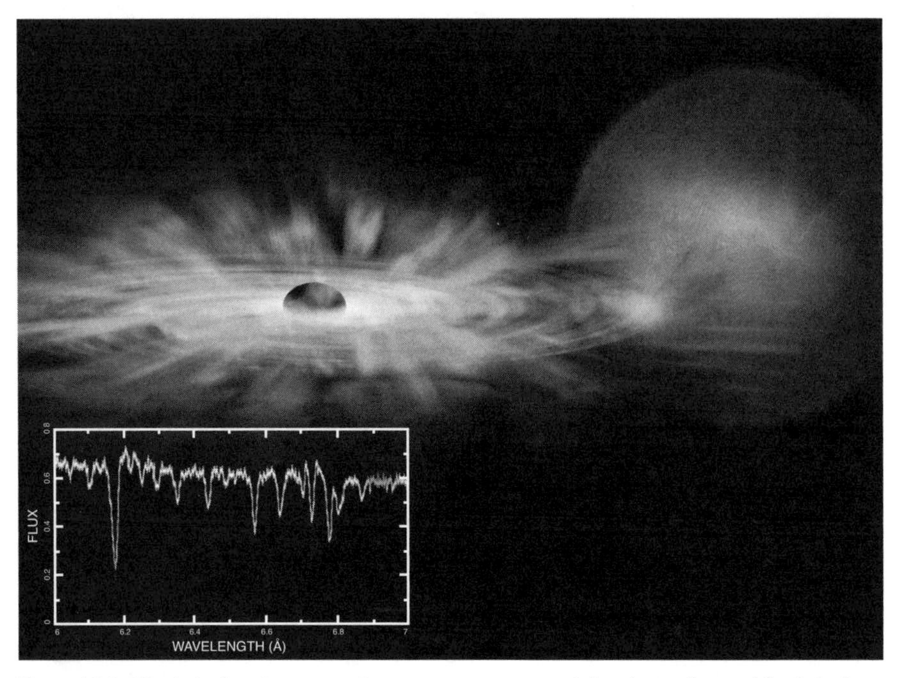

Figure 10.2. Black holes do not emit any energy or particles, but when a black hole is part of a binary system, gas is drawn from the companion onto an accretion disk that glows in X-rays. The binary orbit gives the mass of the black hole. The image is an artist's impression, while the inset shows an X-ray spectrum which diagnoses the temperature of the plasma near the black hole and gives clues to the black hole's properties (NASA/CXC/M. Weiss/J. Miller).

holes with fairly high reliability. The examples studied with X-ray telescopes are the brightest representatives of a population of about 100 million black holes in the Milky Way.[31]

X-ray observations have also pushed the limit of our understanding of black holes. In 2007, a research team used Chandra to discover a black hole in M33, a nearby spiral galaxy. The black hole was sixteen times the mass of the Sun, making it the most massive stellar black hole known.[32] Moreover, it was in a binary orbit with a huge star seventy times the Sun's mass. The formation mechanism of the black hole that placed it in such a tight embrace with its companion is unknown. This is the first black hole in a binary system that shows eclipses, which provides unusually accurate measurements of mass and other properties. The massive companion will also die as a black hole, so future astronomers will be able to gaze on a binary black hole where energy is lost as gravitational radiation and the two black holes dance a death spiral as they coalesce into a single beast.[33]

Stellar Cataclysm

The formation of a black hole is just one half of the story when a massive star dies. As the star reaches the end of its life and all fusion energy sources are exhausted, it suffers a dramatic gravitational collapse. The crushing force of in-falling gas squeezes the core into a black hole, but much of the mass rebounds outward and the outflowing gas meets more in-falling gas and heats it momentarily to billions of degrees. Heavy elements like gold and silver and platinum are created in that thermonuclear blast and they surf into space to become part of future generations of stars.[34] This is a supernova.

For centuries and millennia after the supernova event, outrushing gas interacts with the interstellar medium to form a delicate filigree of nebulosity and filaments of glowing gas. Supernova remnants are most clearly seen in X-rays, and with Chandra their structure can be studied in great detail. The spectrometers on Chandra have mapped the ejection of iron and silicon and oxygen and nitrogen, elements essential for planet-building and for life.[35] Cosmic alchemy is part of our story since every carbon and oxygen atom in every person was once part of a previous generation of stars. The imagers on Chandra have observed the shock-heated filaments in enough detail to model the process that accelerates particles to within a fraction of a percent of the speed of light (plate 18).

Chandra has observed supernova remnants at different stages of their evolution and used them to piece together the expansion history of a typical dying star. This research has an intriguing connection to human history. Tycho's supernova of 1572 in the constellation of Cassiopeia and the Crab Nebula supernova of 1054 in the constellation of Taurus were two of the brightest "guest" stars of the last millennium, visible to the naked eye and in the historical record of many cultures.[36] Recently, Chandra data led to a revised age for the supernova remnant RCW 86, and the detonation date of AD 185 perfectly matches the notation of a guest star in records of the Chinese court astronomers, making this the first supernova in recorded human history.[37]

At the center of a supernova remnant is either a black hole or a neutron star. Neutron stars aren't as exotic as black holes, but they are quite bizarre. The pressure of gravity forces protons to merge with electrons and create pure neutron material; without the electrical repulsion to keep the particles separate they are as close together as marbles in a jar. A neutron star is like a giant atomic nucleus with an atomic number of 10^{57}. Chandra can see neutron stars via the high-energy phenomena spawned by their intense magnetic fields. A neutron star has a magnetic field of about 10^{12} Gauss, equivalent to 10 billion refrigerator magnets. They are spinning at typical speeds of 60 rpm and sometimes as quickly as 45,000 rpm. Imagine something a million times the mass of the Earth that's the size of a city and spinning nearly a thousand times a second! Chandra images of dozens of neutron stars have revealed glowing clouds of magnetized particles blowing away from the neutron star like a wind.[38]

Black Holes: From Speculation to Observation

In 1964, Riccardo Giacconi and his team discovered Cygnus X-1, a source in the constellation known as the Swan that is generally accepted as the first black hole detected. Cygnus X-1 has a mass nine times that of the Sun, with an implied event horizon size of 16 miles, smaller than a city. But in the early 1960s, X-ray astronomy was an emerging science,[39] and a singularity was then considered purely theoretical.

Science writer Dennis Overbye recounts how physicist John Wheeler, based on his students' prompting, eventually came to accept the probabilities of the cataclysmic collapse of a star:

> In 1939, J. Robert Oppenheimer, who would later be a leader in the Manhattan Project, and a student, Hartland Snyder, suggested that Einstein's equations had made an apocalyptic prediction. A dead star of sufficient mass could collapse into a heap so dense that light could not even escape from it. The star would collapse forever while spacetime wrapped around it like a dark cloak. At the center, space would be infinitely curved and matter infinitely dense, an apparent absurdity known as a singularity.
>
> Dr. Wheeler at first resisted this conclusion, leading to a confrontation with Dr. Oppenheimer at a conference in Belgium in 1958, in which Dr. Wheeler said that the collapse theory "does not give an acceptable answer" to the fate of matter in such a star.

However, by 1967, during a presentation in New York, Wheeler had reconsidered and, "seizing on a suggestion shouted from the audience, hit on the name 'black hole' to dramatize this dire possibility for a star and for physics."[40] Mitchell Begelman and Martin Rees likewise note that it was Wheeler who first officially used the term "black hole" in reference to a collapsed star and comment that "the name immediately caught on." They write, "Here was something infinitely more mysterious, and perhaps much more sinister as well—a place where any form of matter or energy could enter, lose its identity, and be lost forever to the Universe."[41] In *Black Holes and Time Warps,* Caltech astrophysicist Kip Thorne similarly observes: "Of all the conceptions of the human mind, from unicorns to gargoyles to the hydrogen bomb, the most fantastic, perhaps, is the black hole . . . a hole with a gravitational force so strong that even light is caught and held in its grip; a hole that curves space and warps time."[42]

In 1970, Stephen Hawking developed theorems that demonstrated black holes would be a standard phenomenon of Einstein's general relativity and of the universe. By the mid-1970s, not surprisingly, black holes were resonating across the realms of fiction, popular music, art, and film. The Canadian rock group Rush released "Cygnus X-1, Book 1: The Voyage" on their album *A Farewell to Kings* (1977). It's an extended rock composition written in homage to the X-ray emission source that Riccardo Giacconi discovered in Cygnus. The lyrics read in part:

In the constellation of Cygnus
There lurks a mysterious, invisible force
The Black Hole
Of Cygnus X-1[. . . .]
Invisible
To telescopic eye
Infinity
The star that would not die
All who dare
To cross her course
Are swallowed by
A fearsome force. . . .

I set a course just east of Lyra
And northwest of Pegasus
Flew into the light of Deneb
Sailed across the Milky Way[. . . .]
The X-ray is her siren song
My ship cannot resist her long
Nearer to my deadly goal
Until the Black Hole—
Gains control. . . .

Sound and fury
Drowns my heart
Every nerve
Is torn apart. . . .[43]

Located approximately 6,000 light-years from Earth, Cygnus X-1 is now thought to be a high-mass X-ray binary system in which gaseous material from the blue supergiant star HDE 226868 is accreting onto a nearby black hole created by the collapse of a star nine times the mass of our Sun.[44] The system emits powerful X-ray radiation as gas from the blue supergiant siphons onto, and is heated by, the black hole. Interest in Cygnus X-1 was further sparked by a public and congenial bet between Stephen Hawking and Kip Thorne regarding the physics of the object. By 1990, having argued that Cygnus X-1 was not a black hole, Hawking admitted to losing the wager.[45]

Science in the Matrix of Culture

No longer the stuff of science fiction, black holes pervade all facets of culture. Wallace and Karen Tucker, in recounting the history of the Chandra X-ray Observatory, note that the telescope has offered "substantial observational evidence for the existence of black holes," which they claim have since "become part of the popular literature as a metaphor for an irretrievable loss of material, time, money, and so on."[46] In the hands of English novelist Martin Amis, in *Night Train* (1988), black holes become a metaphor for death.[47]

But black holes are understood in the public arena as far more substantial than a mere metaphor. In 2008, it was reported that physicists at CERN's Large Hadron Collider could produce miniature black holes that in a runaway scenario might swallow the Earth. In fact, public concern over the possibility of scientists manufacturing microscopic black holes led to the facility posting an information and safety web page that reassures general audiences of the following:

> Speculations about microscopic black holes at the LHC refer to particles produced in the collisions of pairs of protons, each of which has energy comparable to that of a mosquito in flight. Astronomical black holes are much heavier than anything that could be produced at the LHC. According to the well-established properties of gravity, described by Einstein's relativity, it is impossible for microscopic black holes to be produced at the LHC. There are, however, some speculative theories that predict the production of such particles at the LHC. All these theories predict that these particles would disintegrate immediately. Black holes, therefore, would have no time to start accreting matter and to cause macroscopic effects.[48]

A year later J. J. Abrams's film *Star Trek* (2009) depicted Romulans generating, at will, miniature black holes that could devour starships and planets. Actually *Star Trek* and black holes have a history that goes back to the same year Wheeler started using the term. The idea of a collapsed or black star was floated in the first season of the original *Star Trek* series in an episode titled "Tomorrow is Yesterday," first aired in January 1967, in which the starship *Enterprise* encounters a "black star of high gravitational attraction."[49] As Captain Kirk and the crew attempt to escape its gravity well, their ship is slung into a time warp that sends them plunging back through time. The *Enterprise* ends up adrift in Earth orbit in 1968, just days prior to the first manned Moon mission. Though not referred to as a black hole, the black star that triggers their adventure was the product of Einstein's singularity and Oppenheimer and Snyder's collapsed star.

From space artists like David A. Hardy to amateur astronomers like Stephen Cullen, many have speculated what Cygnus X-1 might actually look like.[50] Hardy's rendering depicts powerful X-ray jets shooting from the poles of the binary system's black hole. Cullen actually photographed those jets in May 2009. "Luckily, Cygnus X-1 inhabits a region of space thick with gas that emits specific kinds of radiation when hit by shock waves, energetic particles, and relativistic jets from black holes," writes Cullen, who notes that astronomer David Russell, and a team working with him, had already obtained "visual evidence of a jet-powered nebula caused by the northern jet slamming into the interstellar medium." But Cullen thought there might be a powerful jet emerging from the southern pole, and there was. "Unlike the northern region, the images did not show a well-defined shell of shocked gas," he recalls. "What I did see, however, was a diffuse, fan-shaped glow that traced a path directly back to Cygnus X-1."[51] Having obtained clear images of the previously undetected jet, Cullen subsequently invited Russell's comments and together the astronomers collaborated to publish Cullen's results. The discovery demonstrates the extent to which the Chandra Observatory has prepared astronomers to anticipate what might be explored in wavelengths beyond visible light.

Not only have Chandra's X-ray observations demonstrated profound aspects of the high-energy universe otherwise unavailable to telescopes, but its images have opened the universe for visually impaired readers as well. In 2007, astronomy educator Noreen Grice and astronomers Doris Daou and Simon Steele published *Touch the Invisible Sky* in which Braille text and especially prepared Braille images detail discoveries by Chandra, Spitzer, and other telescopes that reveal the universe in electromagnetic wavelengths not visible to the human eye.[52] In that text and other NASA

Braille books, readers can experience the infrared, ultraviolet, and X-ray universe, through their fingertips.[53]

The popular fascination with black holes seems hinged on the fact that a singularity and its event horizon contort our wildest imaginings about the physical universe. We think we can wrap our minds around just about any natural phenomena, but black holes defy all logic at the point of the singularity and leave unanswered questions regarding the physics at its core. Kip Thorne asserts that black holes have much to tell us about the birth and the future of the universe: "Gravitational-wave detectors will soon bring us observational maps of black holes, and the symphonic sounds of black holes colliding—symphonies filled with rich, new information about how warped space-time behaves when wildly vibrating. Supercomputer simulations will attempt to replicate the symphonies and tell us what they mean, and black holes thereby will become objects of detailed experimental scrutiny."[54] Like the music of the universe discussed in the chapter on WMAP, the universe has more to tell us in the narratives or symphonies we are yet to learn from black holes. One thing at least seems certain. The new understandings of massive black holes that Chandra is unfolding will inevitably be woven into the matrices of human culture and language, poetry and music, and even our dreams.

The Enigmatic Center of Our Galaxy

Twenty years ago, evidence began to accumulate that the center of the Milky Way galaxy contained a dark mass that could not be accounted for as normal stellar remnants.[55] Star motions in the center were so rapid they indicated a huge black hole, about four million times the mass of the Sun. The evidence for this supermassive black hole is now better than for the more conventional black holes that result from the death of a massive star. The mass is not just based on stellar velocities but on tracking the entire stellar orbits as they loop around the central dark object.[56]

Chandra did not contribute the evidence that cemented the case for a large black hole in the galactic center, but it showed that the black hole was unusually anemic, emitting far less high-energy ra-

Figure 10.3. At the center of the Milky Way galaxy, 27,000 light-years away, in the direction of the Sagittarius constellation, is a very dense star cluster. In this X-ray image, spanning 3 light-years, the bright regions represent hot gas from overlapping supernova remnants and to the lower left of the bright region, dynamical evidence indicates a black hole about 4 million times the mass of the Sun. The actual center of the galaxy is marked and is an ultra-compact radio source called Sag A* (NASA/CXC/MIT/F. Baganoff).

diation than other massive black holes. Even as it devours 10^{19} kg of material each second, it radiates energy a million times less efficiently than a set of stars of equivalent mass. Yet, while the black hole in the galactic center is very feeble, it's not boring. Ten years ago, an X-ray flare was seen coming from the vicinity of the black hole and since then, hundreds of flares have been seen, occurring almost daily.[57] They raise the level of X-ray emission tens or hundreds of times above the quiescent state (figure 10.3). Such rapid flares must be created within ten times the event horizon scale, close to the point of no return. The galactic center also harbors a source of energy intense enough to create anti-matter in the form of positrons.[58]

The real mystery is the inactivity of the galactic center. The center of our galaxy harbors a massive star cluster with star densities thousands of times higher than those seen in the solar neighborhood, and there is plenty of gas available within 10–20 light-years. So why is the black hole so quiet? The best guess is that it's currently starved because explosive events have cleared away much of the gas from around it. Chandra has provided evidence for this explanation. There are lobes of X-ray emission that indicate the black hole was more active 5,000 or so years ago, and blobs of plasma emerging from the center argue for quasi-periodic activity. Remarkable observations a few years ago saw a cloud of gas near the black hole brighten and fade in only a few years, responding to an X-ray pulse that had traveled for three hundred years to get there. We can infer that the black hole was a million times brighter three hundred years ago, when Queen Anne had just ascended to the throne in England and the largest town in her American colonies was Boston, with a population of seven thousand.[59] However, the galactic center is about 27,000 light-years from Earth, so all of this action really happened in the late Stone Age before humans settled down into civilizations and the information is just reaching us now.

Supermassive Black Holes

In the 1960s, long before the Milky Way was known to have a supermassive black hole, violent activity was seen in the centers of some distant galaxies. The manifestations were gas motions a hundred times faster than would be seen in a normal stellar system, intense radio emission concentrated in a tiny region of the galactic nucleus, and radiation across the electromagnetic spectrum from radio waves to gamma rays. Some of the strongest X-ray sources in the sky are galaxies hundreds of millions of light-years away, which means they have an X-ray emission thousands of times stronger than a galaxy like the Milky Way. Rapid variability of the radio, optical, and X-ray emission localized the intense activity to regions light-days across, only ten times bigger than the Solar System. No star cluster can pack so much radiation into such a small volume; the only plausible explanation was a gravitational engine: a black hole.[60]

Imagine a city like Los Angeles viewed from high above in a helicopter, where every streetlight, every house light, and every car light is standing in for a star. There are probably 100 million lights in the greater Los Angeles area, so in this analogy each light represents a thousand stars. Los Angeles is like the disk of a spiral galaxy. Now imagine a light one inch across in the city center that emits a hundred times more radiation than all the city lights put together. That's the intensity of an active galactic nucleus. It would be clearly visible from an immense height, long after the individual lights in the city had faded from view. We can see an incredibly bright point of light but the surrounding galaxy is too faint to see. That's a quasar.

Through the 1990s astronomers realized that *every* galaxy contains a black hole, with a mass that scales with the mass of the old stars in the center of the galaxy.[61] Moreover, these black holes are not active most of the time, so are not feeding and spewing out the energy that makes them easy to detect. The Milky Way is a medium-size galaxy with a 4-million-solar-mass black hole that's currently inactive. The brightest quasars harbor black holes a thousand times larger, or a few billion times the mass of the Sun. These extraordinary gravitational engines form and grow quite naturally as galaxies form and grow but steadily consume matter, and by occasional mergers. It's remarkable that nature makes black holes spanning a factor of a billion in mass, from a few times the mass of the Sun to the mass of a small galaxy. Their sizes range by the same factor of a billion, but they all share the properties of having an event horizon and (it is supposed) a central singularity. Strange as it might seem, massive black holes do not represent an extremely dense state of matter. The size of the event horizon is proportional to mass, but the volume increases by the cube of the size, so the more massive a black hole is the less dense it is within the event horizon. The 3-billion-solar-mass beast at the heart of the elliptical galaxy M87 is not much denser than water! Modest density does not, however, reduce their intrinsic strangeness (figure 10.4).

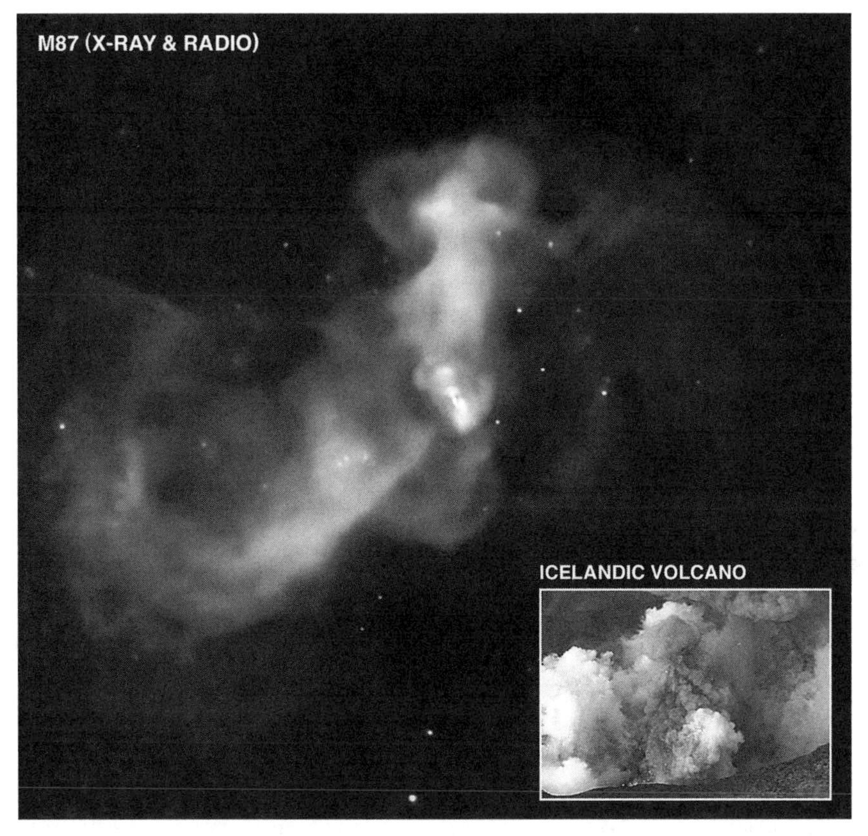

Figure 10.4. M87 is a giant elliptical galaxy with a "super-volcano" of X-ray activity originating from a black hole about 3 billion times the mass of the Sun at its center. The black hole sends out jets of material along the poles of its spin axis that create shock waves and interact with the intergalactic medium in a way analogous to a large volcano interacting with Earth's atmosphere (NASA/CXC/KIPAC/N. Werner/E. Million).

Chandra has played a vital role in telling the story of supermassive black holes and their behavior. In nearby galaxies it has detected vast bubble-like cavities and wavelike ripples in hot gas near the center, signs of "blowback" from the central engine.[62] X-rays provide evidence of repetitive explosive activity, where mass falls onto the black hole, sparking it into activity and triggering jets of high-energy particles, which then evacuate the nearby volume and starve the black hole until new material accumulates. In this way, a massive black hole can be inactive most of the time. Chandra has also seen binary black holes in the centers of merging galaxies.

In one case, two huge black holes are only 3,000 light-years apart and will merge in the next hundred million years, provoking a cataclysm of gravity waves and new activity and yielding a single galaxy with a larger black hole.[63] This is direct evidence for the manner in which galaxies and their embedded black holes have grown by mergers and acquisitions over billions of years.

Black holes and galaxies evolve on cosmic timescales of billions of years, so to figure out how they change the strategy is to make deep surveys that capture the census of active and inactive black holes at a range of distances or redshifts. The Chandra Deep Field was stared at for 2 million seconds, over three weeks, to create the deepest X-ray image of the sky ever made. This deep field, covering a patch of sky smaller than a postage stamp held at arm's length, is combined with a shallower but wider survey to tell the full story.[64] One of us (CDI) has participated in this research, with a survey of active galaxies in a two square degree region that netted more than two thousand supermassive black holes, and showed their evolution in cosmic time. This survey was sensitive enough to catch feeble accretors and black holes similar in size to the one in the Milky Way.

The broad conclusion of these surveys is that star formation and black hole growth in a galaxy are tightly linked. The heaviest black holes, a hundred million times the Sun's mass or larger, ate voraciously in the first few billion years after the big bang, and have been grazing or fasting since then. Black holes ten to a hundred million times the Sun's mass had a more well-regulated diet, and because they consume smaller proportions of their gas and dust early on, they continue to grow even today. X-rays are crucial to telling this story, since they can penetrate regions that would extinguish optical and ultraviolet radiation.[65] The typical size of a galaxy forming a supermassive black hole of a particular mass reduces with cosmic time, so the biggest black holes formed first. Massive galaxies in the far-flung universe misspent their youths by building monstrous black holes. This result is apparently at odds with the hierarchical mode of formation of structure in the universe, where small objects form first and gradually grow into larger objects. This unexpected phenomenon has been called "downsizing."[66]

A by-product of this research is the solution to an old mystery of the X-ray sky. When the first X-ray satellites took their data in the 1960s and 1970s they saw a faint diffuse glow that spanned the entire sky. This X-ray "background" was unexplained. Surveys with Chandra and XMM-Newton show that the background is actually composed of a myriad of pinpoints of X-ray emission from active galaxies 3–8 billion light-years away. Also, a majority of the emission is from galaxies where the nuclear activity is shrouded by dust and would not have been visible to an optical telescope. These surveys have also shown that black holes in the most massive galaxies are "green." By comparing the available fuel with the energy required to evacuate cavities in the central regions, the efficiency can be calculated. It's high because the mass is accreted slowly and smoothly and the energy is extracted very close to the event horizon of the black hole. If a family car was as efficient as one of these supermassive black holes, it would get a billion miles to a gallon of gas.

Hunting Dark Matter and Dark Energy

Chandra has also weighed in on two of the profound mysteries of cosmology: the nature of dark matter and dark energy. These two components of the universe account for 95 percent of its behavior, yet the physical basis for them is not known and is not part of standard physics. Dark matter outweighs normal matter by a factor of six and binds galaxies and clusters and stops them from flying apart, as well as causing the expanding universe to decelerate for most of the first two thirds of its existence. Dark energy dominates dark matter by a factor of three and has caused the cosmic expansion to accelerate in the most recent third of the universe's existence.[67]

Dark matter and dark energy don't interact strongly with normal matter like the atoms in our bodies, so stealth and cunning must be used to ensnare them and measure them. In both cases the laboratories used are rich clusters of galaxies, consisting of thousands of galaxies moving swiftly under the action of gravity and a huge cloud of superhot gas, so energetic that it emits X-rays. In 2006, an object called the Bullet Cluster was used for a convincing demonstration that dark matter actually exists. In the Bullet Cluster, a bullet-shaped cloud of hundred-million-degree gas is produced by a high-speed collision between a large cluster and a smaller one. The hot gas was slowed by the collision, due to a drag force analogous to air resistance. By contrast, the dark matter hardly interacted at all and sailed through during the collision, ending up on either side of the hot gas.[68] This result would not have occurred unless weakly interacting dark matter dominated the mass of both clusters. In particular, alternate theories of gravity, conjured up to avoid needing dark matter, fail to explain the observations. It seems we have to live with dark matter.

Dark energy is even more ephemeral, announcing its presence (and ubiquity) only by the effect it has on the cosmic expansion. The long and hard search for independent evidence of its existence settled on clusters of galaxies. In a study that took nearly a decade to complete, researchers showed that rich clusters suffer from "arrested development."[69] It's more difficult for clusters to grow when space is being stretched. By comparing the size and age of clusters with simulations of how they should grow under different conditions of cosmology, the results cement the interpretation of dark energy as a universal repulsive agent. They also rule out alterations to gravity theory and confirm that general relativity is a good description of the behavior of matter and radiation on large scales. The enigma of dark energy has not been solved, but its status as the biggest challenge in both physics and cosmology has been enhanced. Whether it is a black hole devouring a companion, massive black holes causing mayhem in the centers of galaxies, or clusters being pulled apart by the accelerating expansion of space, Chandra has provided data to illustrate the violence of the universe we inhabit.

A result of these insights is a new sense of the power of gravity. Stars are powered by gravity, and gravity also governs the nature of the most compact and energetic objects we've ever discovered. Dark forces even govern the expansion of the universe. When we think of other worlds, we think of planets illuminated by stars, where any life that exists is beholden to the energy from the star. But in a fundamental sense, starlight is just the inefficient leakage of radiation from mass-energy conversion by fusion. The true source of the starlight is gravity—a star is a gravitational engine. Black holes and neutron stars have no light but they have intense gravity. While they seem alien and utterly different from our world, given suitable protection or adaptation living creatures might be able to live near these compact stellar husks, using gravity to power their dreams.

11 HST

THE UNIVERSE IN SHARP FOCUS

Above all scientific projects, the Hubble Space Telescope encapsulates and recapitulates the human yearning to explore distant worlds, and understand our origins and place in the universe. Its light grasp is 10 billion times better than Galileo's best spyglass, and many innovations were needed for it to be realized: complex yet reliable instruments, the ability for astronauts to service the telescope,[1] and the infrastructure to support the projects of thousands of scientists from around the world. The facility and its supporters experienced failure and heartache as well as eventual success and vindication.

Hubble's legacy has touched every area of astronomy, from the Solar System to the most distant galaxies. In the public eye, it's so well known that many people think it's the *only* world-class astronomy facility. In fact, it operates in a highly competitive landscape with other space facilities and much larger telescopes on the ground. Although it doesn't *own* any field of astronomy, it has made major contributions to all of them. It has contributed to Solar System astronomy and the characterization of exoplanets, it has viewed star birth and death in unprecedented detail, it has paid homage to its namesake with spectacular images of galaxies near and far, and it has cemented important quantities in cosmology, including the size, age, and expansion rate of the universe.[2]

Ranked by size of the mirror, Hubble wouldn't make it into the top fifty largest optical telescopes.[3] Its preeminence is based on three factors associated with its location in Earth orbit. The first is liberation from the blurring and obscuring effects of the Earth's atmosphere. Ground-based telescopes typically make images far larger than their optics would allow because turbulent motion in the upper atmosphere jumbles the light and smears out the images. Hubble gains in the sharpness of its vision by a factor of ten relative to a similar-sized telescope on the ground. Earth orbit also provides a much darker sky, which affects the contrast and depth of an image. The difference you might see in going from a city center to a rural or mountain setting is only part of the story; natural airglow and light pollution affect even the darkest terrestrial skies. A vacuum can't obfuscate. The last feature of a telescope in Earth orbit is its ability to gather wavelengths of radiation that would be partially absorbed or even quenched by the Earth's atmosphere. Hubble has taken advantage of this by working at infrared and ultraviolet wavelengths.

The Hubble Space Telescope (HST) is well into its third decade of operations, and it's easy to take for granted the beautiful images that are released almost weekly. But it was not an effortless journey for NASA's flagship mission.

A Long and Bumpy Road

In 1946, Yale astronomy professor Lyman Spitzer wrote a paper detailing the advantages of an Earth-orbiting telescope for deep observations of the universe.[4] The concept had been floated even earlier, in 1923, by Hermann Oberth, one of the pioneers of modern rocketry. In 1962, the U.S. National Academy of Sciences gave its imprimatur to the idea, and a few years later Spitzer was appointed chair of a committee to flesh out the scientific motivation for a space observatory. The young space agency NASA was to provide the launch vehicle and support for the mission. NASA cut its teeth with the Orbiting Astronomical Observatory missions from 1966 to 1972.[5] They demonstrated the great potential of space astronomy, but also the risks—two of the four missions failed. We've already encountered Spitzer since he gave his name to NASA's infrared Great Observatory. Spitzer worked diligently to convince his colleagues around the country of the benefits of such a risky and expensive undertaking as an orbiting telescope.

After the National Academy of Sciences reiterated its support of a 3-meter telescope in space in 1969, NASA started design studies. But the estimated costs were $400–500 million and Congress balked, denying funding in 1975. Astronomers regrouped, NASA enlisted the European Space Agency as a partner, and the telescope shrunk to 2.4 meters. With these changes, and a price tag of $200 million, Congress approved funding in 1977 and the launch was set for 1983. More delays followed. Making the primary mirror was very challenging and the entire optical assembly wasn't put together until 1984, by which time launch had been pushed back to 1986. The whole project was thrown into limbo by the tragic loss of the *Challenger* Space Shuttle in January 1986. When the shuttle flights finally resumed, there was a logjam of missions so another couple of years slipped by.[6]

Hubble was launched on April 24, 1990, by the shuttle *Discovery*. A few weeks after the systems went live and were checked out, euphoria turned to dismay as scientists examined the first images and saw they were slightly blurred. The telescope could still do science but some of the original goals were compromised. Instead of being focused into a sharp point, some of the light was smeared into a large and ugly halo. This symptom indicated spherical aberration, and further in-flight tests confirmed that the primary mirror had an incorrect shape. It was too flat near the edges by a tiny amount, about one-fiftieth of the width of a human hair. Such was the intended precision of Hubble's optics that this tiny flaw made for poor images.[7] Hubble's mirror was still the most precise mirror ever made, but it was precisely *wrong*.

The spherical aberration problem may be ancient history and in the rearview mirror now, but at the time it was a public relations nightmare for NASA. Its flagship mission could only take blurry images. Commentators and talk show hosts lampooned the telescope and David Letterman presented a Top Ten list of "excuses" for the problem on the *Late Show with David Letterman*.[8] More seriously, the episode became fodder for case studies in business schools around the country. The fundamental error was the result of poor management, not poor engineering. The Space Telescope project had two primary contractors: Perkin-Elmer, who built the optical telescope assembly, and Lockheed, who built the support systems for the telescope. There was also a network of two dozen secondary contractors from the aerospace industry. The mission was jointly executed by Marshall Space Flight Center and Goddard Space Flight Center, whose relationship involved rivalry and was not always harmonious, with overall supervision from NASA Headquarters. Complexity of this degree can be a recipe for disaster without tight and transparent management, and clear communication among the best technical experts.

When the primary mirror was being ground and polished in the lab by Perkin-Elmer, they used a small optical device to test the shape of the mirror. Because two of the elements in this device were mis-positioned by 1.3 millimeters, the mirror was made with the wrong shape. This mistake was then compounded. Two additional tests carried out by Perkin-Elmer gave an indication of the problem, but those results were discounted as being flawed! No completely independent test of the primary mirror was required by NASA, and the entire assembled telescope was not tested before launch, because the project was under budget pressure. Also, NASA managers didn't have their best optical scientists and engineers looking at the test results as they were collected. The agency was embarrassed and humbled by the failure. Their official investigation put it succinctly: "Reliance on a single test method was a process which was clearly vulnerable to simple error."[9] In this way a multi-billion-dollar mission was hamstrung by a millimeter-level mistake and the failure to do some relatively cheap tests. In the old English idiom: penny wise, pound foolish. The propagation of a small problem into a huge one recalls another aphorism from England, where a lost horseshoe stops the transmission of a message and the result affects a critical battle: for the want of a nail, the war was lost.

A Telescope Rejuvenated

In fact, NASA had lost a big battle, but they weren't yet ready to concede the war. With Hubble working at part strength, the agency immediately began planning to diagnose and correct the problem with the optics. Although a backup mirror was available, the cost of bringing the telescope down to Earth and re-launching it would have been exorbitant. It was just as well that the telescope had been built to be visited by the Space Shuttle and serviced by astronauts. The instruments sitting behind the telescope fit snugly into bays like dresser drawers and they could be pulled out and replaced with others of the same size.

It also helped that the mirror flaw was profound but relatively simple, and the challenge was reduced to designing components with exactly the same mistake but in the opposite sense, essentially giving the telescope prescription eyeglasses. The system designed to correct the optics was the brainchild of electrical engineer James Crocker, and it was called COSTAR, or Corrective Optical Space Telescope Axial Replacement.[10] COSTAR was a delicate and complicated apparatus with more than five thousand parts that had to be positioned in the optical path to within a hair's breadth. Installing COSTAR raised the difficulty level of an already challenging first servicing mission that was planned for late 1993. Seven astronauts spent thousands of hours training for the mission, learning to use nearly a hundred tools that had never been used in space before. They did a record five back-to-back space walks, each one grueling and dangerous, during which they replaced two instruments, installed new solar arrays, and replaced four gyros. This last fix was needed because of the disconcerting tendency of Hubble's gyros to fail at a high rate, in a way never seen in lab testing. Without working gyros, the telescope could not point at or lock on a target. Before leaving, the crew boosted Hubble's altitude, since three years of drag in Earth's tenuous upper atmosphere had started to degrade the orbit.

The first servicing mission was a stunning success. It restored the imaging capability to the design spec and added new capabilities with a more modern camera. It also played significantly into the vigorous debate in the astronomy and space science community over the role of humans in space. NASA had always placed a strong bet that the public would be engaged by the idea of space as a place for us to work and eventually live. But after the success of Apollo, public interest and enthusiasm waned. (It even waned during the program; a final three planned Moon landings were scrapped.) Also, most scientists thought that it was cheaper and less dangerous to create automated or robotic missions than to service them with astronauts. NASA's amazingly successful planetary missions from the 1970s to the 1990s were seen as evidence of the primacy of robotic spacecraft. Hubble was of course designed to be serviced by astronauts, but the often-unanticipated problems they were able to solve in orbit, coupled with the positive public response (and high TV ratings during the space walks, at least the early ones), persuaded many that the human presence was essential and inspirational.

More servicing missions followed. Each one rejuvenated the facility and kept it near the cutting edge of astronomy research. The second mission in 1997 installed a sensitive new spectrograph and Hubble's first instrument designed to work in the infrared. Astronauts also upgraded the archaic onboard computers. The third mission in 1999 was moved forward to deal with the vexing problem of failing gyros. Before the mission flew, the telescope lost a fourth gyro, essentially leaving it dead in the water.[11] All six gyros were replaced and the computer was upgraded to one with a blistering speed of 25 MHz and capacious two megabytes of RAM (that's fifty times slower and thousands of times less storage than the average smart phone). The fourth mission in 2002 replaced the last of the original instruments and also removed COSTAR since it was no longer needed—each subsequent instrument has had its optics designed to compensate for the aberration of the primary mirror. The infrared instrument was revived, having run out of coolant two years early. New and better solar arrays were installed, along with a new power system, which caused some anxiety since to install it the telescope had to be completely powered down for the first time since launch. Except for its mirror, Hubble is reminiscent of the ship of Theseus, a story from antiquity where every plank and piece of wood of a ship is replaced as it plies the seas.

The fifth and last servicing mission almost didn't happen. Once again, the mission was affected by a Shuttle disaster, in this case the catastrophic loss of *Columbia* and crew in February 2003. NASA Administrator Sean O'Keefe decided that human repairs of the telescope were too risky and future Shuttle flights could only go to the safe harbor of the International Space Station. He also studied engineering reports and concluded that a robotic servicing mission was so difficult that it would most likely fail. At that point, Hubble was destined to die a natural death as gyros and data links failed.

When in January 2004 it was announced that the Hubble Space Telescope would be scrapped, the public's response was overwhelmingly in support of refurbishing the instrument in orbit. David Devorkin and Robert Smith report that a "'Save the Hubble' movement sprang into life."[12] Robert Zimmerman similarly comments: "The public response . . . was, to put it mildly, loud. Very soon, NASA was getting four hundred emails a day in protest, and over the next few weeks editorials and op-eds in dozens of newspapers across the United States came out against the decision."[13] Zimmerman details the history, development, and deployment of the telescope as well as its more significant findings. He suggests that Hubble, more than any other telescope or major scientific instrument, has allowed humankind to explore what lies at the furthest depths of space and that this is one reason for the widespread public sense of ownership of what some newspapers have called the "people's telescope."[14] Not since Edwin Hubble did his pioneering work on cosmology with the 100-inch reflector at Mount Wilson Observatory has public interest in a particular instrument been so intense or pervasive.[15]

O'Keefe and other NASA officials were taken aback by the public response. They'd expected astronomers to lobby hard for another servicing mission, and they weren't surprised that astronauts weighed in, saying they'd signed up knowing the risks and they wanted to service the telescope. But it turned out that a large number of people felt attached to Hubble through its spectacular pictures and newsworthy discoveries. They felt clear affection for the facility, and since the taxpayer had indeed paid for it, the agency took notice and O'Keefe first reevaluated and then reversed his decision.[16] The fifth servicing mission went off without a hitch in May 2009, leaving the telescope in better shape overall and with certain capabilities a hundred times more effective than its original configuration.[17] Two new instruments were installed, two others were fixed, and many other repairs were carried out since it was agreed by everyone that this was the last time the telescope would be serviced (figure 11.1).

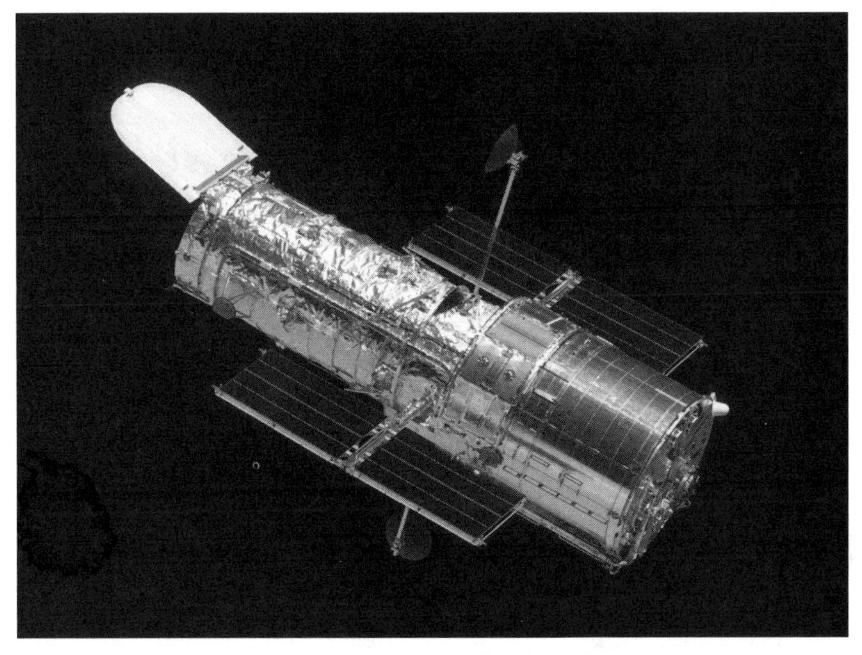

Figure 11.1. The Hubble Space Telescope is still in high demand as astronomy's flagship facility, over two decades after it was launched. In large part, this is due to five servicing missions with the Space Shuttle, where astronauts performed extremely challenging space walks to replace and upgrade vital components, and install new state-of-the-art instruments (NASA/Johnson Space Center).

Hubble's continual rejuvenation is a major part of its scientific impact. The instruments built for the telescope are state-of-the-art, and competition for time on the telescope has consistently been so intense that only one in eight proposals gets approved. All this comes with a hefty price tag. Estimating the cost of Hubble is difficult because of how much to assign to the Shuttle launches and astronaut activities, but a twenty-two-year price tag of $6 billion is probably not far from the mark. For reference, the budget was $400 million when construction started and the cost at launch was $2.5 billion. Compared to slightly larger 4-meter telescopes on the ground, Hubble generates fifteen times as many scientific citations (one crude measure of impact on a field) but costs one hundred times as much to operate and maintain.[18] Regardless of its cost, the facility sets a very high bar on any subsequent space telescope. As Malcolm Longair, Emeritus Professor of Physics at the University of Cambridge and former chairman of the Space Telescope Science Institute Council, has observed: "The Hubble Space Telescope has undoubtedly had a greater public impact than any other space astronomy mission ever." He also notes that this small telescope's "images are not only beautiful, but are full of spectacular new science," much of which was unimagined by the astronomers who conceived and launched the instrument.[19]

The People's Telescope

The Hubble Space Telescope has contributed to the identification of exoplanets, the dark energy that permeates the universe, and massive black holes that lurk in nearby galaxies. Probably no other science facility has left its mark in so many homes or done more to advance the general public's understanding of the structure, age, and size of the universe. Breathtaking photographs of regions of star formation, stunning spiral galaxies, exploding planetary nebula, and the most distant galaxies in the visible universe spill out of coffee table books, adorn the walls of children's bedrooms, and serve as computer screensavers. Why is that? The answer is not simply that the Hubble images pervade popular culture, though of course they do. There are multiple factors that might account for the telescope's tremendous popularity and an increasing public awareness of, and affection for, the telescope.

While there are telescopes with twenty-five times the light-gathering power on Earth, Hubble remains the premier tool of astronomers due to its exquisite sensitivity, and the ubiquity of the Hubble photographs has been unprecedented. From newspapers and magazine covers, to planetarium and museum programs and displays, popular science books, posters, calendars, and postage stamps, Hubble images pervade popular culture. Moreover, the emergence of the Internet has afforded global access to the telescope's photos and scientific results. In July 1994, when Comet Shoemaker-Levy 9 slammed into Jupiter with an estimated explosive force of six hundred times humanity's entire nuclear arsenal, Hubble offered up-close views of the devastating planetary impacts. Astronomer David Levy, co-discoverer of the comet, reported that millions of people around the world watched on television or via the Internet as the comet's line of fragments bombarded the massive planet.[20]

Art historian Elizabeth Kessler offers several less obvious reasons why the Hubble Space Telescope is so cherished worldwide. She contends that Hubble's spectacular images have become "interwoven into our larger visual culture" in part because of their very deliberate construction along the lines of sublime art, which often seeks to evoke grandeur, great height and breadth of field, and an overwhelming awe of the power of nature (plate 19). She points out that many of the Hubble images released by the Space Telescope Science Institute (STScI) and the associated Hubble Heritage Project reflect the aesthetics of nineteenth-century paintings of the American West. "Light streams from above" in the Hubble images as in landscape paintings, explains Kessler,[21] despite the fact that there is no up or down in space. Through such framing strategies Hubble's photos are often configured like landscape paintings or photographs.

Kessler argues that what also endears Hubble to so many is that its photos provide a means for public audiences "to imagine the possibility of seeing such spacescapes with our own eyes."[22] But Kessler is careful to clarify that all published Hubble photos are interpretations, usually composites of multiple images captured at various wavelengths of light. Even if we could travel at many times the speed of light across the galaxy to observe astrophysical objects, because of the rods and cones in the human eye, we would likely be unable to see color in faint light sources. The reason is that the cones in the human eye allow us to see color, but cones need lots of light to do so. The rods require less light, but do not pick up color. That's why photographs of the Milky Way arced across the night sky often include color that we cannot see with the naked eye.[23]

Kessler points out that the Hubble Heritage images are extensively processed and represent not what the human eye would see, "but a careful series of steps that translate numeric data into picture."[24] Hubble collects data in visible light but also supports a suite of instruments that "see" at wavelengths not visible to the human eye, such as infrared and ultraviolet light that can reveal additional details. In actuality, HST's electronic detectors see only intensity, which can be represented in grayscale, a range of shades of black and white, or gray. A set of filters made of colored glass, such as red, green and blue, are rotated in front of the detector as images are captured.[25] With spectroscopy, different wavelengths of light are dispersed with a grating and linearly arrayed on the detector. Once the data have been collected, color is added in processing by combining the images using different filters with the appropriate weights to reproduce "true color"[26] that can be used to distinguish gases within a nebula or their temperatures, or to distinguish young, hot, newly formed stars from older, cooler stars.

The Science in Hubble's Iconic Pictures

Kessler explains that some astronomers were initially dismissive of Hubble's "pretty pictures." They felt such images had less to do with science and that the public would accept the false color photos as what Hubble actually sees. But as Kessler points out, Hubble was the vehicle for changing this perception, particularly in the case of the stunning images of the Eagle Nebula (M16) released in 1995 (plate 20). A group of astronomers at Arizona State University led by Jeff Hester were interested in photo-evaporation, by which they suggested radiation from a massive star in the nebula is evaporating gaseous material away from sites of newly forming stars. Kessler notes that Hester's team "did not plan the observation with the intention of creating a visually impressive picture."[27] However, their remarkable image of the Eagle Nebula awed scientists and general audiences alike. And, as anticipated, in the nebula's 10-trillion-kilometer gaseous columns, Hester's team identified regions of newly forming stars.

Upon release of the Eagle Nebula images, Kessler reports, "*The New York Times*, *Washington Post*, *USA Today*, and other major newspapers printed articles featuring the image, and *Newsweek*, *U.S. News and World Report*, and *Time* all ran stories in the following weeks."[28] Kessler highlights the fact that while the news reports detailed Hester's theory regarding new star formation, the real emphasis of the news coverage was the spectacular beauty of the cloud pillars in the Eagle Nebula, subsequently dubbed the "Pillars of Creation." A report in 1999 on Hubble's popularity by the Space Telescope Science Institute indicates that "web hits at the time of the M16 release were about a half a million a month."[29] The fascination with Hester's image of the Eagle Nebula demonstrated that stunning astronomical photos could capture both important scientific data and amazing spacescapes that readily communicate complex information to general audiences. Kessler points out that NASA quickly realized the tremendous public support for Hubble might be better served through an organization that would regularly request, produce, and release images taken by Hubble. The Hubble Heritage Project emerged out of such interest.

Commenting on what he and members of the Hubble Heritage Project were looking for in the composition of spectacular photos like that of the Eagle Nebula, Jeff Hester explains:

> I want [people] to realize, when they look at images like that, that what they are looking at is humans acting on this urge within us. Going out there and proactively building the technology and the tools that allow us to reach out with our minds and our aesthetic and our imaginations and our sense of wonder and our sense of curiosity. . . . That these [photographs] are not just nicely aesthetic, but . . . testaments to what we as people are able to do. To take images of columns of clouds of gas and dust several thousand light years away in which stars are forming, and to bring them home and to make that part of the universe known to us, to extend the sphere of the human mind and imagination out to encompass those things.[30]

Hester nevertheless understands the difficulty in communicating complex scientific information:

> [Science] demands a rigorous vocabulary, the vocabulary of mathematics. It demands that you learn to think in that vocabulary. That's not a vocabulary that's accessible to a lot of people. . . . [I]t really matters that you find a way to get around the vocabulary problem, that you find a way to bother to think about who it is that you're talking to, what is it that they are familiar with . . . [so] that instead they can start to get their arms around [it] a little bit.[31]

The Hubble Heritage Project's home page provides thumbnails of approximately two hundred images; these are the best of the best from among tens of thousands of images taken by the facility.[32] According to Kessler, the project was the brainchild of planetary scientist Keith Noll and astronomer Howard Bond. Along with Anne Kinney and Carol Christian, these scientists collaborated to create Hubble Heritage images to promote public education and general interest in astronomy and space science. Kessler contends that they have expertly achieved these goals by blurring the boundaries between art and science to contribute to both astronomical investigation as well as popular understanding of cosmology. Keith Noll summarizes the project's objective:

> "[M]y real hope for this is that there are kids that have our pictures on their walls, that maybe spend some time dreaming about what it's like to be in space, what it would be like to travel to these exotic places. . . . But what I really hope is that they all sort of carry this little bit of awe and mystery with them in their lives, so that everything isn't just about getting up and driving in traffic and paying bills. It kind of helps to remember . . . how amazing the universe is.[33]

Hubble's photos easily convey some of the most fascinating aspects of star birth and death, and solar system formation. In many cases it doesn't require specialized or scientific training to immediately grasp the import of some of Hubble's most remarkable images. This is true for the Hubble photos of the Eagle Nebula and its collapsing gases in star birth regions, the newly forming star systems in the Orion Nebula (M42) with its proto-planetary disks the size of our Solar System, the star birth visible in the arms of the Whirlpool galaxy as illustrated in the mosaic of M51 (figure 11.2), or the bizarre and complex cloud formations produced by dying stars at the heart of planetary nebulae.

Perhaps the simplest reason for the Hubble telescope being so cherished is that in a single bound, complex scientific investigation can be conveyed in a handful of amazing images. The tremendous energy radiated as stars blow off their gaseous outer layers is evident at a glance from a single image. While the average person may not have imagined the import of a star's magnetic field in shaping planetary nebulae, the effects of intense radiation and stellar winds are nearly palpable in such photographs. In images with less obvious information, a simple caption can make the meaning immediately comprehensible to the nonspecialist. A short caption regarding the pillars of dust in the Eagle Nebula towering to 10 trillion

Figure 11.2. M51 is a nearby spiral galaxy called the Whirlpool galaxy; this is how the disk of the Milky Way would appear if we could gaze down on it, rather than being embedded in the disk. A nearby companion galaxy is triggering intense star formation in the spiral arms. Hubble Space Telescope can resolve individual bright stars in this galaxy (NASA/STScI/Hubble Heritage Team).

Figure 11.3. The Red Spider Nebula (NGC 6537) has waves of dust and gas extending a hundred billion kilometers. First cataloged by William Herschel over two hundred years ago, the subtle patterns are due to shock waves caused as outflowing material runs into the interstellar medium, a gas more sparse than the best vacuum on Earth (NASA/ESA/STScI/G. Mellema).

kilometers, or 6 trillion miles, or explaining that stellar winds in the Red Spider Nebula (NGC 6537) produce waves of dust and gas a hundred billion kilometers high (figure 11.3), while nearly incomprehensible, has afforded general audiences unprecedented understanding of the incredible energy and matter involved in star birth and death.[34] Hubble images of gravitational lensing caused by large clusters of galaxies, majestic barred spirals whirling through space, and spectacular globular clusters have been indelibly etched into the public imagination. But beyond this popular appeal, the Hubble Space Telescope has generated enough unique data to qualify as the most successful scientific experiment in history.

The Cosmic Distance Scale

What is the science yield from this premier facility, and how has it changed our view of the universe? In the early years of a space telescope, there's often a concerted effort to put a lot of time into a few critical projects, so that major results will be obtained even if the telescope fails. There's understandable tension involved in this decision, since the Hubble has always been oversubscribed by a large factor, and it keeps more astronomers happy if the time is thinly spread among many projects. Riccardo Giacconi, the first director of the Space Telescope Science Institute, was persuaded to set aside substantial time for peer-reviewed "Key Projects." Early in the history of the facility, three of these were approved. Two of them were successful but are not well known—a medium-deep survey of galaxies,[35] and a spectroscopic survey of quasars to probe the hot and diffuse gas in the intergalactic medium[36]—but the third is rightly considered one of the Hubble Space Telescope's finest achievements.

As we saw in the chapter on Hipparcos, distance is fundamental to astronomy, yet it's one of the hardest things to measure. In the everyday world we judge distance based on familiarity with the size of a tree or a person, for example, or the brightness of a lamp or a street light. But in the universe, objects like stars and galaxies range over many orders of magnitude in size and brightness, so their apparent size or brightness is a very unreliable guide to distance. Is that a bright object far away or a dim object nearby? Is that a large object far away or a small object close by? This type of confusion is common. Considered in a different way, poorly determined distances limit our knowledge of how the universe works. Without knowing how far away an object is, we don't know its true size or its true brightness. That also means we don't know its mass or luminosity, which means we can't have a real physical understanding of something that's so remote that we'll never be able to study it in the lab or gather a sample of it. This urge to measure distances accurately becomes acute in the study of the universe as one entity. In an expanding universe, precise distances are needed to answer fundamental questions: what is the size of the universe and how old is it?

Early in the life of the Hubble telescope, hundreds of orbits were devoted to measuring distances to three dozen galaxies. Fittingly, this Key Project used the method that Edwin Hubble had used in the early 1920s to show that the Andromeda nebula was a remote stellar system. Hubble identified examples of a well-known class of variable stars in M31, the Cepheid variables. Cepheids pulsate in a well-understood and well-regulated way, and this behavior manifests as a linear relationship between their luminosity and their period of variation.[37] By finding Cepheids in a remote galaxy and measuring their periods with a series of images taken over hundreds of days, the star's absolute and apparent brightness can be combined to derive the distance. Cepheids are very luminous and can be seen to large distances. But they're buried in dense regions of overlapping star images so the sharp imaging of a telescope in space is needed to pick them out. Wendy Freedman, Rob Kennicutt, and Jeremy Mould led the project, accompanied by twenty-five other astronomers from around the world.

The Key Project had a goal of measuring the distance scale to an accuracy of 10 percent. This might seem unduly modest, but it's a reflection of the difficulty of the experiment. In the decade leading up to the launch of Hubble, astronomers had disagreed by 50 percent or more on the distance scale. In particular, the goal was to measure a quantity called the Hubble constant to within 10 percent. The Hubble constant is the current expansion rate of the universe, measured in units of kilometers per second per Megaparsec. Hubble discovered a linear relation between the distance and the recession velocity of a galaxy. The Hubble constant sets that scaling. So if the Hubble constant is 70, then for every Megaparsec (3.3 million light-years) increase in distance, the recession velocity increases by 70 kilometers per second (157,000 mph). On average, this scaling says that a galaxy a million light-years away is receding from the Milky Way at 47,000 mph, a galaxy 10 million light-years away is receding at 470,000 mph, and so on. Inverting the Hubble constant allows us to project the expansion back to zero separation of all galaxies, and so gives us a good estimate of the age of the universe. By measuring distances out to 20 Megaparsecs (or 66 million light-years), the Key Project would determine the size, expansion rate, and age of the universe.

The formal result of the Key Project was the measurement of a current expansion rate of 73 kilometers per second per Megaparsec, with a random error of 6 and a systematic error of 8, published in the twenty-eighth paper in the series from team members, which indicates just how productive and prolific one Hubble project can be.[38] Synching the distance scale in this way has been a foundation stone in the emergence of "precision cosmology," where overall attributes of the universe are becoming very well-determined. Although the Key Project worked in our cosmic "back yard," or the local 0.01 percent of the visible universe, it leveraged and helped to improve distance measurements both inward and outward. Hubble also was able to measure parallaxes for Cepheids in the Milky Way (along with Hipparcos, which we've previously discussed) and tether the distance scale in direct geometric measures. Knowing Cepheid distances for several dozen galaxies calibrated the use of other types of distance indicators, such as supernovae and the well-defined rotation properties of galaxies, which can be used out into the remote universe.

Dark Force

The Hubble Space Telescope has touched every area of astronomy. But some of its key contributions have profoundly shaped our view of the universe. One of the original hopes for the Hubble was to extend the Key Project, which measured the local or current expansion rate, and measure the entire expansion history of the universe over cosmic time. Hubble's sensitivity and resolution allow it to observe supernovae in distant galaxies. A Type 1a supernova is a white dwarf star that detonates as a result of mass steadily siphoned onto it from a massive companion star. When the white dwarf exceeds the Chandrasekhar limit, which we encountered in the last chapter, it collapses and then explodes as a supernova. This well-regulated process means it's a "standard bomb" with an intrinsic brightness that doesn't vary from one supernova to another by more than 15 percent. At peak brightness, the supernova rivals its parent galaxy in brightness (figure 11.4), so these explosions can be seen to distances of 10 billion light-years or more.[39]

In the mid-1990s, two groups studying supernovae saw something utterly unexpected.[40] In an expanding universe, the effect of normal matter and dark matter is to slow down the expansion rate. By looking back in time with more and more distant supernovae, these researchers had expected to see supernovae appearing slightly brighter than expected for a constant expansion rate. The reasoning is that deceleration reduces the distance between

Figure 11.4. Although the Hubble Space Telescope is relatively small by modern standards, its sharp imaging and sensitive instruments allow it to see supernovas at distances of billions of light-years. When they die, these stars rival the brightness of their surrounding galaxies, enabling the distances to those galaxies to be measured (NASA/STScI/P. Garnavich).

us and a supernova relative to constant expansion, making it appear brighter. Instead, they saw the opposite effect: the distant supernovae were dimmer than expected. The interpretation was the distance to the dying star was larger than expected. Rather than decelerating, the universe has been accelerating! Cosmic acceleration is a very puzzling effect, because it implies a force acting in opposition to gravity. No such force is known in physics, so the cause of the phenomenon was called "dark energy," where that phrase is really just a placeholder for ignorance. Dark energy seems to have the character of the cosmological constant, the term that Einstein added to the solutions of his equations of general relativity to suppress natural expansion (and which he later called the greatest blunder of his life). It's new and fundamental physics.

The discovery of cosmic acceleration was based on images and spectra taken with ground-based telescopes. The Hubble Space Telescope didn't make the initial discovery. But confirming the result, extending the measurements to higher redshift, and putting constraints on the nature of dark energy—all of those have been essential contributions. Hubble's depth has been used to trace the expansion history back over two thirds of the age of the universe. The data show that acceleration reverts to deceleration more than 5 billion years ago.[41] Astronomers can now apportion the two major components of the universe: dark matter and dark energy. Dark energy accounts for 68 percent, dark matter accounts for 27 percent, diffuse and hot gas in intergalactic space is about 4.5 percent, and all the stars in all of the galaxies in the observable universe amount to only 0.4 percent of the cosmic "pie." Dark forces govern the universe.

Imagine the expanding universe with a brake and an accelerator. The "driver" isn't very competent so they press both pedals at the same time. Dark matter is the brake because gravity slows the expansion. Dark energy is the accelerator. In the first two thirds of the expansion history, the dark matter dominates and the expansion slows with time. But the effect of dark matter weakens, because the density and the gravity force go down, while dark energy has a constant strength in both time and space. So the pressure on the brake eases while the accelerator is pressed the same amount, and about 5 billion years ago all galaxies started to separate at ever-increasing rates. Some cosmologists consider it an unexplained coincidence that dark energy and dark matter—two mysterious entities with fundamentally different behavior—happen to have roughly the same strength and crossed paths relatively recently in cosmic time. This is the only time in cosmic history they are close to equal strength; for most of the early history of the universe, dark matter was utterly dominant and forever into the future dark energy will dominate. This coincidence only sharpens the enigma.

Dark Force Redux

Hubble has weighed in on the other major ingredient of the universe: dark matter. The existence of dark matter was first indicated in an observation of the Coma Cluster of galaxies by the maverick Caltech astronomer Fritz Zwicky. Zwicky measured redshifts or radial velocities for galaxies in the cluster and saw velocities that were much higher than anticipated. The galaxies were buzzing around like angry bees at over 500 kilometers per second, over a million miles per hour.[42] If the cluster had the mass indicated by the stars in all the galaxies, it wasn't enough to keep those galaxies bound in one region of space. In fact, visible mass was insufficient by a factor of ten. The Coma Cluster should be flying apart. But it's not, so Zwicky hypothesized an invisible form of matter to hold it together. It had to be a form of matter that exerted gravity but didn't radiate light or even interact with radiation—dark matter.

This observation was so odd and so unexpected that most astronomers simply set it aside. (It didn't help that Zwicky was brilliant but extremely cantankerous, and he made a lot of enemies in the profession with his blunt and often rude comments.) But in the 1970s the rotation of spiral galaxies also indicated unseen forms of matter extending far beyond the visible stars. Astronomers revisited Zwicky's observation and found that it was correct, and that other clusters of galaxies showed the same effect. Dark matter was a ubiquitous feature of the universe, on galactic scales and beyond galaxies in the space between them.

The Hubble Space Telescope has cemented the measurement of dark matter in a very elegant way. Einstein's theory of general relativity says that mass bends light, and this prediction was confirmed in 1919 with starlight bending around the limb of the Sun during a solar eclipse. We've seen that both Cassini and Hipparcos had a hand in showing that relativity was correct. Zwicky realized that a galaxy could bend light by a detectable amount and he urged astronomers to search for the effect. Lensing was finally seen for the first time five years after Zwicky died, when the twin-image mirage of a single quasar was observed in 1979. In the mid-1980s, the phenomenon of lensed arcs was discovered with 4-meter telescopes on the ground. Each lensed arc is an image of a galaxy behind a rich cluster, magnified and distorted by the cluster. Cluster lensing has a very particular signature because the arcs are fragments of concentric circles centered on the cluster core. The beautiful part of the effect is that the gravitational deflection is sensitive to *all* mass, light or dark, so it's a reliable way to weigh a cluster.

Hubble's exquisite imaging has been used to study lensed arcs in dozens of clusters.[43] Each background galaxy that gets imaged is a little experiment in gravitational optics (and a confirmation

Figure 11.5. The cluster of galaxies Abell 2218 acts as a gravitational lens, where its visible and dark matter distort and amplify the light of more distant, background galaxies. They are seen as tiny arcs in this Hubble image, many of which are concentric with the center of the cluster. The lensing effect is an affirmation of general relativity, where mass causes space-time curvature and the universe acts like a gigantic optics experiment (NASA/ESA/SM4 ERO Team).

of general relativity). In some clusters there are hundreds of little arcs, so the mass measurement of the cluster is very reliable—it's like having an optics experiment with hundreds of light rays (figure 11.5). The observations clinch the fact that dark matter exceeds normal matter by a factor of six, or the ratio of the 27 percent to 4.5 percent contributions mentioned earlier. They also allow the dark matter to be mapped in the cluster, and the spatial distribution is critical information in helping decide the physical nature of this universal but mysterious substance.

Hubble has also played a pivotal role in locating a large type of dark object: massive black holes. Since the discovery of quasars in the 1960s, it has been clear that only a gravitational "engine" could generate so much energy from such a small volume. The energy source for quasars is thought to be a supermassive black hole, billions of times more massive than the Sun. As we saw in the last chapter, black holes aren't always black. Nothing can escape from the event horizon, but a black hole will gather hot gas into a rotating disk around it. The disk siphons material into the black hole while the poles of the spinning black hole act as giant particle accelerators. The accretion disk emits huge amounts of ultraviolet and X-ray radiation while the jets emit radio waves. For a long time, astronomers thought that the galaxies surrounding the quasars were special because they housed such a black hole. Over the last fifteen years, Hubble has used its spectrographs to study stellar velocities near the centers of apparently normal galaxies near the Milky Way. Often, the data showed a sharp rise in star velocities near the nucleus of the galaxy.[44] Calculating the density of matter that would generate such high velocities in such a small volume, the only possible explanation was an efficient gravitational engine—a black hole.

But these black holes were surprising in two ways. First, they weren't as massive as the black holes that powered quasars. They ranged from 10 million to a few hundred million times the mass of the Sun. Second, the galaxies otherwise looked completely normal. Apparently, these black holes weren't active even though there was plenty of "food," or gaseous fuel in the center of these galaxies. As the data accumulated, it became clear that *every* galaxy has a black hole, with mass proportional to the mass of old stars in the galaxy, but those black holes must only be active a small fraction of the time and inert the rest of the time.[45]

This result has led to a paradigm shift in extragalactic astronomy. There's no division between "normal" and "active" galaxies—all galaxies are active at some level but not all the time. Black holes are a standard component of a galaxy. There are a few intermediate mass black holes residing in globular clusters and in dwarf galaxies, filling in the mass range from a thousand to a million solar masses. Nature knows how to make black holes spanning a factor of a billion in mass! The low end of the range is collapsed stellar corpses the size of a city and the high end is behemoths in galaxies ten times larger than the Milky Way. Moreover, in their active phases black holes eject mass and quench star formation and generally alter the properties of the surrounding galaxy. The co-evolution of galaxies and black holes is now a major field of research; with its exquisite resolution and sensitivity Hubble is playing a big role. Some time in the first hundred million years or so after the big bang, the first stars and galaxies formed, and black holes began to grow at the same time.

The Deepest Picture Ever Taken

To learn how galaxies formed, astronomers took their best facility and pushed it to the limit. The result was the deepest picture of the sky ever made. The Hubble Ultra Deep Field had its genesis in an earlier project called the Hubble Deep Field, and a bold decision by the second director of the Space Telescope Science Institute, Bob Williams. In 1995, Williams devoted the 10 percent of the observing time that he had at his discretion as director to a very deep multi-color image of a single patch of sky. To see why this was bold, let's take a brief excursion into the culture and sociology of research astronomy.

For astronomers, the Hubble Space Telescope is the best game in town. The Hubble has made more discoveries and generated more papers than any other research facility. Every year, astronomers craft proposals for time, and there are many times more proposals than can get on the telescope. There's a natural tendency to spread the bets. Directors had used their discretionary time similarly, giving most of it to small proposals to ease the over-subscription. Williams decided to put all his eggs in one basket by devoting 150 orbits—a huge allocation of time—to a single deep image. His decision changed the culture of astronomy. He let the research community decide where the telescope should be pointed for those 140 hours, and what color of filters should be used, but he insisted that the data be processed and made public immediately for any astronomer to use. The tiny region chosen in Ursa Major—1/28,000,000 of the sky—contained over three thousand galaxies, and the data paper for the Hubble Deep Field has been cited more than eight hundred times by other research papers.[46] Also, this large investment of a scarce resource in one field persuaded infrared, radio, and X-ray astronomers to follow suit. Other leading telescopes put copious time into complementing the optical images with data across the electromagnetic spectrum, and in many cases those data were also made available quickly. An intriguing mix of competition and altruism spurred the research forward.

But what if the one field you pick isn't typical for some reason? A premise of cosmology is that our location isn't special or unusual. This assumption is called the cosmological principal; in practice it means showing that the universe is homogeneous and isotropic. Homogeneous means roughly the same at all locations, which is hard to prove since we can't travel beyond the galaxy in a space ship. Isotropic means the same in all directions. While there's been no indication that we see different numbers and types of galaxies looking in one direction in the universe compared to any other, astronomers were nervous, so Bob Williams committed additional Hubble time to a small, deep field in the southern sky in 2000. Since then, deep fields have sprouted like mushrooms. When a new sensitive camera was installed during the fourth servicing mission in 2002, the Space Telescope Institute director at the time, Steve Beckwith, upped the ante by putting four hundred orbits, a million seconds of observing time distributed in four colors, into a tiny patch of sky in the direction of the Fornax constellation. That's the Hubble Ultra Deep Field.[47] The most distant light in this image has taken 95 percent of the age of the universe to reach us, so it comes from close to the "dawn" of light. To get a sense of this incredible image, hold a pin out at arm's length; the head of the pin covers as much sky as the image produced by Hubble's CCD camera. Astronomers harvested 10,000 galaxies from this minuscule bit of the sky. The faintest are five billion times fainter than the eye can see, and Hubble can only collect one photon per minute from them—think of trying to see a firefly on the Moon. Surveying the entire sky to this depth would take a million years of uninterrupted observing.

The numbers are staggering, and they can be used to derive some important information about the contents of the universe. Since the Ultra Deep Field covers 1/13,000,000 of the sky, the projected total number of galaxies in all directions is 130 billion. Each galaxy will on average contain 400 billion stars, so there are about 10^{23} stars in the visible universe, or a hundred thousand billion billion. That's a mind-bending number, but the real excitement comes from the implications for life. We've learned that planets are ubiquitous around Sun-like stars and expect to know soon about the abundance of habitable and Earth-like planets. The number of potential biological experiments in the universe may not be very different from the number of stars. What odds would you put on us being alone? When you look at the faint galaxies littering these deep fields, mere smudges of ancient light, it's irresistible to imagine that in many of them or even all of them someone or *something* is looking across the canyons of time and space back at you.

Characterizing Distant Worlds

Closer to home, astronomers have grown confident that there are habitable worlds in our galactic neighborhood as they harvest exoplanets and begin to detect objects close to the Earth in mass. The Hubble Space Telescope wasn't in on the ground floor of the discovery of planets beyond the Solar System; that was the work of ground-based observers patiently working on small telescopes for decades. However, it has played a central role in going beyond detection to begin to characterize the planets.

With more than 850 exoplanets confirmed, and dozens more being discovered every week, the thrill of discovery is not what it was in the late 1990s. Almost all of these planets have been discovered by an indirect method, where the influence of an orbiting planet is seen in a periodic Doppler shift of the parent star. No one doubts that these planets are real, but the evidence of seeing the planet is more direct. So it was very exiting when Hubble provided the first optical image of an exoplanet in 2008. A planet three times Jupiter's mass is orbiting the bright star Fomalhaut, twenty-five light-years away.[48] The planet is embedded in a dust disk of the young star and was spotted in archival images taken for a different purpose at two different epochs. Another research team has developed special imaging processing techniques that may be able to dig as many as a hundred new exoplanets out of the vast trove of Hubble archival images.

One of the many surprises in exoplanet research was the existence of hot Jupiters, planets like one of the giant planets in our Solar System but orbiting closer to their stars than Mercury does to the Sun. These planets present an opportunity because the chances of an alignment that lead to an eclipse as seen from the Earth greatly increase. Over a hundred exoplanets have been seen to eclipse their parent star, dimming it slightly for a few hours. These events are repeatable and Hubble has observed several exoplanet eclipses with its spectrographs.[49] The spectrum of the star shows absorption from a trace amount of heavy elements, but when the transit occurs, some starlight filters through the atmosphere of the gas giant and an *extra* imprint of absorption is added to the spectrum from ingredients in the planet atmosphere. So far, sodium, oxygen, carbon, and hydrogen were detected in the atmosphere of one planet, and carbon dioxide, methane, and water (or rather, steam!) in the atmosphere of another.[50] Both are Jupiter-sized and far too hot to be habitable, but these observations are showing the path for the detection of biomarkers: the chemical imprints of biology in the atmosphere of an exoplanet. This may well be the way we first discover life beyond Earth.

Finally, Hubble has contributed to the statistical understanding of exoplanets. Looking toward the crowded stellar bulge of the Milky Way 26,000 light-years away, the telescope found sixteen exoplanet candidates orbiting a variety of stars.[51] Five of the planets are in an extreme category not yet found by any other search method: super-rapid orbiters that whirl around their stars in less than a day. They were discovered in 2006 using the transit method. Extrapolating to the entire galaxy, the survey projects to 6 billion Jupiter-sized planets in the Milky Way. Theory predicts terrestrial planets in roughly equal or greater numbers than giant planets, so the distant worlds are out there, just waiting for us to reel them in.

Completing Hubble's Work

The Hubble Deep Field and the Ultra Deep Field images would have fascinated Edwin Hubble perhaps most of all, since they plumb the depths of space near the edge of the observable universe and reveal our universe as it was 400 million years after the big bang. In current cosmological models based on the big bang, the universe has an edge in time rather than space. Looking out, we look back in time due to the finite speed of light. At the limits of Hubble's vision, no edge in space has been detected. Rather, we've looked so far back in cosmic time that we're seeing the epoch where the first small galaxies were created and began merging into the larger and mature galaxies we see around us now. The edge in time corresponds to "first light," but beyond that was an early period called the "dark ages" when there were no stars and galaxies. Hubble has been so successful that it has almost completed the fundamental task of seeing the entire population of galaxies in the visible universe over all cosmic time. The Ultra Deep Field images provide new information regarding how highly structured spiral galaxies are formed. Astronomers now understand that large spiral galaxies represent mergers, over eons, of fragments of the young, chaotic galaxies seen at the faintest levels of brightness.[52] "To some degree, we can follow the stages of galaxy formation right into the fog bank of the Dark Ages and beyond, even toward the big bang itself," writes Jeff Kanipe, who describes the Hubble Space Telescope "as both a pathfinder and a plumb line into the vast depths of our cosmic origins."[53] The Deep Field images represent humankind's first glimpse of those origins.

The Hubble Space Telescope is named in recognition of Edwin Hubble, an observational astronomer who in the 1920s and 1930s worked at Mount Wilson Observatory with the Hooker 100-inch telescope, then the largest reflecting telescope in the world. During his lifetime Hubble was accepted as "the world's preeminent observational astronomer,"[54] in part because the 100-inch reflector allowed Hubble to seek answers to some of the then fundamental questions of astronomy—whether there were galaxies beyond the Milky Way, how to create an accurate distance ladder to such galaxies, and what could be known at the limits of the observable universe. In 1923, Hubble determined a distance to the Andromeda galaxy of roughly 800,000 light-years. Though today we know it lies about 2.2–2.3 million light-years from our galaxy, Hubble's calculation settled once and for all the debate over whether anything existed beyond the Milky Way galaxy. It was a monumental find. Hubble accomplished this through Henrietta Leavitt's research into the regular periodicities of Cepheid variable stars. Upon discovering and charting a Cepheid variable in Andromeda, Edwin Hubble could determine its intrinsic brightness to calculate a distance to the galaxy.[55] Later, in 1929, he used the 100-inch reflector, and the inestimable skills of Milton Humason as an observer and photographer, to determine that galaxies are racing away from each other at remarkable speeds. Hubble didn't use the phrase when he published his result, but this marked the discovery of the expanding universe.[56]

Perhaps it was Grace, Edwin Hubble's wife, who dubbed him "the mariner of the nebulae." She was an eloquent writer whose memoirs of Hubble are archived at the Huntington Library in San Marino, California. Gale Christianson's biography, which draws upon Grace's unpublished memoirs of Hubble, vividly recreates the scene of the astronomer at work at his telescope: "Like a captain on the bridge of a great ship, the master mariner barked out his orders—so many hours and so many degrees. Then came the metallic whining of the traverse, a series of loud clicks, a final heavy clang of the Victorian machinery as the 100-inch was clamped" into place for viewing far-flung galaxies, or nebulae as Hubble referred to them. Christianson extends the metaphor, characterizing Hubble as a sentinel mariner "[s]lipping, night after night, silent and alone, past the distant shoals, of the nebulae."[57] Hubble himself wrote of astronomical exploration in the terms of oceanic exploration. He spoke of the 100-inch reflecting telescope as performing reconnaissance at the "dim boundary—the utmost limits of our telescopes," as Hubble in the cold early morning hours would "search among ghostly errors of measurement for landmarks that are scarcely more substantial."[58]

Cosmic Voyages

The characterization of Edwin Hubble maneuvering the 100-inch reflector like a mariner upon a vast sea evokes another celebrated captain who navigated the depths of space in the science fiction television series *Star Trek* launched in September 1966. As noted in the chapter on the Voyager mission, Captain James T. Kirk of the starship *Enterprise*, portrayed by William Shatner, commanded a fictional five-year mission to explore the far reaches of our galaxy.

When Gene Roddenberry began producing *Star Trek*, the world's attention was turned toward space as the final frontier and the dream of traveling vast distances across our galaxy was at a fever pitch. The Apollo missions of the late 1960s and early 1970s captivated audiences around the world, who watched with trepidation as Neil Armstrong, and through him humankind, stepped onto the surface of another world for the first time. We delighted in watching the astronauts skipping and hopping across the lunar surface, and bounding lazily over cratered terrain in the lunar rover. The final episode of the original *Star Trek* series was broadcast in June 1969, just a month before Neil Armstrong and Buzz Aldrin stood on the lunar surface. Many in that generation came away from the truncated Apollo program wistful that the limits of money and politics prevented NASA from further manned lunar exploration.

One effect of *Star Trek* first airing in the runup to the moon landing was that the series inspired the first space-faring generation, including many at NASA. William Shatner writes of the original series that "every time [NASA] launched a manned rocket our ratings went up, meaning people were very interested in space: and when our ratings went up Congress voted more money for the space program." Even if we don't assign causation, this is interesting. The perceived interconnection between *Star Trek* and NASA extended to NASA itself. Shatner recalls:

> NASA officials often invited us to launches, and finally I decided to go to one of them. They treated me as space royalty, eventually allowing me to sit in the LEM, the moon landing module, with an astronaut. I was lying in the hammock like seats . . . looking out of the small windows at the universe displayed as the astronauts would see it. The astronaut, who was teaching me how to fly this craft, told me to look at a certain section of the star system—and as I did, flying beautifully across the entire horizon came the Starship *Enterprise*.[59]

That model of the *Enterprise* was assembled, Shatner points out, by some of the same engineers who built the Apollo spacecraft that landed on the Moon.

So compelling was Roddenberry's vision that in the late 1970s, the space agency recruited Nichelle Nichols, who played communications officer Lieutenant Uhura in the original series, to help in attracting young people to NASA's program. In 1977, NASA had received roughly 1,600 applications for new hires, but fewer than one hundred from women, and far fewer from minorities. Within four months of Nichols's recruiting, applications rocketed up to 8,400, with approximately 1,650 from women and 1,000 from minority candidates.[60] Years later, in 1992, during the STS 47 mission, Space Shuttle astronaut Mae Jemison, the first African American woman in space, made it her practice, in honor of Nichelle Nichols, to begin each of her shifts with "Hailing frequencies open," an often scripted line for Nichols.[61] That same year, in 1992, in recognition of *Star Trek*'s contribution to the popular fascination with space exploration, a portion of Roddenberry's ashes were flown aboard Space Shuttle *Columbia*.

Fred Hoyle predicted over sixty years ago that when "the sheer isolation of the Earth [had] become plain to every man whatever his nationality or creed . . . a new idea as powerful as any in history will be let loose." Hoyle optimistically expected "this not so distant development may well be for the good, as it must increasingly have the effect of exposing the futility of nationalistic strife" and would inevitably reconfigure "the whole organization of society."[62] Hoyle's point about a new understanding of Earth in the context of the vastness of space is in part why *Star Trek* and, later, the Hubble Space Telescope, garnered such global interest. Both NASA and *Star Trek* contributed to a deep cultural narrative about the possibilities of exploring the universe through international collaboration. Only twenty-four men traveled to the Moon, but unimaginable surveys of deep space have been realized via the Hubble Space Telescope. The telescope continues to capture and convey stunning images of galaxies and nebula and inspire a new generation who dreams of a time when not just a handful of carefully selected astronauts, but they themselves can explore worlds within our galaxy and beyond. That powerful cultural dream has been realized as Hubble's cameras extend our vision to the very limits of the observable universe.

When the *Apollo 8* astronauts toured the world upon return from the Moon, command module pilot Michael Collins recalls that regardless of which nations they visited, people congratulated them on the fact that "we," humankind, made it to the Moon.[63]

Collins noted that people around the world embraced the Moon landing as a collective human achievement. The Hubble Space Telescope, an international effort between NASA and the European Space Agency (ESA), is similarly embraced and loved globally as a shared human achievement. Hubble's remarkable images have forever altered our understanding of the cosmos and demonstrate how far humankind has reached across the universe in our as yet most palpable survey of the intergalactic abyss.

12 WMAP

MAPPING THE INFANT UNIVERSE

Awareness of the size and age of the universe is hard-won knowledge that has taxed scientists for the past 2,500 years. To ancient cultures, the sky was a proximate canopy that circled overhead, and there was no sense of the vast distance to the stars, let alone the idea that something might lie beyond those pinpoints of light. The ancient Greeks were the first civilization to spawn a class of philosopher-scientists, who applied logic and mathematics to their observations of the sky.

Cosmology has its root in the Greek idea of "cosmos," or an orderly and harmonious system. In the Greek view, the antithetical concept of "chaos" referred to the initial state of the universe, which was darkness or an abyss.[1] Thus, order emerged from disorder when the universe was born. Pythagoras is believed to be the first to use the term cosmos, and the first to say that the universe was based on mathematics and numbers, although in truth, so little is known about Pythagoras and his followers that direct attribution of these ideas is impossible. Pythagoras is also credited with "harmony of the spheres," a semi-mystical, semi-mathematical idea that simple numerical relationship or harmonics were manifested by celestial bodies, with the overall result having commonality with music. Pythagoreans didn't think the music of the spheres was literally audible.[2]

Aristotle's geocentric cosmology dominated Western thought for nearly two millennia, but Aristarchus developed a heliocentric cosmology that implied large distances to the stars, so as not to observe a parallax shift from one season to another. Mapping the stars in the third dimension didn't become possible until parallax was measured in the nineteenth century, giving William Herschel an inkling of the extent of the system of stars that we inhabit. Twin foundational discoveries by Edwin Hubble early in the twentieth century—the distances to the nebulae, or galaxies, and the universal recession velocities of galaxies—set the stage for modern cosmology. By the mid-twentieth century, the universe was known to be billions of years old and billions of light-years in extent.

A Beginning for the Universe

It was only in the most recent seconds of human existence, comparatively speaking, that the universe began to take shape in the human imagination. For the vast majority of the 200,000 years since the emergence of *Homo sapiens*, the reaches of the depths of space were unfathomable. In just the last hundred years humans began to *discover* the universe and develop a basic understanding of its mass and age. One key to unfolding our current view was detection and mapping of the cosmic microwave background, the remnant light and heat of the big bang. The temperature of the vacuum of space is 2.725 K, a trace above absolute cold, exactly what the universe should have cooled to if it had expanded to its current size from a hot and dense initial state 13.8 billion years ago. NASA's COBE and WMAP spacecraft have mapped this signature of the moment when the abyss of space and everything in it came into being.

Prior to and throughout the first third of the twentieth century, most people, even most astronomers, simply assumed that the universe always existed. Science historian Helge Kragh comments, "The notion of a universe of finite age was rarely considered and never seriously advocated."[3] Astronomers knew very little about the depths of space before the 100-inch telescope at Mount Wilson Observatory near Pasadena became operational. At that time, they intensely debated whether the Milky Way comprised the entire universe, or whether the nebulae might lie far beyond our galaxy. On an October night in 1923, American astronomer Edwin Hubble, working with the 100-inch, then the largest telescope in the world, observed a variable star in M31, the Andromeda Nebula, which ultimately confirmed that it was millions of light-years beyond the Milky Way. The announcement of Hubble's result in 1925 radically altered the scientific and public understanding of our place in space. Just a few years later, in 1929, Hubble and his assistant Milton Humason again rocked the world by reporting that remote galaxies were racing away from the Milky Way at 700 miles per second or faster. Their observations indicated that the universe was expanding, a seemingly preposterous idea that Albert Einstein himself initially refused to believe.

Given that relativity theory recognized space and time as inseparable, the Belgian astronomer and Jesuit priest Georges Lemaître interpreted Hubble and Humason's findings of the "runaway" galaxies as meaning only one thing. The universe itself was expanding, which in turn suggested that the universe must have been smaller, denser, and hotter in the distant past. Among Lemaître's many contributions to cosmology, three of his most simple and yet profound ideas were that the universe had a beginning, that both relativity and quantum theory were needed to explain this origin in terms of expanding space-time, and that Edwin Hubble's receding galaxies were evidence of this cosmic expansion.

Lemaître was the first to suggest that Hubble and Humason's redshifted nebulae indicated the expansion of space-time itself.[4] In 1931, in the journal *Nature*, Lemaître offered a short proposal for what English astronomer Fred Hoyle later derogatorily dubbed the big bang. In that article, Lemaître postulated "the beginning of the universe in the form of a unique atom, the atomic weight of which is the total mass of the universe."[5] He theorized inflation of the cosmos from a "primeval nebula" or "primeval atom" of dense matter. Working from Einstein's theory of general relativity as well as emerging theory in quantum mechanics of elementary atomic particles, Lemaître proposed that the universe inflated from a dense, highly compacted soup of subatomic particles that at a moment of quantum instability resulted in the unfolding of space. "What's remarkable about his *Nature* letter," writes John Farrell, "is that—apart from discussing the idea of a temporal beginning of the cosmos—it marks the first time that a physicist directly tied the notion of the origin of the cosmos to quantum processes."[6] Even before the neutron had been discovered, Lemaître understood that the beginnings of the universe could be explained in part via quantum theory and argued that "all the energy of the universe [was] packed in a few or even in a unique quantum." By Lemaitre's estimation space and time or space-time could "only begin to have a sensible meaning when the original quantum had been divided into a sufficient number of quanta. If this suggestion is correct, the beginning of the [universe] happened a little before the beginning of space and time."[7] Lemaître depicted the early universe as analogous to a "conic cup," the bottom of which represents "the first instant at the bottom of space-time, the now which has no yesterday because, yesterday, there was no space."[8]

In 1934, in "The Evolution of the Universe," Lemaître outlined his "fireworks theory of evolution" in which the stars and galaxies, having evolved over billions of years, were merely "ashes and smoke of bright but very rapid fireworks."[9] He described our situated view from Earth, scanning the night skies, as we look back toward the primordial past: "Standing on a well-chilled cinder, we see the slow fading of the suns, and try to recall the vanished brilliance of the origin of the worlds."[10] Lemaître additionally intuited that a fossil light would be the signature of the universe's beginning. As James Peebles, Lyman Page, and Bruce Partridge point out: "One learns from fossils what the world used to be like. The fossil microwave background radiation is no exception."[11] Lemaître expected evidence of a fossil radiation from the early stages of the universe could be detected. Thinking in 1945 that cosmic rays were the signatures of this fossil light, Lemaitre supposed that these "ultra-penetrating rays" would reveal the "primeval activity of the cosmos" and were "evidence of the super-radioactive age, indeed they are a sort of fossil rays which tell us what happened when the stars first appeared."[12] Just weeks before his death in 1966, Lemaître celebrated learning of the discovery of the cosmic microwave background, the fossil light he had anticipated. Throughout his career, he debated with Einstein whether or not his cosmological constant was a repulsive force that "could be understood as a vacuum energy density."[13] Cosmologists now regard Einstein's cosmological constant as indicative of the effects of dark energy contributing to the universe's expansion (figure 12.1).

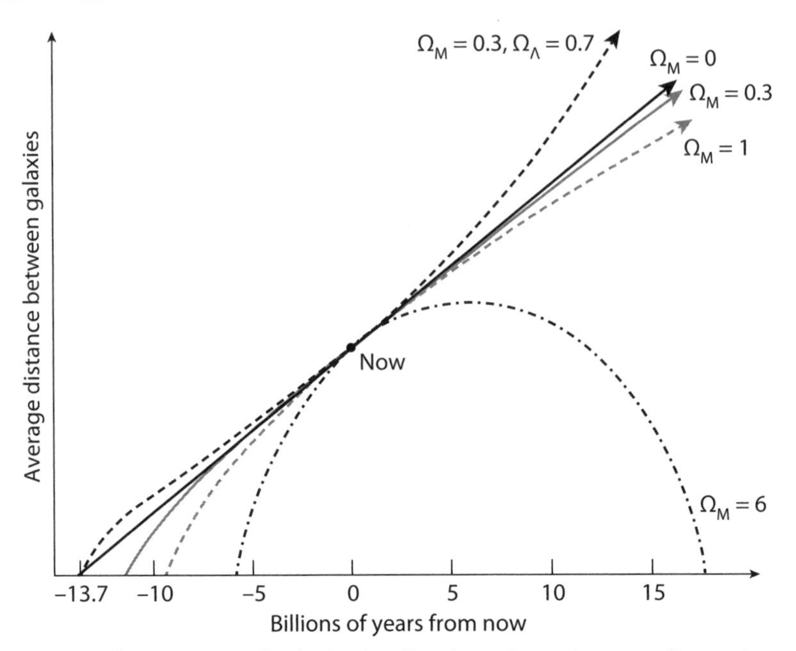

Figure 12.1. The recession of galaxies implies the universe is expanding; using general relativity, the expansion history can be calculated. The curves show the past and future expansion of the universe in terms of matter content (Ωm) and dark energy content (Ωλ). Observations agree with the upwards curving dashed line. The expansion history of the universe was dominated initially by dark matter and more recently by dark energy (Wikimedia Commons/BebRG).

The Hot Big Bang

Pivotal roles in science are often taken by outsiders. The "father" of the big bang model was a part-time lecturer at the Catholic University of Leuven when he published a paper in a journal that was little read outside Belgium. George Lemaître was a Catholic priest who had moved from civil engineering into physics and astronomy. Einstein was initially skeptical, but at a seminar by Lemaître in Princeton in 1935, he expressed his admiration for the beauty of the theory.

The other intellectual parent of the scientific theory of creation was a Russian émigré called George Gamow. This physicist was elected to the Academy of Sciences of the U.S.S.R. at the age of twenty-eight, the youngest person ever to gain that honor. With his student Ralph Alpher, he showed that the early hot universe could have produced the amount of helium we see, which is too much to have been produced by stars. Gamow jokingly had his colleague Hans Bethe listed on the paper so it would read Alpher-Bethe-Gamow, in a pun on the first three letters of the Greek alphabet; Bethe had no other role in the paper.[14] In 1948, the same year, Gamow and Robert Herman predicted that the afterglow should have cooled down after billions of years, filling the universe with microwave radiation at a temperature of five degrees above absolute zero, or 5 K.[15] However, nobody pursued the prediction, due in part to a lack of widespread awareness of the theory, and in part to the primitive state of microwave technology at the time.

The new theory was given its catchy name by Fred Hoyle in a BBC radio broadcast in 1949.[16] Even though he advocated the rival "steady state" theory, which didn't involve a hot and dense early phase for the universe, he claimed the label was descriptive and not pejorative. As Hoyle noted, with typically sardonic wit, the big bang was an audacious theory: the entire universe, holding enough matter to yield more than a trillion trillion stars in many billions of galaxies, somehow emerged instantaneously and without any precedent from an iota of space-time! Steady state theory called for the gradual creation of matter in the vacuum of space between the receding galaxies. Although spontaneous creation of matter was ad hoc physics, it seemed like a more modest proposition.

For fifteen years, the status of the theory remained tentative. The universal recession of galaxies certainly pointed to a time when the universe was smaller, denser, and hotter. The explanation of the fact that the universe is a quarter helium by mass (and 10 percent by number of atoms) was a success, but the difficulty of measuring cosmic abundances of other light elements meant no further progress could be made on testing this idea, called big bang nucleosynthesis. Counts of extragalactic radio sources indicated the population was evolving and argued against the steady state theory. The missing ingredient was a decisive observation that favored the hot big bang. It came by accident in 1964 when two engineers working at Bell Labs in Holmdel, New Jersey, detected a microwave signal of equal intensity in every direction in the sky (figure 12.2). NASA's Wilkinson Microwave Anisotropy Probe, or WMAP, is the illustrious descendant of this pioneering experiment.

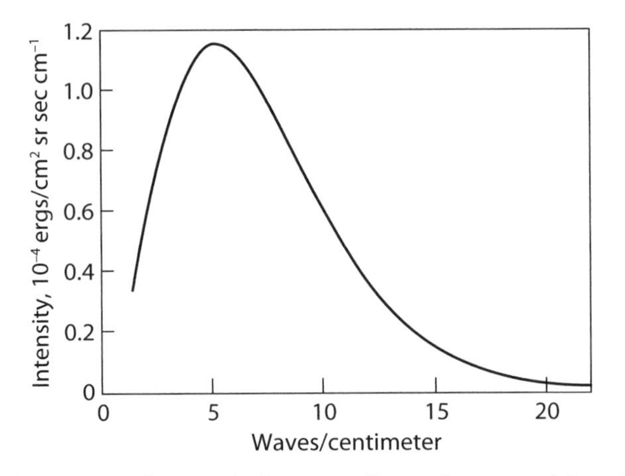

Figure 12.2. Theorists working with the expanding universe model predicted that the universe should be filled with relic radiation from the big bang, diluted and cooled by the expansion to just under 3 K. NASA's COBE satellite measured the spectrum and shows the radiation had exactly the predicted temperature; the data fit the model so well that the error bars are smaller than the thickness of the curve (NASA/COBE/FIRAS Science Team).

Cosmic Microwave Background Radiation

Scientific discovery rarely unfolds smoothly or predictably. Eight years before the theoretical prediction of relic radiation, Andrew McKellar measured the spectra of stars and discovered interstellar material that was excited to a temperature of 2.3 K.[17] He had no explanation for the excitation, which is caused by the radiation from the big bang.[18] While Gamow's prediction of a universe bathed in cold, microwave radiation sat in the literature, several experimenters had the detection of the radiation within their grasp but either did not control systematic errors well enough or were not aware of the importance of the observation. Robert Dicke at Princeton was, and by 1964 he and his team were hot on the trail of the big bang signature. But as they were preparing a radiometer on the roof of the physics building they got a call from Bell Labs. "Boys, we've been scooped," was Dicke's memorable response.[19] The Nobel Prize was awarded in 1978 to Arno Penzias and Robert Wilson of Bell Labs for their discovery.

What was the nature of the radiation that Penzias and Wilson detected? To understand this involves rewinding the history of the universe to its very early epochs. The Hubble expansion is a linear relationship between distance and recession velocity: more distant galaxies are moving away from us quicker. Although at first glance this seems to imply that we have a privileged location in the universe, a hypothetical observer in another galaxy would see exactly the same relationship that Hubble saw. In a uniform three-dimensional expansion each observer thinks they are the center of the universe. Since all cannot be, none are. Nor can we see an edge to the universe, so we can't place ourselves with respect to a boundary. The Copernican principle holds. There's no discernable center to the universe.

Reversing the expansion projects to a time when all galaxies were on top of each other: the big bang. But a simple backward extrapolation overestimates the age of the universe because matter tugs on other matter, so the expansion rate has slowed since the earliest epochs. In an expanding universe model the major observable is redshift, a stretching of the radiation from distant galaxies due to the expansion of space-time itself. Redshift is simply related to the factor by which the universe has expanded since the radiation was emitted. One plus the redshift is the expansion factor. Ironically, the universe is easier to understand in early times. Before structure forms, the universe behaves just like a simple gas, where the temperature and the average density both increase going back toward the big bang. Once gas starts to collapse by gravity, the physics is very complex. Stars and galaxies started forming when the universe was about ten times smaller than it is now, about 13 billion years ago.[20]

Extrapolating further backward, there was a time when the universe was much denser than it is now and hot enough that atoms were ionized. Electrons liberated from atomic nuclei interacted with radiation and stopped the photons from traveling freely. It was as if the universe was shrouded in an impenetrable fog. As the universe expanded and thinned out and cooled, it became transparent and radiation could travel without interruption. This special epoch is the earliest time we can "see" into the universe. In a big bang model, the background radiation comes from a time when the universe was a thousand times smaller than it is today, and a thousand times hotter. Infrared photons from 380,000 years after the big bang, when the temperature was about 3000 K, have been stretched a thousand-fold to become microwave photons in a vast and frigid universe with a temperature a little below 3 K.

The picture of the sky in microwaves is an extraordinary baby picture of the universe. Imagine as an adult that you were shown a picture of yourself a few hours old. Since those waves are from the universe as a whole, they permeate space and they travel in every direction through expanding space. There are trillions of relic photons from the big bang in any volume like that of one breath. However, their radiant intensity coming from any direction in space is only 0.00001 Watt or a ten-millionth of a light bulb.[21] If you can find an old-fashioned image tube TV and tune it between stations so you see only static, about 1 percent of the white specks on the screen are interactions of the dots of phosphor with those microwaves.[22] The big bang is all around us.

Fingerprints of the Creator

Observations of the cosmic microwave background rapidly improved, and it was soon found that the radiation had a spectrum almost perfectly consistent with one temperature, a type of radiation with what is called a thermal spectrum. This was additional support for its interpretation as relic radiation, because such a smooth spectrum only results from radiation that's in equilibrium with its surroundings. Since in this case the surroundings are the universe itself, thermal radiation is expected. Its temperature is 2.725 K and it is the most accurately measured temperature in nature.[23]

By the early 1970s, theorists had predicted that the microwave radiation should not be perfectly smooth. That's because a slightly uneven distribution of matter causes very small variations in temperature, with denser regions hotter. The subtle variations in density act as the seeds for later structure formation. Theories of galaxy formation could not generate large lumps of matter without a little lumpiness with which to start. The initial variations are not really like "lumps" since they are physically extremely large and extremely shallow. In a purely metaphorical sense, the mighty oak trees that are present-day galaxies grew from the tiny acorns of anisotropy in the background radiation.

NASA's Cosmic Background Explorer (COBE) was launched in 1989 to make more precise measurements of the microwave radiation than could be made from the ground or from high altitude balloons. COBE was cheap by modern standards, about $150 million, and extraordinarily successful. It confirmed the exquisite thermal nature of the spectrum, ruling out the last few remaining potential explanations other than a big bang. With only four years of data, the satellite was able to detect minute variations from smoothness; the radiation deviated from a constant temperature from one part of the sky to another by one part in a hundred thousand.[24] These were the long-sought seeds of structure formation. Commentators and media pundits breathlessly embraced the story when project leader George Smoot talked about having discovered the "fingerprints of God."[25] Smoot and his colleague John Mather shared the 2006 Nobel Prize in Physics for their heroic work in advancing cosmology with the detection of these tiny fluctuations.

But there's an extraordinary twist to this story. The smoothness of the microwaves and their perfectly thermal spectrum are difficult to explain in the standard big bang model because the universe was expanding so quickly early on. At the time the microwaves were released, two points in space were receding at nearly sixty times the speed of light. Under these conditions, there's no way disparate parts of the universe could come into equilibrium so adjacent patches of the sky shouldn't be at exactly the same temperature.[26] A related puzzle is the near-flatness of space. General relativity is based on curved space-time and it was expected that the vast mass of the universe would give an imprint of curvature. The cosmic background microwaves have traveled across the entire universe so should reveal if the space they've traveled through is curved. It's not. To explain the smoothness of the radiation and the flatness of space, cosmologists have hypothesized a fantastically early time, only 10^{-35} seconds after the big bang, when the entire universe expanded exponentially due to physics involved with the unification of three fundamental forces of nature. This event is called inflation.

Inflation modifies the big bang theory by positing that all we can see to the limit of vision of our telescopes—called the observable universe—is a small bubble of space-time that inflated to become large, smooth, and flat. The totality of space-time is very much larger, perhaps infinitely larger. Moreover, the variations in radiation that will grow to become galaxies are quantum fluctuations from a tiny fraction of a second after the big bang.[27] It's an extraordinary hypothesis.

Wilkinson Microwave Anisotropy Probe

Enter the Wilkinson Microwave Anisotropy Probe. WMAP was conceived as a way of pushing to a new level of precision and a new set of tests of the big bang theory. Most of those tests involve looking at anisotropies in the radiation, small variations in temperature from one part of the sky to another.

The all-sky map of microwave radiation has to have foreground emission from the Milky Way removed before it can be interpreted. There is a temperature gradient across the sky caused by the motion of the Solar System relative to the universe as a whole at a speed of about 360 kilometers per second. The microwave sky is 0.00335 K warmer toward the direction of our motion and the same amount cooler in the direction opposite to our motion. That small signal is also modeled and subtracted out.[28] What's left is a mottled pattern of very low-level variations. COBE had enough sensitivity to detect the variations statistically, but with an angular resolution of 7 degrees (the angle of your outstretched fingers at arm's length) it could not say much about the detailed structure of the radiation.

COBE was a small satellite that traveled in a 900-kilometer high orbit of the Earth. The instrument that measured temperature variations had two horn receivers pointing in different directions, with the satellite rotating every 70 seconds so they could sweep across the sky. WMAP was a much larger and more sophisticated satellite, even though it was a third the mass of COBE. It collected microwaves with a pair of 1.5-m dishes and its receivers detected the radiation in five frequency bands. It rotated every 130 seconds and made a complete map of the sky every six months. The satellite was launched in 2001 and sent to a Lagrange point (where the gravity of the Earth and Moon balance) 1.5 million kilometers from Earth, where the contaminating radiation is much lower than in low Earth orbit. As a result, WMAP was forty-five times more sensitive than COBE and it was able to resolve thirty-five times smaller regions on the sky, or two times smaller than the angle of the full Moon in the sky (plate 21). The large difference is comparable to the gain of the Hubble Space Telescope over a one-foot diameter telescope of the ground.[29]

WMAP operated flawlessly for ten years, and the exciting results of COBE and WMAP generated the momentum for a third-generation microwave satellite called Planck, named after the Nobel Prize–winning German physicist. Planck is primarily a European mission. Launched by ESA in 2009 to a location at the same Lagrange point as WMAP (L2), it improves on WMAP in both sensitivity and angular resolution.[30]

Piper at the Gates of Dawn

Variations in the microwave radiation capture important information about conditions in the early universe. To describe these variations, astronomers like to think about the power spectrum of the variations, or how much of the variation is on a particular angular scale. In this formalism, l is the angular frequency of variations. For example, $l = 2$ corresponds to two cycles across the sky or variations over 100 degrees, showing the dipole of the Local Group motion mentioned earlier. The 7-degree limit of COBE corresponded to $l = 30$ and the much better 0.3-degree resolution of WMAP reaches almost to $l = 1000$. Think of l as the number of waves to go all around the sky in a circle and l^2 as the number of "tiles" needed to cover the sky. Having more tiles means each one covers a smaller area of sky. The shape of the angular power spec-

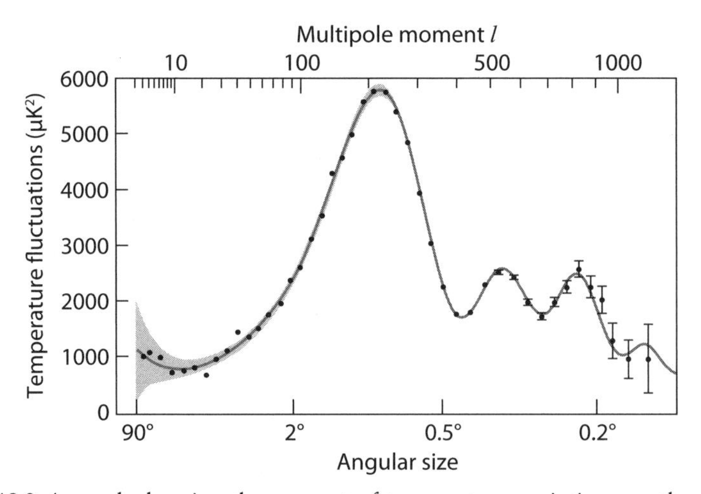

Figure 12.3. A graph showing the amount of temperature variations on the vertical scale, versus the angular scale of those variations on the horizontal scale. The quantity l is like a harmonic so it represents how many waves can fit across the sky, so larger l means smaller feature on the microwave map of the sky. The peak in microwave power on scales of a degree dates from 400,000 years after the big bang (NASA/WMAP Science Team/S. Larson).

trum is compared to predictions from the big bang model (figure 12.3). It's fair to think of l as characterizing the "harmonics" of variation of the radiation.[31]

The physics of the early universe is esoteric, but we can gain insight by the analogy with sound as long as we don't stretch the analogy to the breaking point. Before 380,000 years, while radiation was still trapped by matter, the electrons and photons acted like a gas, with the photons ricocheting off the electrons like little bullets. As in any gas, density disturbances moved at the speed of sound as a wave, a series of compressions and rarefactions. Compression would heat the gas and rarefaction would cool it, so the sound waves would manifest as a shifting series of temperature fluctuations. When electrons combined with protons to become neutral atoms, the photons from slightly hotter and cooler regions traveled unimpeded through the universe. So the temperature variations that we see now are a "frozen" record of fluctuations from that time.

If there was a "piper at the gates of dawn," then who or what was the piper?[32] Inflation is presumed to be the mechanism in the very early universe that rendered space flat and smooth, so it also must have been the source of the initial tiny variations. Those variations from inflation were hugely expanded quantum fluctuations, with the special property that the strengths of disturbances on all different scales were about equal. The disturbances are all produced at once from the very moment of creation, so they make sound waves that are synchronized. The result is sound waves with a series of harmonics or overtones, like the sounds from a flute with holes at regular intervals. Other models for the origin of the disturbances tend to be more chaotic or random, so they predict sound waves like those from a flute with holes at irregular or random intervals. Inflation is the music of the spheres dreamed of by Pythagoras.

In the flute analogy, the fundamental tone is a wave with its largest amplitude at either end of the tube and its minimum amplitude in the middle. The overtones are whole number fractions of the fundamental tone, so one half its wavelength, one third, one quarter, and so on. In the early universe, however, the waves are oscillating in time as well as space, so the waves originate in the first iota of time at inflation and they end at the time the universe becomes transparent about 380,000 years later. The fundamental tone is a wave that has maximum positive displacement (or equivalently, maximum temperature) at inflation and has oscillated to maximum negative displacement (or minimum temperature) at the time of transparency. The overtones oscillate two, three, four, or more times faster and so cause successively smaller regions of space to reach their maximum amplitude 380,000 years later.

Thus, we have all the ingredients to interpret the graph of amount of temperature variation versus angular frequency, l, measured by WMAP. There's one more subtlety. Inflation predicts that all the harmonics should have the same strength. But sound with very short waves dissipates, because it's carried by the collisions between particles and when the wavelength is less than the distance a particle travels before hitting another particle, the wave dissipates. In the air, this is just 10^{-5} cm. But in the "empty space" of the universe before recombination, photons travel 10,000 light-years before colliding. So the high harmonics are reduced or damped out. After a thousand-fold expansion, those scales are now 10 million light-years. Thus, we don't expect to see significant structure in the local universe on scales much more than ten times that size, and the clustering of galaxies is indeed weak on scales that large, chalking up another success for the big bang model. The piper at the gates of dawn is playing a strong fundamental note, with faint echoes from the higher harmonics, and a steady descent into high frequency "hiss."

Space Music

While astronomers were tuning their high-fidelity radio antennas to outer space, electronic artists began turning electromagnetic radiation into music.[33] Penzias and Wilson accidentally discovered the cosmic microwave background in 1965 just as high-fidelity equipment necessary for radio astronomy, as well as synthesizers and electronic instruments suitable for rock music, were being developed. Both fields were poised for an unprecedented exploration of the cosmos.

Bell Laboratories was the perfect place for this seemingly unrelated exploration of outer space. Having produced the first high-fidelity phonograph recording in 1925, Bell Labs advanced the technology used in radio astronomy and in generating electronic music. Karl Jansky, while investigating at Bell Labs the reasons for static in long-distance shortwave communications, in 1931 laid the foundations for radio astronomy by detecting naturally occurring radio sources at the center of the Milky Way. A quarter of a century later, Max Mathews, also at Bell Labs, began using a computer to synthesize sounds. Often credited as the father of computer music, Mathews wrote the code for the computer program Music 1, which in later iterations became a widely used programming language for computer-generated music.[34] The lab's simultaneous interest in radio technologies and electronic music is hardly surprising, since from its founding by Alexander Graham Bell, the company that became Bell Labs was dedicated to developing electronic media such as the phonograph, telephone, radio, and other communication technologies.

Music historian Pietro Scaruffi contends that it was not the electric guitar, as one might expect, that would "revolutionize rock music down to the deepest fiber of its nature," but the emergence of electronic synthesizers and computer programs designed for musical composition. By 1966, Robert Moog had developed the synthesizer, "the first instrument that could play more than one 'voice' and even imitate the voices of all the other instruments."[35] Within four years, Moog was marketing the Minimoog, a portable version that allowed for live performance with the synthesizer. And in the decade that followed, high fidelity or hi-fi technology emerged as turntables, synthesizers, oscillators, and other electronic devices became more widely used precision instruments.

Even as Penzias and Wilson at Bell Laboratories were measuring the exact temperature of the cosmic microwave background, avant-garde electronic artists like the German group Tangerine Dream began exploring the new music genres made possible by electronic instruments, keyboards, and synthesizers. The group is credited with launching what was referred to as *kosmische musik*, cosmic or space music that later evolved into disco, ambient, techno, trance, and other new age genres.[36] Their album *Alpha Centauri* (1971) is purportedly the first electronic rock space album in history. Scaruffi writes, "Tangerine Dream's music is the perfect soundtrack for the mythology of the space age. . . . They were contemporaries with the moon landing. The world was caught in a collective dream of the infinite. Tangerine Dream gave that dream a sound."[37] In 1972, they produced the album *Zeit* (Time) that included among other tracks "Birth of Liquid Plejades" and "Nebulous Dawn." Other space-themed selections by the group were titled "Sunrise in the Third System," "Astral Voyager," and "Abyss." Edgar Froese, one of the group's founding musicians, wrote "NGC 891" for his solo album *Aqua* (1974) in reference to the crisp, nearly edge-on galaxy in Andromeda oriented so that its dust lanes sharply highlight the galaxy's outer spiral arms. English astronomer James Jeans included an image of NGC 891 in his popular astronomy books of the 1930s, and astronomers repeatedly cited this seemingly perfect spiral galaxy to illustrate what the Milky Way would look like from 30 million light-years away.[38]

In 1920, Leon Theremin invented an electronic instrument that produced electronic sound by moving one's hands near two antennas that controlled pitch and volume. The theremin's eerie tones often were used to generate sequences meant to evoke space-like themes, and the instrument served as a mainstay in both avant-garde and rock, particularly in the Beach Boys' selection "Good Vibrations." Such electronic instruments, along with synthesizers, modulators, and amplifiers, became invaluable to sound designers working in the film industry. Ben Burtt, for instance, widely known for the "synthesized sound worlds" of the *Star Wars* films, actually developed the now easily recognized laser-gun sounds by recording and manipulating the twang of a guy-wire from a radio tower after accidentally getting his backpack hooked on one.[39]

Just as the first generation of astronauts was walking in space and on the lunar surface, the related genre of "space rock" was likewise exploring the cosmos. The genre emerged as an art form during the late 1960s via the British psychedelic movement. The title track to David Bowie's album *Space Oddity* (1969), which opened with the memorable phrase "Ground Control to Major Tom," shaped much of space rock to come. Pink Floyd's *The Dark Side of the Moon* (1973), with its closing track "Eclipse," is one of the best-selling rock albums in history. American songwriter Gary Wright's "Dream Weaver" (1975) was composed using only keyboards, synthesizers, and drums and according to Wright was intended as a kind of fantastical train ride through the cosmos. Innovators like Brian Eno and Steve Roach contributed to spaced-themed ambient music in their evocative soundscapes. Eno's *Apollo: Atmospheres & Soundtracks* (1983), the soundtrack for Al Reinert's space documentary *For all Mankind*, was intended to capture "the grandeur and strangeness" of the Apollo missions. Eno characterizes the soundtrack as evoking the astronauts' somewhat disorienting experience of "looking back to a little blue planet drifting alone in space, looking out into the endless darkness beyond, and finally stepping onto another planet." He adds that the score was an attempt to extend "the vocabulary of human feeling just as those missions had expanded the boundaries of our universe."[40]

But the intersection of music and cosmogony goes back to prehistory, when nomadic people would literally "sing the place" to recreate and remember a physical landscape in the form of song. Songlines in Australia provide the Aborigines with unerring navigation over harsh terrain that can extend for thousands of miles.[41] Language may have started as song and the aboriginal Dreamtime sings the world into existence. Bringing the concept into the digital age, eclectic singer-songwriter Bjork released an album in 2011 titled *Biophilia*, where each track is a sensory experience rooted in sound but extending into an iPhone app.[42] With songs ranging from "Cosmogony" to "Dark Matter," Bjork explores inner and outer space with her ethereal, electronic, sonic environments.

Celestial Harmonies

The correlation between music and exploration of outer space is deeply rooted in the human psyche. The Greek mathematician Pythagoras was purportedly the first to show that musical notes, octaves, chords, and their harmonics were the product of simple mathematical ratios and whole numbers. The Pythagoreans realized that a string attached to an instrument produces a fundamental tone when plucked. Halve the string and the tone will be an octave higher. Other harmonics, or overtones, occur at the third, fourth, fifth, seventh, and ninth divisions or intervals of the string. The Pythagoreans further posited that the universe itself could be understood in terms of this basic mathematics of music, and that the planets, for instance, would be distanced from the Sun in intervals similar to the intervals along a string that produce harmonics.[43] By the early 1600s, German astronomer Johannes Kepler was also captivated by Pythagoras's notion of a "music of the spheres." He theorized in *Harmonices Munde* (The Harmony of the World) that musical harmonies could be seen in the motion of planetary bodies. As physicist Amedeo Balbi explains, "Building on ideas by Pythagoras and Plato . . . Kepler was trying to give a scientific foundation to the concept of the 'music of the spheres', the idea according to which each planet moving around the Sun produces a definite sound."[44] Kepler suggested their orbits would be determined by the mathematics of musical harmonics.

Stars, galaxies, even planets naturally produce electromagnetic energy, or low-energy radio emission, as they rotate and move through space. Several planets in our Solar System generate strong electromagnetic emissions from their powerful magnetospheres. In the 1970s, the Voyager spacecraft recorded Jupiter's radio emissions. Of course, the human ear cannot detect sound in the empty vacuum of space. However, we can "hear" the whine of Earth and the eerie sounds of Saturn's radio emissions via satellite recordings translated by computer programs that render measured frequencies as audible sound.[45]

Far more impressive are the acoustic tones of the Sun, as we learned in the chapter on SOHO. Astrophysicists are learning a great deal about the interior composition and physics of the Sun via helioseismology, by which astronomers track acoustic sound waves that travel through the entire body of the Sun. Amazingly, the Sun resonates with myriad sound waves that can be measured like musical notes that are directly analogous to fundamental tones and complex harmonic overtones. Turbulence at the top of the Sun's convective zone segments the photosphere into granules, thousands of kilometers across, that pulse up and down. These pulses send sound waves careening through the body of the Sun and reveal a characteristic acoustic imprint at various internal boundaries like that between the convective and radiative zones.[46] By recording the acoustic waves traveling through our Sun, astrophysicists have a far better understanding of the internal structure of stars, the abundance of helium in our Sun revealing its age, the Sun's rate of rotation, and how temperatures at the core compare to its million-degree corona. These internal acoustic waves also are associated with storms on the solar surface and are being used to predict future sunspots by detecting disturbances inside the Sun.[47]

Synesthesia is the experience of one sense being rendered as another so that you can taste a color or hear visual phenomena. Just as musicians were experimenting with electronic and psychedelic explorations of outer space, astronomers were documenting the Sun's acoustic waves and experiencing their own synesthesia. In the case of the tonal resonances generated by the Sun's pulsing surface, scientists can hear what they visually detect of this resonance. The first phase of helioseismology, launched from 1960 to 1979, interestingly emerged during the same period as the rise of electronic space music.

Grooves in the Cosmic Pond

As early as 1937, James Jeans observed that "the tendency of modern physics is to resolve the whole material universe into waves, and nothing but waves. These waves are of two kinds: bottled-up waves, which we call matter, and unbottled waves, which we call radiation or light."[48] WMAP's image of the early universe would seem to confirm Jeans's observation that the universe might be understood in terms of waves. Even as sound waves traveling through the Sun's plasma reveal information about its structural composition, fluctuations in the plasma that comprised the very early universe reveal critical data about its matter density and structure. Amedeo Balbi, a member of ESA's Planck mission, the successor to WMAP, writes, "The primordial plasma resonated like an enormous bell, and the mechanism which started the vibrations could plausibly only be one: a period of inflation that occurred a tiny fraction of second after the big bang."[49]

Balbi points out that galaxies are clustered throughout space and they have a very slight preference for spacings that can be thought of as corresponding to sound waves. These galaxy separations designate grooves or "acoustic peaks," as in a wave, and represent gravity and temperature variations on the cosmic microwave background, which evolved into a subtle imprint on the distribution of galaxies. One wave that's clearly detected in the galaxy distribution has a scale of 300 million light-years.[50] Astrophysicist Jean-Pierre Luminet likewise claims that the early universe "'rang' like a musical instrument." He explains that the variations or peaks and troughs we see in the cosmic microwave background reveal details of the universe's mass and density in much the same way acoustic waves reverberating through a drum reveal its structural properties. "If you sprinkle sand on the surface of a drum," writes Luminet, "and then gently vibrate the drum skin, the grains of sand will assemble into characteristic patterns" that reveal data regarding "the size and shape of the drum" or "the physical nature of the drum skin."[51] Similarly, in analyzing the variation in clusters and superclusters of galaxies, we are gathering details about the matter density of the primeval universe.

University of Virginia astronomer Mark Whittle likewise characterizes the variations in the microwave background as similar to the crest and trough of sound waves: "The waves are actually very long; they're many thousands of light years, and so they correspond to frequencies, pitch which is very, very low by human standards, roughly 50 octaves below human ears."[52] Whittle has produced a sound file simulating what the big bang would sound like if modulated for the human ear. Describing his recreation of the acoustic peaks in the cosmic microwave background as a "sort of a raw, deep roaring sound," Whittle is quick to note that there's "actually musicality present."[53] He describes WMAP's image of the microwave background as "a microscope . . . a telescope . . . a time machine, all rolled into one, and stored in it is enough information to kind of diagnose what the character of the universe is today, what its future will be and what its birth was." Of this "extraordinary document," Whittle reminds us that embedded in the image is the primeval narrative of the universe "written by nature in nature's own language."[54] COBE and WMAP (and now Planck) have given nature a voice.

Amedeo Balbi observes in *The Music of the Big Bang*, "Ripples in the matter density of the early Universe had to leave a permanent imprint in the ancient cosmic light—as a seal impressed into wax."[55] Cosmologists theorize that etched into the primordial plasma at the beginning of time were density and heat variations that became the galaxy clusters we observe today. Balbi's metaphor of matter density fluctuations imprinted on the cosmic microwave background as indentations in wax brings to mind an early method of creating phonograph records by etching grooves in wax. Early sound recordings involved "engraving" data into a wax overlay of a zinc record. Music or voice data were recorded by tracing with a sharp stylus a spiral onto beeswax coating the zinc record. The disc was then treated with chemicals that preserved the grooves where the stylus had removed the wax, after which the phonograph record could be played.

As early as 1931, Georges Lemaître apparently wondered whether information about the universe had been recorded in the primordial quantum, even as information is preserved in the grooves of a phonograph record. Lemaître observed that "the whole story of the [universe] need not have been written down in the first quantum like the song on the disc of a phonograph."[56] However, the WMAP results suggest that, in fact, it was. From Lemaître to Balbi, the technology of the grooved record and the shape of an acoustic wave served as apt metaphors for peaks in the cosmic microwave background.[57] Had we some other technology for recording sound in the time of Lemaître, perhaps astronomers would have characterized these acoustic peaks in other terms. On the other hand, scientists find that acoustic and electromagnetic waves are means by which the universe, the stars, and planets tell their story. Astronomers and cosmologists, since Pythagoras, have never given up on finding music in the cosmos. And now they have.

Precision Cosmology

WMAP has not only put us "in tune" with the cosmos; it has refined and sharpened our view of the extraordinary event that created all matter and energy 13.8 billion years ago.

WMAP has taken quantities that were poorly known or only hinted at and turned them into well-measured cosmological parameters.[58] The temperature is measured to a precision of a thousandth of a degree. Since space can be curved according to general relativity, the universe can act as a gigantic lens. To do this vast optics experiment we look at the microwaves that have been traveling across space for billions of years. The fundamental harmonic of the microwave radiation sets the size of the "spot." Radiation from that typical spot size travels through space and the angular size can be magnified or de-magnified depending on whether space has positive or negative curvature, which is like the universe acting as either a convex or a concave lens. WMAP has shown that the spot size doesn't change so the universe is behaving like a smooth sheet of glass. The inference is that space is not curved; the universe is flat to a precision of 1 percent.[59] This is just as expected from inflation.

WMAP has measured the mass and energy of the universe with unprecedented precision. The ratio of fundamental to first harmonic powers depends on the baryonic or normal matter content of the universe. Ordinary atoms only make up 4.6 percent of the universe, to within 0.1 percent. The strength of the second harmonic is sensitive to dark matter, and shows that dark matter is 23 percent of the universe, to within 1 percent.[60] Knowing the mass and energy content of the universe, General Relativity can be used to calculate the current age. It's 13.73 billion years with a precision of 1 percent or 120 million years, though that precision depends on assuming that space is exactly flat (which is a good assumption). These numbers have been refined and slightly altered by Planck. The cosmic "pie" has dark slices and just a sliver of visible stuff (figure 12.4).

We've also been able to learn about the epoch when the first stars and galaxies formed using the WMAP data. If stars formed soon enough after recombination, they would have ionized the still-diffusing gas. That would have recreated the conditions where photons bounce off electrons, as was routinely the case before recombination. This late scattering imprints polarization on the microwave radiation, analogous to sunlight being polarized when it bounces off a water surface. Polarization requires a special direction, and that can only be seen if photons travel relatively freely before interacting. Before recombination no polarization is expected. WMAP saw a polarization indicating that 20 percent of the photons were scattered by a sparse fog of ionized gas a couple of hundred million years after the big bang. This is surprising because astronomers didn't expect the first stars to form that quickly.[61]

The sum of all WMAP's improvements leaves little doubt that the universe began with a hot big bang. The model is described quite precisely and any competing idea would have to clear some very high bars of evidence to be viable. It's extraordinary that we can know some attributes of the overall universe better than we know attributes of the Earth.

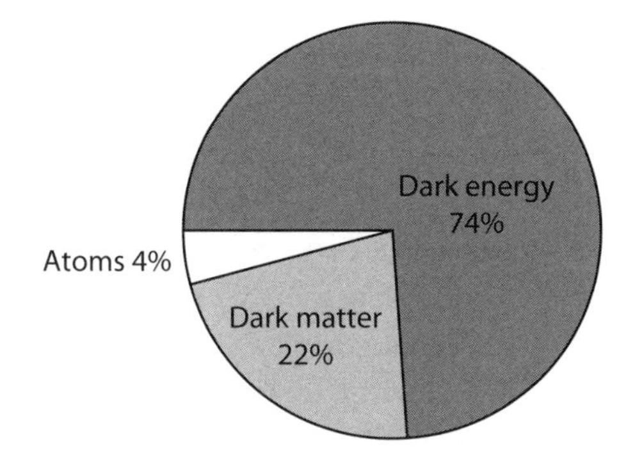

Figure 12.4. Observations of the microwave background radiation combined with ground-based observations determine the contents of the universe. Most of the universe is in the form of enigmatic dark matter and dark energy, with just a small component of the normal matter that comprises all stars, planets, and people (NASA/WMAP Science Team).

Probing the Very Early Universe

The landscape in cosmology has shifted. Scientists no longer worry about the validity of the big bang model. It rests on a sturdy tripod of evidence. One leg is the expansion of the universe as traced by the recession of galaxies. Another leg is the cosmic abundance of light elements in the first three minutes, when the temperature was 10 million degrees. The third is the microwave background radiation. WMAP has advanced the level of diagnostic power of the third piece of evidence to a level that is unprecedented. The big bang model has so far passed all tests with flying colors.

The frontier of cosmology involves gaining better physical understanding of dark matter and dark energy, and pushing tests of the big bang to earlier eras. WMAP has "weighed" dark matter with better accuracy than ever before, and it has shown that dark energy is an inherent property of space-time itself (as with Einstein's "cosmological constant) rather than being a particle or field existing in space-time. The current frontier is the epoch of inflation, an incredible trillion trillion trillionth of a second after the big bang. Inflation was motivated by the unexpected flatness and smoothness of the universe, plus the lack of space-time glitches surviving to the present day, like monopoles and strings.

Inflation got early support from WMAP data showing that the strength of temperature variations was independent of scale. But in current theory, "inflation" is not one thing; it's an umbrella for a bewildering array of ideas about the infant universe. Since it refers to a time when all forces of nature except gravity were unified, inflation models involve speculation as to the fundamental nature of matter. Superstrings are one such concept for matter, but almost all the theories have to incorporate gravity in some way, so the theory of the very early universe is proximate to the search for a "theory of everything."

In 2003, the WMAP team announced a new measurement resulting from the polarization measurements and increasingly refined temperature maps. The temperature variations are generally equal in strength on all scales, but as data improved, it became clear they were not exactly equal on all scales. The sense of this deviation was exactly as predicted by the favored inflation models, and different from predictions by rival theories for the early expansion (exotic "cold big bang" models were also ruled out). In 2010, using WMAP's seven-year data, the team even managed to use the data to confirm the helium abundance from the big bang, and they put constraints on the number of neutrino species and other exotic particles.[62] Most gratifyingly, the data confirmed the fundamental correctness of the model of temperature variations as resulting from acoustic oscillations. The piper's tune is better understood than ever before (plate 22).

Inflation is required by the data, and cosmologists are homing in on the correct model. Part of the inflation landscape is the fact that the physical universe—all that there is—is much larger than the visible universe—all we can see. Inflation also motivates the idea of the multiverse: parallel universes with wildly different properties that emerge from the quantum substrate that preceded the big bang.[63] Our dreams of other worlds should now be expanded to encompass other universes filled with worlds. The ambition and scope of these theories is extraordinary, and it undoubtedly would have amazed Pythagoras to know how far we've taken his ideas of a universe based on mathematics and harmony.

13 Conclusion

NEW HORIZONS, NEW WORLDS

AT THE BEGINNING OF THIS BOOK we encountered the Greek philosophers who let their imaginations roam beyond the visible, everyday world. One was Democritus, who was forty years younger than Anaxagoras; apparently, they knew each other. Democritus was known as the "laughing philosopher" for his habit of seeing the lighter side of life and mocking human frailties. He developed an original idea of Leucippus into the atomic theory. According to Democritus, the physical world was made of microscopic, indivisible entities called atoms.[1] The atoms were in constant motion and they could take up an infinite number of different arrangements; sensory attributes like hot, smooth, bitter, and acrid emerged from assemblages of atoms but were not properties of the atoms themselves.[2] It's a strikingly modern view of matter.[3] By analogy, a beach looks smooth from a distance, but close up we see it's actually composed of particles ranging from tiny sand grains to pebbles and boulders.

Democritus speculated in a similar way about the Milky Way, where its smoothness conceals the fact that it is composed of the combined light of myriad stars. Several centuries later, Archytas made a logical argument that there was space beyond the whirling crystalline spheres.[4] Arriving at the edge of heaven, he imagined, and extending your arm or a staff, either it meets resistance and a physical boundary, or it extends beyond the edge. If the universe is contained, it must be contained within something larger. If we can reach beyond the edge, we define a new edge and that must continue without limit. In this way he argued that space must be infinite.[5] The Greek imagination leapt downward to the invisibly small and upward to the unknowably large. By the late twentieth century, scientists knew that the span of scales from the size of a proton to the size of the observable universe was 42 powers of ten.

Human imagination, however, is not so neatly bounded. We can easily imagine what *is*—a phenomenon that's allowed by the laws of nature—as well as what *isn't*—a phenomenon that violates laws of physics as we know them. There's even a third, limbic category—a situation that doesn't violate any laws of nature but which doesn't occur in our universe.[6] In the early twenty-first century, scientists are once again pushing the boundaries of physical explanation, small and large. So particle theorists dream of nine-dimensional vibrating strings deep within every subatomic particle—worlds within the world—and cosmologists dream of an ensemble of space-times with wildly different properties, of which our universe is just one example—worlds beyond the world.

On Earth, the horizon is the farthest you can see because of the curvature of the surface of the planet. The distance is surprisingly small; standing in a small boat on a large body of water you could only see three miles in any direction. Sailors in antiquity were familiar with the slow disappearance of a ship as it sailed away, until only the tip of the mast was visible.[7] This led them to speculate and search for exotic lands that might lie "beyond the horizon." Astronomy also has the concept of a horizon as a limit to our view or to our knowledge. The event horizon emerges from the general theory of relativity. Einstein's formulation of gravity is geometric; instead of the linear and absolute space and time of Newton's theory, Einstein made an explicit connection between space and time and posited that mass curves space. A compact enough object will distort space-time sufficiently to form a black hole and the event horizon is the surface that seals a black hole off from the rest of the universe.[8] It's not a physical barrier. Rather, it's an information membrane; radiation and matter can pass inward but not outward. Black holes are enigmatic because the region inside the event horizon lies beyond the scrutiny of observations. Cosmology has its own version of this idea, which is complicated by the expansion of the universe. The "particle" horizon for the universe divides events into those we can or can't see at the present time. The universe had an origin, so there are regions from which light has not had time to reach us in the 13.8 billion years since the big bang. If we're patient, light from ever more distant regions will reach us, so the observable universe grows with time. Meanwhile, the "event" horizon of the universe is the boundary between events that are visible at some time or another and events that aren't visible at any time.[9] Standard big bang cosmology predicts the existence of space-time beyond our horizon, possibly containing many star systems in addition to the 10^{23} visible through our telescopes. There are many worlds beyond view, perhaps more than we can imagine.

Observational astronomy is a young activity. After tens of thousands of years of naked-eye observing, the telescope is just four hundred years old. Space astronomy is only fifty years into development, and the cost of launching a big glass into orbit means that the Hubble Space Telescope isn't even among the fifty largest optical telescopes.[10] In what follows we summarize what the future might hold for space science and astronomy, in terms of the missions recently launched or those on the near horizon, a few years from launch. The mid-horizon is five or ten years from now, where the uncertain funding landscape and the high cost of missions make the view very indistinct. The far horizon can depend on technologies not yet developed or perfected and is beyond view. Any projection of more than a few decades is in the realm of speculation and imagination. To close this book, we consider three pairs of missions that promise to advance our knowledge of other worlds—two within the Solar System, two looking at nearby stellar systems, and two studying the distant universe.

The heir to Viking and the Mars Exploration Rovers is the Mars Science Laboratory, or MSL.[11] Following the tradition they established with Spirit and Opportunity, NASA asked students to submit essays to give MSL a name. The winner, from among over nine thousand entries, was sixth grader Clara Ma, who wrote: "Curiosity is an everlasting flame that burns in everyone's mind. It makes me get out of bed in the morning and wonder what surprises life will throw at me that day."[12] And so the most sophisticated robot ever built was named Curiosity. Clara Ma got to inscribe her name on the metallic skin of the rover before it was bundled up for launch. Over a million people worldwide are part of the mission in a smaller way, by having their names added digitally to a microchip onboard. The public is clearly engaged in this Mars rover. In the months leading up to launch, Curiosity had over 30,000 followers on Twitter (plate 23).

The Mars Science Laboratory is a mission with a high degree of difficulty. Technical problems forced it to slip from a launch window in 2009 to the next one in late 2011, and meanwhile the budget ballooned to over $2.5 billion by the time of launch on November 26, 2011. Choosing a single place to land on Mars and answer profound questions about the entire planet was difficult; mission planners had a series of five workshops to whittle sixty possible sites down to four and they let the decision float as long as they could until selecting Gale Crater. It's like playing roulette and "betting the farm" on a single number and a single spin of the wheel. The stakes are enormous—imagine trying to identify a single place on Earth that would be representative of all terrestrial geology and biology. Site selection was made a lot easier by maps from the Mars Reconnaissance Orbiter that shows features and surface rocks as small as a sofa. It hinged on engineering and safety constraints like having a navigable terrain and avoiding high latitudes where less energy is available to run the rover. Beyond that, the key landing site requirement was habitability: evidence for presence of surface water in the past.[13] It's hoped that Gale Crater will show evidence of once having been a shallow lake bed.

Pathfinder, which crawled over Mars in 1997, was little bigger than a child's radio-controlled toy car. Spirit and Opportunity were the size of golf carts. Curiosity is like a small SUV. It's ten feet long, nine feet wide, seven feet tall, and it weighs a ton.[14] With that large mass, NASA couldn't use the "bouncing airbag" landing method of previous rovers, where retro rockets slowed the lander and the payload sprouted airbags on all sides to protect it as it bounced to a lazy halt in the gentle Martian gravity. MSL used a procedure challenging enough to have flight engineers reaching for the Tums when the spacecraft started its final maneuvers 150 million miles from Earth with no real-time control possible. It steered through a series of S-shaped curves to lose speed, similar to those used by astronauts in landing the Space Shuttle. The target landing area is twelve miles across; that sounds large but it's five times smaller than for any previous lander. A parachute slowed its descent for three minutes, then jettisoned its heat shield, leaving the rover exposed inside an aeroshell, attached to a "sky crane" mechanism. Retro rockets on the upper rim of the aeroshell further slowed the descent and then the aeroshell and parachute were jettisoned. At that point, the sky crane became the descent vehicle, with its own retro rockets guiding it toward the surface. About a hundred meters above the surface the sky crane slowed to a hover, and it lowered the rover to the surface on three cables and an electrical umbilical cord. The rover made a soft "wheels-down" landing and the connections were released (and severed as a fallback if the release mechanism were to fail). The sky crane flew off to crash land at a safe distance.[15] Landing was set for August 6, 2012. The rover then readied itself for two years of exploring the red planet. Whew.

That was the plan. And the outcome: everything worked flawlessly. All the scary outcomes and potential disasters were avoided and the spacecraft touched down gently near the edge of Gale Crater on August 6, 2012. Hundreds of engineers cheered, and the millions who watched online shared the pride of the team in their technical feat. In fact, the landing worked so well that NASA has decided to use the technology again to launch another rover in 2020, with an entirely different set of scientific instruments onboard.

Curiosity has a science payload five times heavier than any previous Mars mission. There are ten instruments; eight have U.S. investigators, and one each comes from Russia and Spain. Curiosity has a titanium robotic arm with two joints at the shoulder, one at the elbow, and two at the wrist. The arm can extend seven feet from the rover, and it's powerful enough to pulverize rocks while being delicate enough to drop an aspirin tablet into a thimble. The arm is versatile.[16] It has a hand lens imager that can see details smaller than the width of a human hair, it has a spectrometer that can identify elements, it has a geologist's rock brush, and it has tools for scooping, sieving, and delivering rock and soil samples to instruments within the rover. A set of three instruments called "Sample Analysis at Mars" will analyze the atmosphere and samples collected by the arm, with an emphasis on the detection of organic molecules and isotopes that trace the history of water in the rocks.[17] The arm will deliver samples to another instrument that uses X-ray spectroscopy and fluorescence methods to analyze the complex mixture of minerals in a typical rock or sample of soil. Engineers took instruments that would fill a living room on Earth and squeezed them into a space the size of a microwave oven on Curiosity.

Film director James Cameron lobbied hard for a high-resolution 3D zoom camera to be included in the mission. The option for 3D imaging had been cut for budgetary reasons in 2007, but NASA was savvy enough to recognize a compelling public relations angle, so Cameron was appointed as a Mastcam co-investigator and informally as a "public engagement co-investigator" and the option was reinstated. However, in early 2011 the idea was shelved because there wasn't enough time to fully test the cameras before launch.[18] Curiosity lost the potential for cinematic 3D imaging with a director's eye, but retained its fixed focal length cameras, which should return crisp 3D images for the duration of the mission. The mast cameras will view Mars from eye level, giving the public a sense of roaming on a distant world themselves. Nearly compensating for the loss of Cameron's cinematography is a hi-tech instrument called ChemCam, which uses laser pulses to vaporize rocks at a distance of up to seven meters.[19] It then performs a chemical analysis of the vapor, returning results within seconds. A "Star Wars" capability like that will undoubtedly capture the public imagination.

If Curiosity's landing site is likened to a crime scene where damage was caused by the action of water, Curiosity is trying to identify the culprit long after they have left the scene. It will tell us beyond any reasonable doubt whether at least one part of this small, arid world was once wet and hospitable for life. Curiosity isn't designed to find life. It's not designed to detect DNA or other essential biological molecules, and it can't look for fossils in the rocks it studies.[20] Rather, it's designed to detect organic compounds and provide an inventory of life's essential elements—carbon, hydrogen, nitrogen, oxygen, sulfur, and phosphorus. Viking's tantalizing verdict on life of "not proven" resonates among planetary scientists so they're cautious in setting expectations for Curiosity, but the improvements will be dramatic. Unlike the Vikings, Curiosity will be at a site almost certain to have been wet in the past. It can roam widely for its samples; the Vikings could only gather soil their arms could reach. Curiosity will do analyses of the interiors of rocks and it will be extraordinarily sensitive, able to detect one part in a billion of organic compounds. It will test the hypothesis that perchlorate can mask the detection of organics. It will decide if low levels of Martian methane are more likely to be geological or biological in origin. It will even study the pattern of carbon atoms in any organic material it detects. Dominance of even or odd numbers of carbon atoms, as opposed to random mixtures of each, would be evidence of the repetitious subunits seen in biological assembly.[21] Early results indicate that the Martian soil is chemically complex and potentially life-bearing.

And then we'll have to wait. The middle horizon for Solar System exploration is cloudy.[22] As impressive as Curiosity is, there's only so much that can be learned from a small payload of instruments shipped to the Martian surface. If chunks of Mars could be brought back to Earth they could be analyzed in exhaustive detail, molecule by molecule. So sample return has long been the "Holy Grail" of Mars exploration.[23] Unfortunately it will be difficult and expensive. The current plan is for two rovers to go to the same site on Mars and drill down as much as six feet for samples, then "cache" them in a sealed storage unit. Then a rocket would be required to boost the samples into Mars orbit, and finally a third mission would ferry them to Earth. Just the first part of this procedure will cost at least $2.5 billion and we wouldn't have the samples in our hands until 2022 or even later.

The alternative is equally exciting: a study of "water worlds" in the outer Solar System. NASA and ESA combined forces to draw up plans for a flagship mission to interesting targets far from the Earth. There were two concepts: an orbiter to study the Jovian moons Europa and Ganymede and another to study Saturn's moon Titan. After a "shootout" in 2009, it was announced that the Jupiter mission would go first. The Europa Jupiter System Mission,[24] or Laplace, will launch no earlier than 2020 and would reach the Jovian system in 2028, and it will consist of two orbiters: one built by NASA to study Europa and Io and another built by ESA to study Ganymede and Callisto. Just NASA's part of this project has a price tag of $4.7 billion so funding is not yet guaranteed.[25] If all goes well, 420 years after Galileo first noted Jupiter's largest moons, they will be observed in unprecedented detail by robotic spacecraft.

These missions could cement the idea that "habitable worlds" don't just include Earth-like planets, but also large moons of giant planets. Mars lies beyond the edge of the traditional habitable zone, where water can be liquid on the surface of a planet. The outer Solar System planets are miniature versions of the Sun in chemical composition, and so uninteresting for potential biology. However, large moons like Europa and Ganymede almost certainly have watery oceans under crusts of ice and rock. Tidal and geological heating provide a localized energy source so all the ingredients for biology are there. NASA's orbiter will measure three-dimensional distribution of ice and water on Europa and determine whether the icy crust has compounds relevant to prebiotic chemistry.[26] ESA's orbiter will do the same for the enigmatic moon Ganymede, the largest in the Solar System.[27] The biological "real estate" of the cosmos could easily be dominated by moons like these, languishing far from their stars' warming rays.

The second pair of missions is exploring new worlds farther from home, across swaths of the Milky Way galaxy. When the first planet beyond the Solar System was first discovered in 1995, it marked the dramatic end of decades of searching and frustration. A planet like Jupiter reflects hundreds of millions of times less light than is emitted by its parent star, and from afar it will appear very close to the star, like a firefly lost in a floodlight's glare. Astronomers therefore used stealth to discover exoplanets. They monitored the spectra of stars like the Sun and looked for Doppler shifts in the spectra caused by giant planets tugging on the star.[28] This method has been an unqualified success; the number of exoplanets doubles every eighteen months, and more than three thousand have been discovered.[29] It's a major step in the continu-

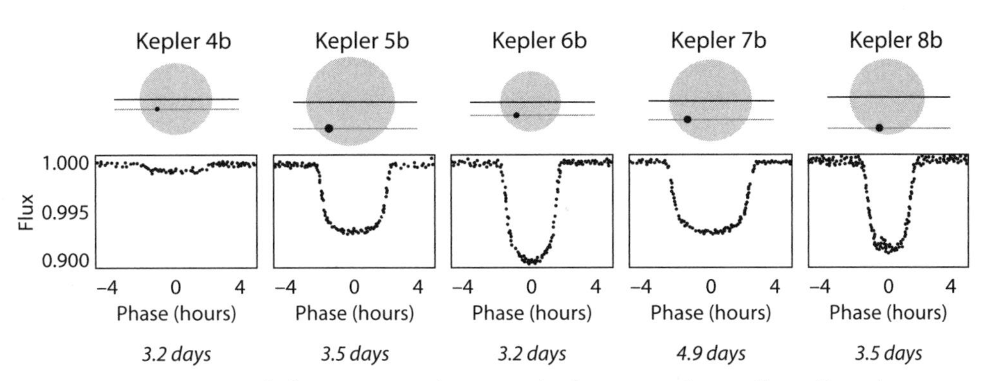

Figure 13.1. Transit light curves. Kepler can only detect exoplanets if our line of sight lies near the plane of their orbits, by measuring the brief eclipses as the planet passes in front of its parent star. The vertical axis is light intensity from a star and each temporary dip is a partial eclipse by a massive planet. Jupiter would dim the Sun by 1 percent and the Earth would dim the Sun by just 0.01 percent, so Kepler needs great precision to detect Earths elsewhere (NASA/Kepler Science Team).

ing Copernican Revolution—planets form throughout the galaxy as a natural consequence of star formation.

Timing is everything in science. It's often fruitless to be too far ahead of your time, but being slow to innovate means being left in the dust. Bill Borucki hit the sweet spot with his idea of a different way to find exoplanets. Working at NASA Ames Research Center, he realized that a planet transiting across its star would dim it slightly, by an amount equal to the ratio of the area of the planet to the area of the star. It was a statistical method since even if all stars had planets, only a small fraction of them would be oriented so that the planet passed in front of the star. He realized that with the stability of the space environment, it would be possible to not only detect the 1 percent dimming caused by a Jupiter transit but to detect the 0.01 percent dimming of an Earth transit (figure 13.1). This was the chance to find other worlds like ours in distant space.[30]

Borucki and his team pitched the project to NASA Headquarters in 1992, but it was rejected as being technically too difficult. In 1994 they tried again, but this time it was rejected as being too expensive. In 1996, and then again in 1998, the proposal was rejected on technical grounds, even though lab work had been done on the proof of concept. By the time the project was finally given the go ahead as a NASA Discovery class mission in 2001, exoplanets had been discovered and the first transits had been detected from the ground.[31] Persistence had paid off.

Kepler was launched in March 2009 and almost immediately showed it had the sensitivity to detect Earths. It is using a one-meter mirror to "stare" at about 150,000 stars in the direction of the Cygnus constellation. Designed to gather data for 3.5 years, its mission was extended in early 2012 through 2016, which is long enough for it to succeed in its core mission: conducting a census of Earth-like planets transiting Sun-like stars. Early in 2013, the Kepler team announced results from the first two years of data. The results were spectacular: over 2,700 high-probability candidates (plate 24), more than quadrupling the number of known exoplanets.[32] Nearly 350 of these were similar to the Earth in size, and fifty-four were in the habitable zones of their stars, ten of which were Earth-sized. The team projected that 15 percent of stars would host Earths and 43 percent would host multiple planets. The estimated number of habitable worlds in the Milky Way is a hundred million. The imagination struggles to contemplate what might exist on so many potentially living worlds.

Kepler is looking at stars within a few hundred light-years, our celestial backyard. Meanwhile, astronomers have their sights set on mapping larger swaths of our stellar system. Due for a launch by the European Space Agency in 2013, Gaia is the heir to Hipparcos, measuring all stars in the Milky Way down to a level a million times fainter than the naked eye can see. The heart of Gaia is a pair of telescopes and a two-ton optical camera, feeding a focal plane tiled with over a hundred electronic detectors. Kepler has a camera with 100 million pixels; Gaia's camera has a billion.[33] Hipparcos studied 100,000 stars; Gaia will chart the positions, colors, and brightness levels of a billion stars. Not only that, it will measure each star a hundred times, for a total of a hundred billion observations. Like WMAP, Gaia will be at the L2 Lagrange point a million miles from Earth, gently spinning to scan the sky during its five-year mission.[34] The result will be a stereoscopic 3D map of our corner of the Milky Way galaxy.

Gaia will detect planets, although that's not its primary objective. It will carry out a full census of Jupiters within 150 light-years, not by looking for a Doppler shift or an eclipse of the parent star but by actually seeing the star wobble as it is tugged to and fro by the planet. This is a truly minuscule effect—the Sun wobbles around its edge due to the influence of Jupiter, but Gaia will be looking for a similar motion on stars hundreds of trillions of miles away. It's like trying to see a penny pirouetting about its edge on the surface of the Moon. As many as 50,000 exoplanets may be discovered this way, compared to the current census of a few thousand candidates. Gaia will also detect tens of thousands of dark and warm worlds called brown dwarfs, which represent the "twilight" state of stars not massive enough to initiate nuclear fusion. It will also be able to detect dwarf planets like Pluto in the remote reaches of the Solar System. Such planets are distinctive and interesting geological worlds in their own right.

Gaia represents our "coming of age" in the Milky Way. It also provides a bridge to the last pair of missions we will consider, since the motions of the oldest stars in our galaxy were imparted at birth, so those motions represent a "frozen record" of how the galaxy formed. Current theories suggest that large galaxies were assembled from the mergers of smaller galaxies. Most of these mergers occurred billions of years ago, but traces remain in the form of streams of stars threading the halo of the galaxy like strands of spaghetti.[35] This type of galactic archaeology is a vital part of understanding how the universe turned from a smooth and hot gas to a cold void sprinkled with "island universes" of stars.

Cosmology is a vibrant field, spurred by new observational capabilities, computer simulations, and increasingly tried-and-tested theories of gravity and the big bang origin of the universe. The mid-horizon in cosmology is defined by the James Webb Space Telescope (JWST), which is an ambitious successor to the Hubble Space Telescope.[36] Hubble's mirror is 2.4 meters in diameter, limited by the size of payload the Space Shuttle could heft into a low Earth orbit. JWST will gather seven times more light with its 6.5-meter mirror, even though it will be launched on an Ariane 5 rocket, which has a maximum payload diameter of 4 meters

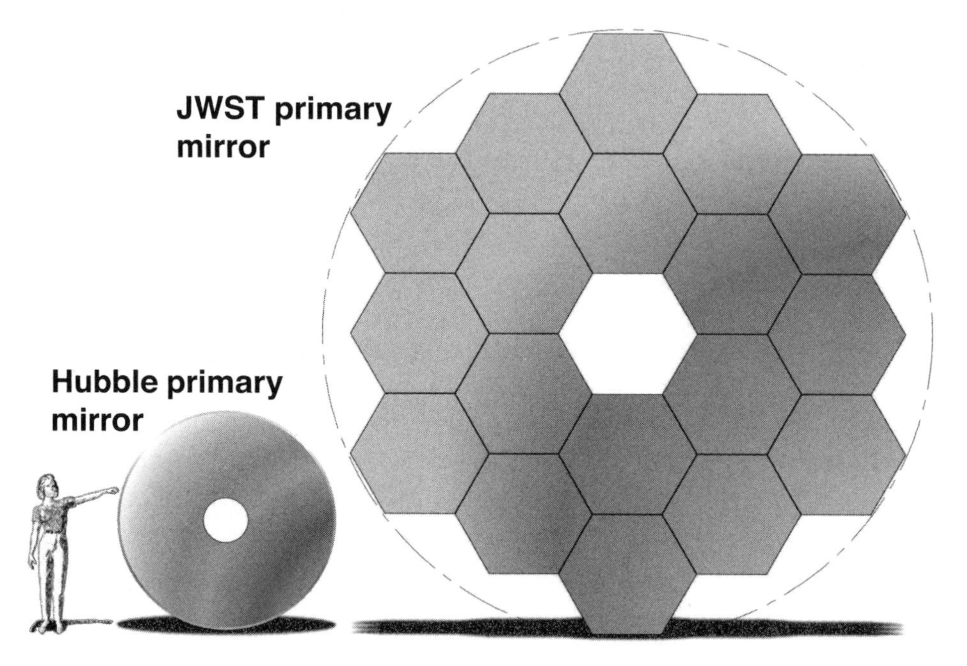

Figure 13.2. The James Webb Space Telescope is the successor to the Hubble Space Telescope, due for launch in 2018. It will be located a million miles from the Earth and will unfold in orbit to become the largest telescope ever deployed in space. Its primary goal is to detect "first light" in the universe, when the first stars and galaxies formed, over 13 billion years ago (NASA/JWST).

(figure 13.2). This clever trick is accomplished by a design where the segments of the telescope unfold after it is deployed, like flower petals, adding greatly to the engineering challenge. JWST will also be located at L2, the increasingly crowded "watering hole" of many astrophysics missions. The telescope and its instruments will be too far away to be serviced by astronauts, so the stakes are high and everyone involved in the project feels the pressure. JWST's cost has ballooned to over $8 billion and the launch date has slipped several times, with a current estimate of 2018. Although the project has broad support from the astronomical community, the high cost and repeated delays have brought it close to cancellation. NASA can only afford one flagship planetary science mission like MSL at a time, and it can only afford one flagship astronomy mission at a time. Until JWST is operating, no other big concept can get traction and a funding start. Astronomers struggle to maintain solidarity under such conditions.

Since JWST is a general-purpose telescope, it will contribute to the study of exoplanets as well as the more remote universe. One of its instruments can take spectra of the feeble light reflected by a planet next to a remote star. By observing in the infrared, the exoplanet will be ten to twenty times brighter relative to the star than it would be at visible wavelengths. Astronomers will take the most Earth-like planets and search for spectral features due to oxygen and ozone. These two gases are telltale indicators of life on Earth, and if they're seen in abundance in the atmospheres of exoplanets, it will be evidence of photosynthetic organisms elsewhere. The observations are extremely challenging, requiring a hundred hours or more per target.[37] But if they succeed, the detection of biomarkers will be a dramatic step in our quest to find a world like our own.

However, the core mission of JWST is to detect "first light." First light is the time in the youthful universe when gravity has had time to congeal the first structures from the rapidly expanding and cooling gas. The big bang was unimaginably hot—a cauldron intense enough to melt the four forces of nature into one super-force. By 400,000 years after the big bang, the temperature had dropped to 3000 K and the universe was glowing dull red as the fog lifted and the photons of the radiation from the big bang began to travel freely through space. The "Dark Ages" began.[38] Theorists are not sure of the exact timing, but they think that it took another hundred million years or so for small variations in density to grow enough for pockets of gas to collapse. This would have been when the universe was at room temperature, a hundred times smaller, and a million times denser than it is today. At this point the Dark Ages ended. It's likely that stars formed first and went through a number of cycles of birth and death before they agglomerated into galaxies. The first stars were probably massive, 80–100 times the mass of the Sun, and they died quickly and violently, leaving behind black holes. In this strange universe no worlds like the Earth existed. The universe was made of hydrogen and helium; it would take generations of stars to live and die before enough heavy elements were created for planets and biology to be possible.

Even JWST will probably not be able to detect the first stars, but it should be able to detect the first galaxies. Its superior capabilities are required because the Hubble Space Telescope has almost exhausted its light grasp as it looks for the first light. Less than 500 million years after the big bang, light travels so far and for so long to reach us that cosmic expansion stretches visible photons to near infrared wavelengths. Distant light slides off the red end of the spectrum and becomes invisible. In a sense, JWST is really looking for "first heat."

To take us beyond the horizon, we look to a mission that is currently taking data. Planck is the successor to WMAP, launched by the European Space Agency in 2009. It has a primary mirror about the size of a dining room table made of carbon fiber reinforced plastic coated with a thin layer of aluminum, weighing only twenty-eight pounds. The detectors are cooled to just above absolute zero. Planck improves on WMAP by a factor of three in angular resolution, a factor of ten in wavelength coverage, and a factor of ten in sensitivity. All of this means Planck will extract fifteen times more information from the cosmic microwave background than WMAP. Nobody doubts the big bang anymore, so Planck is taking the theory to its limits, testing inflation, the notion that the universe underwent a phase of exponential expansion an amazing 10^{-35} seconds after the big bang.[39] By making measurements with a precision of a millionth of a degree, the Planck mission may be able to determine how long inflation lasted and detect the signature of primordial gravitational waves imprinted on the microwaves. Or it may not provide any support for inflation, driving scientists back to the drawing board.

If inflation occurred, quantum fluctuations from the dawn of time became the seeds for galaxy formation. If inflation occurred, there are regions of space and time far beyond the reach of our best telescopes. If inflation occurred, one quantum fluctuation grew to become the cold and old universe around us. The initial conditions of the big bang can accommodate many quantum fluctuations, each of which gives rise to a universe with randomly different physical properties. In most of those universes the properties would be unlikely to support long-lived stars and the formation of heavy elements and life. Inflation therefore supports the "multiverse" concept, where there are myriad other universes with different properties, unobservable by us, and we happen to live in one of the rare universes hospitable for life and intelligent observers.[40]

In fact, the multiverse offers so many possibilities that everything that could possibly happen actually *will* happen somewhere. There might not just be clones of Earth beyond the horizon, but clones of *us* as well. Quantum genesis and the multiverse take us to a point where science becomes as wild and bold as any science fiction—a vision of worlds without end.

Notes

1. Introduction

1. From the characterizations assembled in Diogenes Laërtius in his *Lives of the Eminent Philosophers*, translated by Robert Drew Hicks (1925), Loeb Classical Library.

2. *The Presocratic Philosophers* by J. Barnes (1982), London: Routledge.

3. Much of the evidence of the astronomical theories of Anaxagoras refers to a single long reference in Hippolytus from his book *A Refutation of All Heresies* in the second century BCE. As examples of the paucity of source material for influential thinkers of antiquity, consider that there is no direct and unaltered evidence of the writings of Thales and Pythagoras; the former was the first scientist-philosopher and the latter was the founder of mathematics.

4. Anaxagoras had as a core metaphysical principle that "in everything there is a share of everything." There are big challenges in interpretation when every original ingredient is a part of everything, over all time, and without limit, large or small, as explored by Mansfield (1980), "Anaxagoras' Other World," *Phronesis* 25: 1–4. The idea of a cosmology without any preferred scale has echoes in the modern idea of fractals, see P. V. Grujic (2001), "The Concept of Fractal Cosmos. I. Anaxagoras' Cosmology," *Serbian Astronomical Journal* 163: 21–34.

5. *Anaxagoras of Clazomenae: Fragments and Testimonia* by P. Curd (2007), Toronto: University of Toronto Press.

6. Anaxagoras' troubles may have stemmed as much from politics as from going against religious traditions. He was the teacher, friend, and confidant of Pericles, the influential Athenian leader. Pericles had rivals and antagonists who would have been happy to attack him via his influential mentor.

7. See "Ancient Atomists on the Plurality of Worlds," by J. Warren (2004), *Classical Quarterly* 54, no. 2: 354–65. The language of the Ancient Greeks does not map cleanly onto modern scientific terminology. The atomists held that the Earth had formed from a chance association of atoms, and that an infinite number of other worlds, existing simultaneously and each being transient, had formed the same way. These worlds occupied self-contained universes, or *kosmoi*, each

of which was separate and inaccessible to observations. Since Democritus talked about some *kosmoi* where there is no Sun or Moon, and some where there are no living creatures, we infer that a *kosmos* was a world-system, akin to the observable universe of modern cosmology.

8. *Plurality of Worlds: The Origins of the Extraterrestrial Life Debate from Democritus to Kant* by S. Dick (1982), Cambridge: Cambridge University Press.

9. Ronald Huntingdon, in "Mything the Point: ETIs in a Hindu/Buddhist Context," from *Extraterrestrial Intelligence: The First Encounter*, edited by James L. Christian (1976), Buffalo, NY: Prometheus Books.

10. Contained in *The Works of Lucian of Samosata*, translated by H. W. Fowler and F. G. Fowler (1905), Oxford: Clarendon Press. For the case that Lucian's story represents science fiction, see S. C. Fredericks (1976), "Lucian's True History as Science Fiction," *Science Fiction Studies* 3, no. 1: 49–60. Lucian wrote another science-fictional piece called *Icaromenippus, an Aerial Expedition.*

11. See "Nicolas of Cusa," from vol. 9 of *Dictionary of the Middle Ages*, edited by Joseph R. Strayer (1987), New York: Charles Scribner's Sons.

12. *Giordano Bruno: Philosopher/Heretic* by I. Rowland (2008), New York: Farrar, Straus, and Giroux.

13. Giordano Bruno was in England for a few years, where he might have become acquainted with the work of Thomas Digges, who in 1576 had published a translation of Copernicus's epochal book where he presented the stars as extending through infinite space.

14. The first translation of *Somnium*, by Joseph Lane in 1947, was never published. Most interpretations have been based on the translation by Edward Rosen (2003), Dover Publications.

15. *Kepler's Somnium: Science Fiction and the Renaissance Scientist* by G. Christianson (1976), in *Science Fiction Studies* 3, no. 1: 76–90.

16. Summarized in *Discoveries and Opinions of Galileo*, translated by Stillman Drake (1957), New York: Anchor. Galileo was called in front of the Inquisition for a stern warning, but never paid the ultimate price for his profession of Copernicanism.

17. *The Extraterrestrial Life Debate: 1750–1900* by M. Crowe (1999), London: Courier Dover.

18. Plenitude was invoked by the atomists, and was used in a religious context by Nicolas of Cusa; the term derives from *The Great Chain of Being* by Arthur Lovejoy (1936), Cambridge, MA: Harvard University Press. This quote from Ralph Cudworth's *The True Intellectual System of the Universe* (1678) is typical of the argument: "It is not reasonable to think that all this immense vastness should lie waste, desert, and uninhabited, and have nothing in it that could praise the Creator thereof, save only this one small spot of Earth." Teleology has often been invoked, controversially, in biology throughout the past three centuries. Immanuel Kant promoted both principles in his philosophical writing.

19. *The Last Frontier: Imagining Other Worlds from the Copernican Revolution to Modern Science Fiction* by K. Guthke (1990), Ithaca, NY: Cornell University Press.

20. *Conversations on the Plurality of Worlds* by B. Fontanelle (1990), Berkeley and Los Angeles: University of California Press.

21. Fontanelle was a Renaissance man, elected to both the French Academy and the French Academy of Sciences, and a gourmand who attributed his longevity and good health to a regular intake of strawberries. Soon afterward, Christian Huygens elaborated on Fontanelle's ideas in a more scientific treatment called *The Celestial Worlds Discovered* (1698).

22. Surprisingly, and regrettably, Flammarion's book is out of print; French language editions can occasionally be bought from antiquarian booksellers online.

23. According to David Kyle's *A Pictorial History of Science Fiction* (1976), London: Hamlyn Publishing Group, Huge Gernsback coined the term "science fiction" in 1929, and thereafter scientific and fictional accounts of other worlds established separate traditions, which do, however, continue to enrich each other.

24. Subrahmanyan Chandrasekhar spent over five years reworking the *Principia* from first principles, resulting in *Newton's Principia for the Common Reader* (1995), Oxford: Clarendon Press. Chandrasekhar was a Nobel Prize winner in Physics and one of the best theoretical astrophysicists of the twentieth century, but despite over three hundred years of advances in mathematical methods to draw on, he felt he could not improve on the original.

25. See *Remembering the Space Age* by S. Dick, editor (2008), Washington, DC: NASA (NASA SP-2008-4703). The history of space exploration is quite unlike the history of astronomy. Observational astronomy grew through four centuries of maturing telescope and detector technology. The earliest astronomers were "citizen scientists" with erratic support from royalty and wealthy patrons; formal government support in the United States did not begin until the 1950s. The space program had the singular driver of superpower rivalry in the depths of the Cold War (with an earlier military precursor when Werner Von Braun came to the United States to continue the development of rocket technology that had been used by the Germans in World War II). NASA was founded a year after the launch of Sputnik, as a civilian counterpart to the aggressive program to launch nuclear weapons. Outside the United States and Russia, the space program got a later start and has always been funded by governments as a purely civilian activity.

26. Modern cosmology has not resolved the ancient issue of whether or not there are infinite worlds. Certainly, the principle of plenitude is supported, since the potential for planets and life among 100,000 billion billion stars is substantial. However, our view of the expanding universe is limited by time, not space; there are physical regions where information has not had time to reach us since the big bang. The standard big bang model predicts the existence of regions of space that have not yet come into view and suggests that those regions might dwarf the observable universe.

27. "A Jupiter-mass Companion to a Solar-type Star," by M. Mayor and D. Queloz (1995), *Nature* 378: 355–59.

28. Some time during the past decade, the fraction of refereed publications in astronomy with first authors from the United States dropped below 50 percent for the first time ever; see Schulman et al. (1996), "Trends in Astronomical Publication Between 1975 and 1996," *Publications of the Astronomical Society of the Pacific*109: 1278–84. Meanwhile, almost all space missions with price tags above a billion dollars involve the space agencies of more than one nation.

2. Viking—Discovering the Red Planet

1. *Martian Metamorphoses: The Planet Mars in Ancient Myth and Religion* by E. Cochrane (1997), Ames, IA: Aeon Publishing.

2. In an early version of *Beauty and the Beast*, Ares was the lover of Aphrodite, and she bore him six children, who were all minor gods. Among them, Phobos and Deimos were his primary attendants, so naturally the red planet's two small moons were named after them. The most massive known dwarf planet in the Solar System was named after Ares' sister Eris. Fittingly, Eris lives at the periphery of the Solar System with Pluto, given a similar cold shoulder as the god of the underworld.

3. In Rome, a priesthood called the Arval Brothers regularly used to call on Mars to drive off "rust," with its double meaning of wheat fungus and the oxides that corroded farm implements and weaponry. This is ironic because the blood-red tinge of Mars, which was always connected to war and violence, is now known to be caused by oxidation or rusting of the iron-rich soil.

4. Mars's (and other outer planets') retrograde motion is a consequence of the slower velocity of their orbits according to Kepler's laws. As viewed from the Earth toward a backdrop of the stars, Mars appears to move from East to West on successive nights for most of its orbit. But every few years, when Earth makes its closest approach, it overtakes the red planet "on the inside" and so by the relative motions Mars moves backward or West to East on the stellar backdrop, and then a month or so later resumes its East to West motion. This behavior was counterintuitive and deeply mysterious to ancient people.

5. Through the furor that erupted after the broadcast, Welles maintained his innocence of any intention to cause a panic. His demurral was not credible to most people in the entertainment business because Welles was famous for being equal parts precocious and mischievous. There were disclaimers near the beginning, middle, and end of the show, but Welles knew that many people would join the broadcast from a more popular show on another station and so miss the first one. By the time of the second, much later in the show, the cross-cutting reportage had blurred the line between fact and fiction so much that credulous listeners were convinced that the Earth was being invaded. His reflections are captured in the *Orson Welles Interviews* by Mark Estrin and Orson Welles (2002), Jackson: University of Mississippi Press.

6. *The Invasion from Mars: A Study in the Psychology of Panic* by H. Cantril et al. (1982), Princeton, NJ: Princeton University Press.

7. Sensational press accounts and banner headlines in the following days created the impression that the show had caused mass panic. Welles and his company were censured by the broadcaster. In fact, commentators disagree on the impact of the hoax. Nonetheless, the episode served to burnish Welles's growing legend.

8. *War of the Worlds: From Wells to Spielberg* by J. Flynn (2005), New York: Galactic Books.

9. Mars has an opposition, or a time when it lines up with the Earth on the same side of the Sun, every twenty-six months. At its opposition in 2003, Mars was closer than it had been for any time in 60,000 years, though it was only 15 percent closer than the previous opposition. NASA launched the Phoenix mission

in 2003 to take advantage of the opposition, and it launched the Mars Science Laboratory in late 2011 to take advantage of the next opposition in 2012.

10. W. T. Peters (1984), "The Appearance of Venus and Mars in 1610," *Journal of the History of Astronomy* 15: 211–14.

11. Cassini also made the crucial advance of determining a distance to Mars for the first time. He used observations of Mars made from different points of the Earth's surface at different times. Between the two locations, Mars shows a slight parallax shift with respect to the background stars, and this gives a trigonometric distance to the planet.

12. *The Planet Mars: A History of Observation and Discovery* by W. Sheehan (1996), Tucson: University of Arizona Press.

13. As quoted in *Herschel* by H. C. McPherson (1919), London: Macmillan.

14. For Secchi's many contributions to astronomy, see *In the Service of Nine Popes: A Hundred Years of the Vatican Observatory* by S. Maffeo, J.J., translated by G. Coyne, S.J. (1991), Vatican City, Rome: Specola Vaticana.

15. The vagaries of seeing are familiar to all amateur astronomers but increasingly less familiar to professionals since there are no eyepieces attached to large telescopes and the imaging is done by electronic cameras that average out the seeing variations or compensate for them in real time, a technique called adaptive optics.

16. The nature of the illusion that might lead someone to see linear features where none exist has been described by Cecil Adams in *More of the Straight Dope* (1988), New York: Random House. A map of Mars featuring all of its agreed-upon features, but excluding the canals, was pinned up at the front of a classroom and the students were all asked to copy it. The children near the front, who could see the map clearly, reproduced it accurately. But children near the back to whom the fine features of the map were just a blur tended to make maps with lines connecting the smaller features. This represents a strong psychological tendency to "connect the dots" and convert incomplete or incomprehensible information into a coherent whole.

17. *Water and the Search for Life on Mars* by D. Harland (2005), Dordrecht, Netherlands: Springer Press.

18. *The Planet Mars: A History of Observation and Discovery* by W. Sheehan.

19. *La Planète Mars et ses Conditions d'Habitabilité* by C. Flammarion (1892), Paris: Gauthier-Villars et Fils.

20. *Lowell and Mars* by W. Hoyt (1996), Tucson: University of Arizona Press.

21. The landmark publications were *Mars* by P. Lowell (1896), Boston, MA: Houghton, Mifflin; *War of the Worlds* by H. G. Wells (1898), New York: Harper; *Mars and Its Canals* by P. Lowell (1906), New York: Macmillan; *Is Mars Habitable?* by A. R. Wallace (1907), London: Macmillan; and *A Princess of Mars* by E. R. Burroughs (1912), New York: Del Ray.

22. "Decline and Fall of the Martian Empire," by K. Zahnle (2001), *Nature* 412: 209–13.

23. As quoted in Sheehan (1996), from "Our Solar System" by P. Lowell (1916), *Popular Astronomy* 24: 427.

24. *The Superpower Space Race: An Explosive Rivalry Through the Solar System* by R. Reeves (1994), Dordrecht, Netherlands: Plenum Press.

25. Bill Momsen, the JPL engineer responsible for imaging science on the Mariner missions, has posted his personal recollections on the Internet, at http://home.earthlink.net/~nbrass1/mariner/miv.htm.

26. Mariner surveyed just 1 percent of the surface and its trajectory happened to include some of the most heavily cratered terrain on Mars, most of it 4 billion years old. Large life-forms seemed to be convincingly ruled out, allowing only the possibility of extremophile microbes, although the researchers noted that similarly poor resolution images of the Earth might not show signs of intelligent life, "A Search for Life on Earth at Kilometer Resolution" by S. D. Kilston, R. R. Drummond, and C. Sagan (1966), *Icarus* 5: 79–98.

27. Prior to launch, each lander, encased in its aeroshell heat shield, had been subjected to a week of heat sterilization at a temperature of 250°F (121°C). They were then wrapped in a "bioshield" that was jettisoned after the spacecraft left Earth orbit. This procedure was designed to prevent contamination of the Martian surface by terrestrial microbes and it established the norm for planetary protection by all future NASA missions. Ironically, since the Vikings were launched, we have discovered microbes on Earth that can survive temperatures above 121°C and others that can survive the vacuum of space, so the requirements for sterilization have increased since then.

28. In 1981, the first man-made object to soft land on another world was renamed the Thomas A. Mutch Memorial Station, after the former leader of the Viking Lander Imaging Science Team. Mutch disappeared the previous year while climbing in the Himalayas.

29. Strictly speaking, the Russian probe Mars 3 managed a soft landing on December 2, 1971, but it only sent back data for 15 seconds before permanently ceasing transmissions. The reason is unknown.

30. NASA's mission archive for Viking 1 and 2 is at http://nssdc.gsfc.nasa.gov/planetary/viking.html, curated by David Williams at Goddard Space Flight Center. The website includes detailed information on each orbiter and lander, archives of all raw and processed images. The primary reference for the spacecraft, scientific experiments, and data returned is "Scientific Results of the Viking Project," edited by E. A. Flinn (1977), *Journal of Geophysical Research* 82: 735.

31. Quoted from an article by Gentry Lee for SPACE.com and posted on July 20, 2001. His reminiscences in video form are on the JPL website at http://www.jpl.nasa.gov/videos/mars/viking-062206. Gentry Lee is chief engineer for the Solar System Mission Directorate at the Jet Propulsion Laboratory. In addition to his leadership of many NASA missions, he was Carl Sagan's partner in the development and creation of the *COSMOS* television series. Lee co-authored four science fiction novels with Arthur C. Clarke and has written three additional solo novels.

32. *Water and the Search for Life on Mars* by D. Harland (2005), Dordrecht, Netherlands: Springer Press, p. 62.

33. The cameras were equipped to make digital images through three visible filters—blue, green, and red—and three infrared filters. Due to calibration uncertainties, balancing the different channels to produce "true" color was more of an art than a science. The sky in most of the images appears pink looking up due to small dust particles in the lower atmosphere. Early image releases had soil colors

that were more orange than is geologically reasonable, so those colors were later adjusted to be reddish-brown. Most of the published images are seamless mosaics stitched together from individual exposures since the field of view of the camera was small.

34. *Robotic Exploration of the Solar System: Part 1. The Golden Age 1957–1982* by P. Ulivi and D. Harland (2007), Chichester, UK: Springer Praxis, p. 231.

35. Quoted in a NASA Jet Propulsion Laboratory news release associated with the thirtieth anniversary celebration of the Viking 1 landing, NASA Press Release 2006-091.

36. In addition to protean work on planetary science, Viking contributed to the esoteric subject of general relativity. There are four classical tests of Einstein's theory of gravity. One is the gravitational time delay effect, or "Shapiro effect," where a radar or light signal passing near a massive object takes slightly longer to travel than if the massive object were not there. The signal is retarded by the gravitational potential. Careful timing of signals sent to and from the Viking orbiter showed a delay that agreed with the general theory of relativity: "Viking Relativity Experiment: Verification of Signal Retardation by Solar Gravity," by R. D. Reasenburg et al. (1979), *Astrophysical Journal Letters* 234: 219–21.

37. *On Mars: Exploration of the Red Planet 1958–1978*, NASA SP-4212 by E. Ezell and L. Ezell (1984), Washington, DC: National Aeronautics and Space Administration.

38. Our continually improving ability to image the Martian surface was impressively demonstrated in 2006 when the Mars Reconnaissance Orbiter spotted both Viking landers and their nearby heat shields and backshells. These recent images are so sharp it is possible to identify individual rocks photographed by Viking from the surface in 1976.

39. Mutch wrote the forward for a commentary on the orbiter images, "Viking Orbiter Views of Mars," by M. H. Carr et al. (1980), NASA Scientific and Technical Information Branch, NASA SP-441.

40. *Mapping Mars* by O. Morton (2002), New York: Picador.

41. Regolith is a layer of loose, heterogeneous material overlying sold rock. It is distinct from soil because soil is made of rock broken down by weathering and erosion. On Mars, the fine dust and loose particles are caused by meteoric impacts as well as by erosion in the more distant past.

42. "Scientific Results of the Viking Missions," by G. Soffen (1976), *Science* 194: 1274–76.

43. Since the radioactive carbon might be taken in by a purely chemical mechanism, its detection would not be proof of biology. So other samples were sterilized before the carbon source was added, to act as controls; see "The Viking Carbon Assimilation Experiments: Interim Report," by N. Horowitz et al. (1976), *Science* 194: 1321–22. See also *Life on Mars: The Complete Story* by P. Chambers (1999), London: Blandford.

44. "Viking Labeled Release Biology Experiment: Interim Results," by G. Levin and P. Straaf (1976), *Science* 194: 1322–29.

45. The rusting due to the oxidation of iron minerals happened several billion years ago, when Mars was warmer and wetter. Rusting on Mars in its current

cold, dry state probably involves reactions catalyzed by superoxide minerals in the action of strong ultraviolet radiation, see "Evidence that the Reactivity of the Martian Soil is Due to the Action of Superoxide Ions" by A. S. Yen et al. (2000), *Science* 289: 1909–12.

46. Levin has had difficulties getting published in the peer-reviewed literature, but his arguments are summarized in "Modern Myths Concerning Life on Mars," by G. V. Levin (2006), *Electroneurobiologia* 14: 3–15, and in the book *Mars: The Living Planet* by B. Gregorio (1997), Berkeley, CA: North Atlantic Books.

47. "Mars-like Soils in the Atacama Desert, Chile, and the Dry Limit of Microbial Life," by R. Navarro-Gonzales et al. (2003), *Science* 302: 1018–21.

48. The detailed argument is that reactions with iron oxides or salts strongly attenuates the detection of organics using heat treatment and gas chromatography, such that the Martian soil could have several orders of magnitude more organic material than the Viking limit. See "The Limitations on Organic Detection in Mars-like Soils by Thermal Volatization-Gas Chromatography-MS and their Implications for the Viking Results," by R. Navarro-Gonzalez et al. (2006), *Proceedings of the National Academy of Sciences* 103: 16089–94.

49. "A Possible Biogenic Origin for Hydrogen Peroxide on Mars," by D. Schulze-Makuch and D. M. Houtkooper (2007), *International Journal of Astrobiology* 6: 147–54.

50. Quoted in a press release from the American Astronomical Society meeting on January 7, 2007, in Seattle, Washington, available online at http://researchnews.wsu.edu/physical/157.html.

51. Phoenix also refuted the idea that no liquids can exist on the surface of Mars. Viscous briny drops were seen forming on the leg of the lander, probably involving a mixture of perchlorate and water; see "Detection of Perchlorate and the Soluble Chemistry of Martian Soil at the Phoenix Mars Lander Site," by M. H. Hecht et al. (2009), *Science* 325: 64–67.

52. See "Kinetics of Perchlorate- and Chlorate-Respiring Bacteria" by B. E. Logan et al. (2001), *Applied and Environmental Microbiology* 67: 2499–2506.

53. "Reanalysis of the Viking Results Suggest Perchlorate and Organics at Mid-Latitudes on Mars," by R. Navarro-Gonzales et al. (2010), *Journal of Geophysical Research—Planets* 115: 2010–21.

54. See "UV-Resistant Bacteria Isolated from the Upper Troposphere and Lower Stratosphere," by Y. Yang et al. (2008), *Biological and Space Science* 22: 18–25.

55. *The Blue Planet: An Introduction to Earth System Science*, 2nd ed., by B. Skinner and B. Murck (1999) New York: John Wiley & Sons, p. v.

56. "A Physical Basis for Life Detection Experiments," by J. E. Lovelock (1965), *Nature* 207, no. 4997: 568–70.

57. *Gaia: A New Look at Life on Earth* by J. E. Lovelock (1979), Oxford: Oxford University Press, p. 1.

58. Ibid., p. 5.

59. *Dazzle Gradually: Reflections on the Nature of Nature* by L. Margulis and D. Sagan (2007), White River Junction: Chelsea Green Publishing, p. 154.

60. "Life Detection by Atmospheric Analysis," by J. E. Lovelock and D. Hitchcock (1967), *Icarus: International Journal of the Solar System* 2: 149–59.

61. *The Vanishing Face of Gaia: A Final Warning* by J. E. Lovelock (2009), New York: Basic Books, p. 163.

62. *James Lovelock: In Search of Gaia*, by J. Gribbin and M. Gribbin (2009), Princeton and Oxford: Princeton University Press, p. 143. Also see "Lynn Margulis (1938–2011)" by James A. Lake (December 22, 2011) *Nature* 480, no. 7378, p. 458.

63. "Gaia by Any Other Name," by L. Margulis (2004) in *Scientists Debate Gaia*, eds. S. H. Schneider and J. Miller, Cambridge, MA and London: MIT Press, pp. 8–9.

64. "Reflections on Gaia" by J. Lovelock (2004) in *Scientists Debate Gaia*, eds. S. H. Schneider and J. Miller, Cambridge, MA and London: MIT Press, p 2.

65. *Gaia: A New Look at Life on Earth* by J. E. Lovelock, p. 7.

66. "The Atmosphere, Gaia's Circulatory System" by L. Margulis and J. E. Lovelock Reprinted in *Dazzle Gradually: Reflections on the Nature of Nature* by L. Margulis and D. Sagan, pp. 157–71. The *Quarterly* provided a forum for both serious science topics as well as issues in the humanities. Lovelock has always had an iconoclastic streak and it was fitting that his most important scientific idea appeared in a journal that was embraced by the counterculture movement in the 1960s and 1970s.

67. One deceptively simple idea played an important role in gaining acceptance for the theory. "Daisyworld" was a mathematical model to show how organisms interacting with their environment can lead to temperature regulation. In the original version there are two types of daisy, black and white, such that black daisies absorb sunlight and heat the planet while white daisies reflect sunlight and so cool the planet. The simulation tracks the daisies as radiation from the host star increases. Competition between the daisies leads to a balance of populations where the planet's temperature stays close to the optimum for daisy growth. A later modification to the simulation showed that having many species improved the regulatory effects on temperature, indicating that biodiversity was beneficial. For the original model, see "Biological Homeostasis of the Global Environment: The Parable of Daisyworld" by A. J. Watson and J. E. Lovelock (1983), *Tellus B* 35: 286–89. For the improved version, see "Daisyworld Revisited: Quantifying Biological Effects on Planetary Self-Regulation," by T. M. Lenton and J. E. Lovelock (2001), *Tellus B* 52: 288–305.

68. *Earth System Science: From Biogeochemical Cycles to Global Change* by M. Jacobson, R. Charlson, H. Rodhe, and G. Orians (2000), San Diego and London: Academic Press, pp. 5, 6.

69. *The Vanishing Face of Gaia: A Final Warning* by J. E. Lovelock, p. 180.

70. "How Earth's Atmosphere Evolved to an Oxic State: A Status Report," by D. C. Catling and M. W. Claire (2005), *Earth and Planetary Science Letters* 237, p. 16.

71. *Pale Blue Dot: A Vision of the Human Future in Space* by C. Sagan (1974), New York: Random House, p. 228.

72. "The Prophet" by J. Goddell (November 1, 2007), *Rolling Stone*, Issue 1038, pp. 58–98. 7p. Academic Search Premier. Web. August 17, 2012.

73. *The Living Universe: NASA and the Development of Astrobiology* by S. J. Dick and J. E. Strick (2004), New Brunswick: Rutgers University Press, p. 83.

74. *The Ages of Gaia: A Biography of Our Living Earth*, J. E. Lovelock (1988), New York and London: W. W. Norton, p. 150.

75. "Foreword," L. Margulis, *The Ice Chronicles: The Quest to Understand Global Climate Change*, in P. Mayewski and F. White (2002), Hanover and London: University Press of New England, p. 10.

76. *The Ice Chronicles: The Quest to Understand Global Climate Change*. P. Mayewski and F. White (2002), Hanover and London: University Press of New England, pp. 151, 167, 166.

77. University of Delaware website, "Greenland glacier calves island 4 times the size of Manhattan, UD scientist reports," online at http://www.udel.edu/udaily/2011/aug/greenland080610.html.

78. The "2001 Amsterdam Declaration on Global Change" is the collective work of four organizations: the International Geosphere-Biosphere Programme (IGBP), the International Human Dimensions Programme on Global Environmental Change (IHDP), the World Climate Research Programme (WCRP), and the international biodiversity programme DIVERSITAS. The Declaration is available online at the IGBP website at http://www.igbp.net/4.1b8ae20512db692f2a680001312.html.

79. See "Landsat 1," at http://landsat.gsfc.nasa.gov/about/landsat1.html.

80. *The Ages of Gaia: A Biography of Our Living Earth*, by J. E. Lovelock, pp. xix, 19.

81. See "Groundwater Depletion is Detected from Space" by Felicity Barringer (May 30, 2011), New York Times.com online at http://www.nytimes.com/2011/05/31/science/31water.html?_r=2&ref=science.

82. "Revisiting the Viking Missions on Mars," *Talk of the Nation*, NPR Radio Interview by I. Flatow (July 21, 2006), NPR.org online at http://www.npr.org/templates/story/story.php?storyId=5573659.

3. MER—The Little Rovers That Could

1. NASA News Release 2003-081 from June 8, 2003, just three weeks before the rover launch date, available online at http://marsprogram.jpl.nasa.gov/mer/newsroom/pressreleases/20030608a.html.

2. See the resolution at the Library of Congress website, http://thomas.loc.gov/cgi-bin/query/z?c111:H.RES.67.EH.

3. Several authors have noted that detailed robotic survey has given us such a sense of familiarity with Mars that, as Andrew Chaikin writes, we can trace our "fingertips over the entire planet." See *A Passion for Mars: Intrepid Explorers of the Red Planet* by A. Chaikin (2008), New York: Abrams, pp. 231–32.

4. *Solar System Voyage* by S. Brunier (2002), trans. Storm Dunlop. Cambridge: Cambridge University Press, p. 10.

5. *Geography and Vision: Seeing, Imagining and Representing the World* by D. Cosgrove (2008), New York and London: I. B. Taurus & Co., p. 47.

6. In February 2012, NASA ended its involvement in the ExoMars mission in order to pay for cost overruns on the James Webb Space Telescope, one of several adverse consequences to space and planetary science from the huge projected cost

of that flagship mission. The Europeans and Russians are continuing, with an orbiter and static lander planned for 2016, and a rover planned for 2018.

7. One aspect of this work is the creation of an enclosed and self-contained habitat that simulates as closely as possible what astronauts would experience in a manned Mars mission. That's the goal of the Mars Society project, initially envisaged as a worldwide network of research stations but limited by funding to the Utah installation. The habitat provides living and working space for a six-person crew. Crew members can't leave without wearing a space suit. They can communicate with the outside with a built-in time delay that makes direct communication impossible, and they do their work with tools and equipment appropriate to a Mars mission. The second aspect involves locating sites that are Mars analogs in terms of geology or extreme aridity. For an example from Spain that mimics one of the rover sites, see "The Tinto River, and Extreme Acidic Environment under Control of Iron, as an Analog of the Terra Meridiani Hematite Site of Mars," by D. Fernandez-Remolar et al. (2004), *Planetary and Space Science* 52: 239–48.

8. *Postcards from Mars: The First Photographer on the Red Planet,* by J. Bell (2006), New York: Dutton, p. 76.

9. By modern standards, the Pancam does not seem that impressive: 2 megapixels, fewer than you would find on the camera of a typical cell phone. But they are extraordinarily robust, designed to work over a temperature range of 160°C. Each camera has a field of view of 17 degrees, so nine images would have to be mosaicked to mimic what we see with our eyes in a single view, and hundreds were carefully stitched together to get a single panorama.

10. Showing realistic and scientifically accurate colors is not trivial in an environment with a much thinner atmosphere than the Earth, and with airborne dust altering the color palette. Each rover has a combined sundial and calibration target mounted on it, within each view of the pancam. There are four different mineral targets, each mounted in silicone rubber and each a different color, which the pancams use to calibrate their color table. See "Chromaticity of the Martian Sky as Observed by the Mars Exploration Rovers Pancam Instruments," by J. F. Bell, D. Savransky, and M. J. Wolff (2006), *Journal of Geophysical Research* 111: 1–15.

11. "Pancam Multispectral Imaging Results from the Spirit Rover at Gusev Crater," by J. F. Bell et al. (2004), *Science* 305: 800–6, and "Pancam Multispectral Imaging Results from the Opportunity Rover at Meridiani Planum," by J. F. Bell et al. (2004), *Science* 305: 1703–9.

12. *Postcards from Mars: The First Photographer on the Red Planet*, by J. Bell, pp. 70, 70–71, 111.

13. Ibid., pp. 109, 113. For an example of a unique astronomical observation made from the surface of Mars, see "Solar Eclipses of Phobos and Deimos Observed from the Surface of Mars," by J. F. Bell et al. (2005), *Nature* 436: 55–57.

14. *Roving Mars: Spirit, Opportunity and the Exploration of the Red Planet,* by S. Squyres (2005), New York: Hyperion, pp. 2, 3.

15. The combination of low temperature and low pressure means that water cannot exist on the Martian surface; only the frozen and gaseous states are pos-

sible. If a cup of water was teleported (à la *Star Trek*) to the surface, it would explosively evaporate. Actually, the water is boiling since the pressure is so low it reduces the boiling point of water to below the surface temperature.

16. "An Integrated View of the Chemistry and Mineralogy of Martian Soils," by A. S. Chen (2005), *Nature* 436: 49–54.

17. "Mars Exploration Rover Mission," by J. A. Crisp et al. (2003), *Journal of Geophysical Research* 108, no. 8061: 17.

18. Technology developed for the rovers is summarized at http://marsrovers.nasa.gov/technology/.

19. The Mission was NASA's Mars Pathfinder, later renamed the Carl Sagan Memorial Station. Its rover was called Sojourner. See "Overview of the Mars Pathfinder Mission and Assessment of the Landing Site Predictions," by M. Golombek et al. (1997), *Science* 278: 1743–48.

20. "Rock Abrasion Tool: Mars Exploration Rover Mission," by S. P. Gorevan et al. (2003), *Journal of Geophysical Research* 108, no. 8068: 8.

21. First results from Spirit and Opportunity appeared in a special issue of *Science* magazine on August 6, 2004, vol. 305, pp. 737–900.

22. Suggestions of water flowing episodically on Mars even today have proved controversial. There are formations that suggest recent water activity and new gullies that have been carved out in the years that orbiters have imaged the surface, but the interpretation of these features is ambiguous. The controversy is embodied by the two papers "Present Day Impact Cratering Rate and Contemporary Gully Formation on Mars," by M. C. Malin et al. (2006), *Science* 314: 1573–77, and "Modeling the Formation of Bright Slope Deposits Associated with Gullies in Hale Crater, Mars: Implications for Recent Water Activity," by K. J. Kolb et al. (2010), *Icarus* 205: 113–37. The evidence for widespread, nonpolar glaciers is compelling; see "Radar Sounding Evidence for Buried Glaciers in the Southern Mid-Latitudes of Mars," by J. W. Holt et al. (2008), *Science* 322: 1235–38.

23. "In Situ Evidence for an Ancient Aqueous Environment at Meridiani Planum, Mars," by S. W. Squyres et al. (2004), *Science* 306: 1709–14.

24. "Hematite Spherules in Basaltic Tephra Altered under Aqueous, Acid-sulfate Conditions on Mauna Kea Volcano, Hawaii: Possible Clues for the Occurrence of Hematite-rich Spherules in the Burns Formation at Meridiani Planum, Mars," by R. V. Morris et al. (2005), *Earth and Planetary Science Letters* 240: 168–78.

25. "Jarosite and Hematite at Meridiani Planum from Opportunity's Moessbauer Spectrometer," by G. Klingelhoefer et al. (2004), *Science* 306: 1740–45.

26. "Water Alteration of Rocks and Soils on Mars at the Spirit Rover Site in Gusev Crater," by L. A. Haskin et al. (2005), *Nature* 436: 66–69.

27. "Sedimentary Rocks at Meridiani Planum: Origin, Diagensis, and Implications for Life," by S. W. Squyres and A. H. Knoll (2005), *Earth and Planetary Science Letters* 240: 1–10.

28. "Early Mars Climate Models," by R. M. Aberle (1998), *Journal of Geophysical Research* 103: 467–28, 489; and "Martian Surface Paleotemperatures from Thermochronology of Meteorites," by D. L. Schuster and B. P. Weiss (2005), *Science* 309: 594–600.

29. Over the long course of the mission, NASA has posted an update for each rover about once a week, for a total of nearly 800 reports from the surface of Mars. They can be seen at http://marsrovers.jpl.nasa.gov/mission/status.html.

30. From the NASA/JPL press release marking the beginning of the ninth year of Opportunity's work on Mars, see http://marsrover.nasa.gov/newsroom/pressreleases/20120124a.html.

31. Scott Maxwell has kept a blog detailing his work as a rover driver, called "Mars and Me," which can be found at http://marsandme.blogspot.com/. His arch sense of humor is on full display in his bio at the JPL website, where it says: "By day he works at the world-renowned Jet Propulsion Laboratory, to all appearances a mild-mannered computer programmer. But by night, as The Midnight Rover, he roams the crime-ridden L.A. streets, bringing truth, justice, and liberty into the lives of the ordinary citizens who need him most."

32. See her profile and interview at http://solarsystem.nasa.gov/people/profile.cfm?Code=StroupeA.

33. A description of the MER Student Astronaut Program and links to their individual journals can be found on the Planetary Society website at http://www.planetary.org/press-room/releases/2000/1002_Student_Scientists_from_Around_the.html.

34. The "Red Rover Goes to Mars" project is described at http://lego.marshall.edu/.

35. See http://www.marsdaily.com/reports/Astrobot_Biff_Starling_Prepares_for_Mars_Landing.html.

36. *A Traveler's Guide to Mars: The Mysterious Landscapes of the Red Planet*, by W. K. Hartmann (2003), New York: Workman Publishing, pp. 5, 6.

37. For a description of Google Mars, see http://www.google.com/mars/about.html, and also the online *National Geographic* article, "New Google Mars Reveals the Planet in 3D" by Virginia Jaggard at the news section of their website, http://news.nationalgeographic.com/news/2009/02/090204-google-mars.html.

38. "Mars for the Rest of Us," by J. Romero (June 2009), *IEEE Spectrum* at the journal site online at http://spectrum.ieee.org/aerospace/robotic-exploration/mars-for-the-rest-of-us/0.

39. See Sec. 407. "Participatory Exploration" of the Full Text of H.R.6063 National Aeronautics and Space Administration Authorization Act of 2008 at Spaceref.com online at http://www.spaceref.com/news/viewsr.html?pid=27997.

40. Quoted in *The Biological Universe: The Twentieth Century Extraterrestrial Life Debate and the Limits of Science*, by S. J. Dick (2006), Cambridge: Cambridge University Press, p. 401.

41. Ibid., p. 404.

42. *The Lure of the Red Planet,* by W. Sheehan and S. O'Meara (2001), Amherst: Prometheus Books, p. 202.

43. *Radio's America: The Great Depression and the Rise of Modern Mass Culture*, by B. Lenthall (2007), Chicago: University of Chicago Press, p. 12.

44. "The Dry Salvages," *Four Quartets* in *Collected Poems 1909–1962* by T. S. Eliot (1991), Orlando: Harcourt, p. 198.

45. *The Diary of Virginia Woolf. Volume III, 1925–1930*, by V. Woolf, eds. Anne Olivier Bell assisted by Andrew McNeillie (1980), New York and London: Harcourt Brace Jovanovich, p. 147.

46. *A Passionate Apprentice: The Early Journals, 1897–1909*, by V. Woolf, ed. Mitchell A. Leaska (1990), New York and London: Harcourt Brace Jovanovich, p. 399. As she never intended them to be published, Woolf did not use contractions in the informal writing of her diaries.

47. *The Letters of Vita Sackville-West to Virginia Woolf*, eds. L. DeSalvo and M. A. Leaska (1985), New York: William Morrow, pp. 93, 181. Ceres was first identified in 1801 by Giuseppe Piazzi of Palermo, Sicily, but our first closeup view of this likely dwarf planet will occur when NASA's Dawn mission encounters Ceres in 2015.

48. *The Diary of Virginia Woolf. Volume III, 1925–1930*, by V. Woolf, eds. Anne Olivier Bell and Andrew McNeillie (1980), New York and London: Harcourt Brace Jovanovich, p. 153.

49. "A Whiff of Mystery on Mars," by K. Sanderson (2010), *Nature* 463: 420–21. Technical articles on this topic include "Some Problems Related to the Origin of Methane on Mars," by V. A. Krasnopolsky (2005), *Icarus* 180: 359–67, and "Methane and Related Trace Species on Mars: Origin, Loss, Implications for Life, and Habitability," by S. K. Atreya, P. R. Mahaffy, and A.-S. Wong (2007), *Planetary and Space Science* 55: 358–69.

50. "Scientists Find Signs Water Is Flowing on Mars" by K. Chang (August 5, 2011), New York Times.com online at http://www.nytimes.com/2011/08/05/science/space/05mars.html. "Oldest Microfossils Raise Hopes for Life on Mars" by B. Vastag (August 21, 2011), Washington Post.com online at http://www.washingtonpost.com/national/health-science/oldest-microfossils-hail-from-34-billion-years-ago-raise-hopes-for-life-on-mars/2011/08/19/gIQAHK8UUJ_story.html. See also "The Dirt on Mars' Soil: More Suitable For Life Than Thought," by M. Wall (August 22, 2011), Space.com online at http://www.space.com/12695-mars-soil-life-support-study.html.

51. "Foreword," by Ray Bradbury in *Imagining Space: Achievements, Predictions, Possibilities, 1950–2050*, by R. D. Launius and H. E. McCurdy (2001), San Francisco: Chronicle Books, p. 15.

52. *From Sea to Space* by B. Finney (1992), Palmerston North, New Zealand: Massey University, p. 119.

53. *By Airship to the North Pole: An Archaeology of Human Exploration,* by P. J. Capelotti (1999), New Brunswick: Rutgers University Press, pp. 1747–5, and the online article "Space: The Final [Archaeological] Frontier," found at http://www.archaeology.org/0411/etc/space.html.

54. From *Mars* by L. T. Elkins-Tanton (2007), New York: Chelsea House, p. 97, with data reported from MOLA (Mars Orbital Laser Altimeter).

55. Quoted in *Worlds Beyond: The Thrill of Planetary Exploration*, by S. A. Stern (2002), Cambridge: Cambridge University Press, pp. 119–20.

56. *Dying Planet: Mars in Science and the Imagination*, by R. Markley (2005), Durham and London: Duke University Press, p. 344.

57. *Roving Mars: Spirit, Opportunity and the Exploration of the Red Planet,* by S. Squyres (2005), New York: Hyperion, pp. 377–78.

58. "Moon Landing," by W. H. Auden in *Selected Poems*, ed. E. Mendelson (2007), New York: Vintage, pp. 307–8.

4. Voyager—Grand Tour of the Solar System

1. See JPL's "Voyager: The Interstellar Mission" web page online at voyager .jpl.nasa.gov. The distance of each spacecraft from the Sun always increases, but sharp-eyed observers may note that the distance to the Earth sometimes decreases. That's because the Earth moves faster in its orbit of the Sun than the Voyagers are moving away from the Solar System, so at some times in the year the Earth approaches them. However, on average, they're receding from the Earth at the same rate they're receding from the Sun. As of August 2011, Voyager 1 was traveling at 38,153 mph relative to the Sun, while Voyager 2 had a velocity of 34,588 mph relative to the Sun.

2. Of the twenty-four astronauts who traveled to the Moon, twelve landed and set foot on the surface, all between July 1969 and December 1972. John Young and Eugene Cernan each went to the Moon twice, setting foot on it with their second trip, while Jim Lovell flew twice to the Moon without landing on it. This is perhaps the most extraordinary and exclusive club in history.

3. *De Caelo et Mundo, Lib. I, Trac. III, Cap. I*, by Saint Albertus Magnus, as quoted in "A Brief History of the Extraterrestrial Intelligence Concept" by J. F. Tipler (1981), *Quarterly Journal of the Royal Astronomical Society* 22: 133.

4. *Giant Planets of Our Solar System: Atmospheres, Composition, and Structure* by P.G.J. Irwin (2003), New York: Springer-Verlag.

5. "Particles, Environments, and Possible Ecologies in the Jovian Atmosphere," by C. Sagan and E. E. Salpeter (1976), *Astrophysical Journal Supplement* 32: 737–55.

6. The *Apollo 13* story has been well told in Lowell and Kluger (1994) and in the 1995 movie *Apollo 13*, directed by Ron Howard and based on the book. After its gravity assist, the trajectory of the spacecraft was so finely tuned that the astronauts were prohibited from jettisoning their urine for the reminder of the flight. This resulted in Fred Haise suffering a serious urinary tract infection because he didn't drink enough water. Luckily, this was the only medical problem suffered by any of the *Apollo 13* astronauts during their eventful trip.

7. Russians were pioneers in many aspects of the theory of space travel, and NASA has translated and archived the most important papers. Kondratyuk's 1919 manuscript "To Whoever will Read this Paper to Build an Interplanetary Rocket" is archived as NASA Technical Translation F-9285 (1965); Zander's 1925 paper "Problems of Flight by Jet Propulsion: Interplanetary Travel" is archived as NASA Technical Translation F-147 (1964); and Minovitch's work "A Method for Determining Interplanetary Free-Fall Reconnaissance Trajectories" is JPL Technical Memorandum TM-312-130. Minovitch also maintains a website on gravity assist at http://www.gravityassist.com/. Sometimes the term gravitational "slingshot" is used, but this isn't a good physical analogy; in a slingshot centrifugal force is turned into propulsive force.

8. For more on the art of sending spacecraft around the Solar System with precision and grace, see *Fly Me to the Moon: An Insider's Guide to the New Science of Space Travel* by Edward Bulbruno and Neil Tyson (2007), Princeton: Princeton University Press, and its more technical cousin *Capture Dynamics and*

Chaotic Motions in Celestial Mechanics: With Applications to the Construction of Low Energy Transfers by Edward Belbruno (2004), Princeton: Princeton University Press. NASA's Jet Propulsion Lab maintains an online tutorial on the Basics of Space Flight at http://www2.jpl.nasa.gov/basics/index.php.

9. *NASA's Voyager Missions: Exploring the Outer Solar System and Beyond* by B. Evans and D. M. Harland (2003), London: Springer-Praxis.

10. Technical details of the spacecraft and the scientific instruments are archived at NASA's National Space Science Data Center, online at http://nssdc.gsfc.nasa.gov/nmc/spacecraftDisplay.do?id=1977-084A.

11. Predictably, the use of radioactive material to power spacecraft has been controversial, although these small units do not have a self-sustaining nuclear reaction so they are not actually "reactors." The probability of dispersal of the radioactive material in Earth's atmosphere due to the explosive loss of the spacecraft on launch has been estimated at one in a million; see "The Cassini Mission Risk Assessment Framework and Application Techniques" by S. Guarro et al. (1995), *Reliability Engineering and System Safety* 49: 293–302. Congress nearly pulled the plug on the Galileo and Ulysses missions over this concern. The robustness of the RTG canisters was illustrated when the *Apollo 13* Lunar Module jettisoned its RTG energy source on re-entry in 1970. It fell intact to the bottom of the 20,000-foot Tonga Trench in the Pacific Ocean. The corrosion-resistant materials of the canister are expected to keep it intact for nearly 900 years. On the other hand, in the former Soviet Union, over 1000 RTGs have been deployed over the years to power remote lighthouses and navigation beacons and they've become a significant human hazard and environmental headache.

12. Quoted from "Space Explorers," an interview of Edward Stone by Kurt Streeter, published in the "Time Out" section of *Gulf Times*, April 27, 2011.

13. "The Voyager Missions to the Outer Solar System," by E. C. Stone (1977), *Space Science Reviews* 21: 75, the introduction to a special issue of the journal on the Voyagers.

14. *By Jupiter: Odysseys to a Giant* by E. Burgess (1982), New York: Columbia University Press.

15. "The Jupiter System Through the Eyes of Voyager 1," by B. A. Smith et al. (1979), *Science* 204: 951–57.

16. Jupiter has sixty-four moons, most of which are only a few kilometers across. The four moons discovered by Galileo—Ganymede, Callisto, Io, and Europa—each dwarf all the others combined in terms of mass. Thebe and Metis are the seventh and eighth most massive moons of Jupiter, respectively.

17. "The Mountains of Io: Global and Geological Perspectives from Voyager and Galileo," by P. Schenk et al. (2001), *Journal of Geophysical Research* 106: 33201–22.

18. "Hydrated Salt Minerals on Ganymede's Surface: Evidence of an Ocean Below," by T. B. McCord et al. (2001), *Science* 292: 1523–25.

19. *Unmasking Europa* by R. Greenberg (2008), Berlin: Springer Praxis.

20. This and other useful scientific morsels on the Voyager missions are contained in the NASA document "Voyage to the Outer Planets," NASA Facts 2007-12-6, produced by the Jet Propulsion Lab.

21. "Atmospheric Dynamics of the Outer Planets," by A. P. Ingersoll (1990), *Science* 248: 308, and "Dynamics of Triton's Atmosphere," by A. P. Ingersoll (199), *Nature* 344: 315.

22. NASA's Deep Space network consists of three large radio dishes: one in California's Mojave Desert, one outside Madrid, and a third near Canberra in Australia. Separated by roughly 120 degrees in longitude, they give twenty-four-hour coverage for signals from any direction in space. NASA operates the network but share its capabilities with space agencies from other countries. In addition to the Voyagers, the Deep Space Network is following the Opportunity rover and the Mars Reconnaissance Observer, Cassini, and the Dawn mission to the asteroid belt.

23. Technically, the outer limits of the Solar System are defined by the spherical Oort Cloud of comets. They make periodic excursions into the inner Solar System but for most of the time they are at the outer extremity of their orbits 50,000 to 100,000 A.U. from the Sun. The Oort Cloud is hypothesized, but the cold reservoir of trillions of comets at huge distances has never been observed. It's thought that they were originally among the planets but got ejected early in the history of the Solar System.

24. "An Explanation of the Voyager Paradox: Particle Acceleration at a Blunt Termination Shock," by D. J. McComas and N. A. Shwadron (2006), *Geophysical Research Letters* 33, L04102, and "Voyager 1 Explores the Termination Shock Region and the Heliosheath Beyond," by E. C. Stone et al. (2005), *Nature* 309: 2017–20.

25. "The Stellar Destiny of Pharaoh and the So-Called Air Shafts of Cheops Pyramid," by A. Badawy (1964), *MIDAWB Band X*, pp. 189–206, and "Astronomical Investigations Concerning the So-Called Air Shafts of Cheops," by V. Trimble (1964), ibid., pp. 183–87.

26. *Solar System Voyage* by S. Brunier (2002), trans. by S. Dunlop. Cambridge: Cambridge University Press, p. 10.

27. *Voyager's Grand Tour: To the Outer Planets and Beyond* by H. E. Dethloff and R. A. Schorn (2003), Washington, DC: Smithsonian Institution, p. 231.

28. *Journey Into Space: The First Three Decades of Space Exploration* by Bruce Murray (1989), New York and London: W. W. Norton, pp. 167, 168, 156. See also "Live Pictures of Fly-By Are to Be on TV," *New York Times*, August 20, 1989, 1:32. Also see "Many To See Photos From Voyager During Its Flight Past Neptune," Associated Press, on August 2, 1989.

29. Dave Itzkoff reports that the *Cosmos* series has been "viewed by 400 million people in 60 countries, making it public television's most-watched short-form series until [the] Ken Burns documentary 'The Civil War.'" Fox Broadcasting plans in 2013 to launch *Cosmos: A Space-Time Odyssey*, hosted by astrophysicist Neil deGrasse Tyson. Ann Druyan, who co-wrote with Sagan the original series titled *Cosmos: A Personal Voyage,* will reportedly be executive producer and a writer for the new series. See "'Family Guy' Creator Part of 'Cosmos' Update," D. Itzkoff (August 5, 2011), New York Times.com online at http://www.nytimes.com/2011/08/05/arts/television/fox-plans-new-cosmos-with-seth-macfarlane-as-a-producer.html?ref=science&pagewanted=print.

30. *Voyager Tales: Personal Views of the Grand Tour* by D. W. Swift (1997), Reston, VA: AIAA (American Institute of Aeronautics and Astronautics), pp. 324, 325.

31. Ibid., p. 405.

32. *The Dream of Spaceflight: Essays on the Near Edge of Infinity* by W. Wachhorst (2000), New York: Basic Books, p. 78.

33. See *Apollo 8 Onboard Voice Transcription* (1969), NASA Johnson Space Center Mission Transcripts: Apollo 8, p. 183, online at http://www.jsc.nasa.gov/history/mission_trans/apollo8.htm.

34. Casani's note is quoted in *Voyager's Grand Tour: To the Outer Planets and Beyond* by H. E. Dethloff and R. A. Schorn (2003), Washington, DC: Smithsonian Institution, p. 89, and note 25, p. 247.

35. *Murmurs of Earth: The Voyager Interstellar Record* by C. Sagan et al. (1978), New York: Random House, p. 11. The Record Committee comprised Frank Drake; MIT Physics professor Philip Morrison, who was among the first to suggest scanning for possible extraterrestrial radio communication; Harvard astronomer Alastair G. W. Cameron, who helped theorize the formation of Earth's moon; chemical evolution scientist Leslie Orgel, who along with colleague Francis Crick of DNA fame realized the importance of RNA; vice president for Hewlett-Packard B. M. Oliver, who published on the possibility of galactic civilizations; and philosophy of science professor Steven Toulmin. In 1980, Sagan, Bruce Murray, and astronomer Louis D. Friedman founded The Planetary Society, the leading mandate of which is SETI, the Search for Extraterrestrial Intelligence.

36. *Cosmic Journey: The Voyager Interstellar Mission and Message*, written and directed by P. Geller (2003), Executive Producer A. Druyan. Cosmos Studios. DVD.

37. *Murmurs of Earth: The Voyager Interstellar Record* by C. Sagan et al., pp. 154, 154–55, 156–57, 160.

38. Ibid., p. 120.

39. Timothy Ferris writes, "Beethoven remarked he could move himself to tears simply by thinking about the Cavatina," in *Murmurs of Earth: The Voyager Interstellar Record* by C. Sagan et al. (1978), New York: Random House, p. 203. See also *Thayer's Life of Beethoven by A. W. Thayer* Vol. 2, ed. Elliot Forbes (1964), Princeton: Princeton University Press, pp. 1007–8.

40. *Cosmic Journey: The Voyager Interstellar Mission and Message*, written and directed by P. Geller (2003).

41. *Murmurs of Earth: The Voyager Interstellar Record* by C. Sagan et al., p. 27.

42. "Humpback Whale Songs Spread Eastward like the Latest Pop Tune." Science Daily.com, April 15, 2011 online at http://www.sciencedaily.com/releases/2011/04/110414131444.htm.

43. *Murmurs of Earth: The Voyager Interstellar Record* by C. Sagan et al., p. 151.

44. Ibid., pp. 135, 137, 139, 141, 143.

45. *Voyager: Seeking Newer Worlds in the Third Great Age of Discovery* by S. J. Pyne (2010), New York: Viking, pp. 350, 323, 322.

46. *Murmurs of Earth: The Voyager Interstellar Record* by C. Sagan et al., p. 167.

47. Ibid., p. 42.

48. Holly Henry wishes to thank linguist and California State University colleague Parastou Feiz for suggesting this point to me.

49. *On the Origin of Stories: Evolution, Cognition, and Fiction* by B. Boyd (2009), Cambridge and London: Belknap Press of Harvard University Press, p. 295.

50. President Carter's statement is reprinted in *Murmurs of Earth: The Voyager Interstellar Record* by C. Sagan et al., p. 28.

51. "Gettysburg and Now," by C. Sagan and A. Druyan (1997), *Billions and Billions: Thoughts on Life and Death at the Brink of the Millennium*, New York: Random House, pp. 199, 200. Sagan presented this text in slightly revised form to a crowd of approximately 30,000 people gathered to commemorate the battle and rededicate the Eternal Light Peace Memorial located in the Gettysburg National Military Park.

52. *Pale Blue Dot: A Vision of the Human Future in Space* by C. Sagan (1994), New York: Random House, p. 153.

53. *NASA/TREK: Popular Science and Sex in America* by C. Penley (1997), New York and London: Verso, pp. 16, 19. The *Enterprise* prototype shuttle was used to test flight performance in the atmosphere but was not equipped for space flight and reentry. With retirement of the shuttle program, the *Enterprise* shuttle is on display at the Intrepid Sea, Air and Space Museum in New York City.

54. *Journey Into Space: The First Three Decades of Space Exploration* by Bruce Murray (1989), New York and London: W. W. Norton, p. 193.

55. *Up Till Now: The Autobiography* by W. Shatner and D. Fischer (2008), New York: Thomas Dunne Books/St. Martin's, p. 114.

56. See the Kennedy Space Center site online at http://www.kennedyspace center.com/sci-fi-summer.aspx.

57. *Voyager's Grand Tour: To the Outer Planets and Beyond* by H. E. Dethloff and R. A. Schorn (2003), p. 109. Also, the television series *Star Trek: Voyager* was one of several spin-offs from the 1960s series.

58. Ed Stone's interview with Ira Flatow in June 2011 is from *Talk of the Nation: Science Friday*. The interview titled "Voyager 1 Probing Solar System's Outer Edge" is online at NPR's website at http://www.npr.org/2011/06/17/137250831 /voyager-1-probing-solar-systems-distant-edge.

59. "Cold War Pop Culture and the Image of U.S. Foreign Policy: The Perspective of the Original *Star Trek* Series," by N. E. Sarantakes (2005), *Journal of Cold War Studies* 7, no. 4: 78.

60. *Deep Space and Sacred Time:* Star Trek *in the American Mythos* by J. Wagner and J. Lundeen (1998), Westport, CT: Praeger, pp. 3, 136.

61. *NASA/TREK: Popular Science and Sex in America*. C. Penley (1997), New York and London: Verso, pp. 17, 20.

62. *Pale Blue Dot: A Vision of the Human Future in Space* by C. Sagan, pp. 82; 8–9.

63. See the NASA/JPL web page on the Interstellar Mission at http://voyager .jpl.nasa.gov/mission/interstellar.html.

64. To make the comparison between chemical energy and fusion energy, consider a Big Mac. The chemical energy in a 100 gram hamburger patty delivers 2.5 million Joules of energy to your body. But if we heated the patty up enough to fuse its nuclei, we would get about 1 percent of its mass energy, 10^{14} Joules. That enormous factor is why fusion is still considered a much better way to power our civilization than by releasing chemical energy contained in fossil fuels.

65. To put the idea of an interstellar fusion spacecraft in perspective, the Shuttle has a mass of 2,000 tons and it would need to harness the energy from 500,000 kg of fuel. The largest experimental fusion reactor, the Joint European Torus, weighs about 4,000 tons and has only ever sustained a reaction for half a second, producing 5 million Joules. Ignoring the fact that the reactor itself is more massive than the Shuttle, the energy output would have to increase by a factor of 100 trillion to power a trip to a nearby star.

66. *Pale Blue Dot: A Vision of the Human Future in Space* by C. Sagan, pp. 84, 395.

67. *Frontiers of Propulsion Science (Progress in Astronautics and Aeronautics)* by M. Millis and E. Davis (2009), Reston, VA: American Institute of Astronautics and Aeronautics.

68. The Breakthrough Propulsion Physics website is at http://www.grc.nasa.gov/WWW/bpp/index.html. A companion site for the general public called "Warp Drive: When?" describes graphically why new technology will be needed to crack the interstellar travel problem; see http://www.nasa.gov/centers/glenn/technology/warp/warp.html. Thwarted by NASA's tepid support for his program, which evaporated completely in 2003, Marc Millis founded the Tau Zero Foundation, a nonprofit organization designed to educate the public and gather support for a serious commitment for interstellar travel. The website amusingly riffs off a famous saying of Confucius: "A journey of a thousand light years begins with a single step." Since 2004, Paul Gilster has written an excellent blog that serves as the news forum for the Tau Zero Foundation, called *Centauri Dreams*, see http://www.centauri-dreams.org/.

69. *Entering Space: Creating a Spacefaring Civilization* by R. Zubrin (1999), New York: Tarcher Putnam.

70. See DARPA's website on the 100 Year Starship study at http://www.100yss.org.

71. For Ion propulsion and other *Star Trek* innovations, see *How William Shatner Changed the World* (2005), directed by Julian Jones. Allumination Filmworks, DVD.

72. See the X Prize Foundation at http://www.xprize.org as well as the Qualcomm Tricorder Prize at http://www.qualcommtricorderxprize.org. Founded by Peter Diamandis, the X Prize Foundation supports innovation in space exploration, ocean conservation, and preservation of Earth's ecosystems.

73. "Print Your Own Space Station—in Orbit" by M. Wall, posted on November 11, 2010, online at Space.com at http://www.space.com/9516-print-space-station-orbit.html. "NASA Selects Visionary Advanced Technology Concepts for Study" by D. Steitz, posted on August 8, 2011, Press Release online at http://www.nasa.gov/home/hqnews/2011/aug/HQ_11-260_NIAC_Selections.txt.

Information regarding Kurzweil and Diamandis's Singularity University is online at http://singularityu.org.

74. "MIT Scientists Develop a Drug to Fight any Viral Infection" by M. Melnick, August 11, 2011, Time.com online at http://healthland.time.com/2011/08/11/mit-scientists-develop-a-drug-to-fight-any-viral-infection/?hpt=hp_bn6.

75. *The Singularity Is Near: When Humans Transcend Biology* by R. Kurzweil (2005), New York: Viking, p. 28.

76. "New Planet in Neighborhood, Astronomically Speaking," by Dennis Overbye, posted October 16, 2012 on the *New York Times* website at http://www.nytimes.com/2012/10/17/science/space/new-planet-found-in-alpha-centauri.html?_r=0.

77. Quoted in the article "The Long Shot" by Lee Billings, posted May 19, 2009 on the Seed Magazine website at http://seedmagazine.com/content/print/the_long_shot/.

5. Cassini—Bright Rings and Icy Worlds

1. "Captain's Log," by Carolyn Porco (September 21, 2009), the leader of the Imaging Team on the Cassini Mission, online at http://www.ciclops.org/index/5830/Le_Sacre_du_Printemps?js=1. This is an entry from Porco's eloquent commentary regarding Cassini's ongoing mission. Archived entries are available at the CICLOPS (Cassini Imaging Central Laboratory for Operations) website. A spectacular image of Saturn in equinox can be viewed on the Ciclops page titled "The Rite of Spring" at http://www.ciclops.org/view.php?id=5773.

2. The rings are composed of particles ranging from microscopic to those the size of a house, with a composition that is mostly icy and partially rocky. For a general description, see "Saturn: Rings" at NASA's Solar System Exploration website http://solarsystem.nasa.gov/planets/profile.cfm?Object=Saturn&Display=Rings. For more technical background, see "The Scientific Significance of Planetary Ring Systems," by E. D. Miner, R. R. Wesson, and J. N. Cuzzi (2007), in *Planetary Ring Systems*, New York: Springer-Praxis, pp. 1–16.

3. "Captain's Log," by C. Porco (September 21, 2009), online at CICLOPS at http://www.ciclops.org/index/5830/Le_Sacre_du_Printemps?js=1. As Porco occasionally uses ellipses to emphasize a point, ellipses in brackets indicate text deleted from her entries.

4. Galileo Galilei observed the rings in 1610, but with the imperfect optics of his 20-power magnification telescope they looked like "blobs" on either side of Saturn, so he labeled them as moons. Two years later, they seemed to have disappeared—Galileo had inadvertently observed a plane crossing. Then in 1655, Huygens's superior optics and 50-power magnification telescope let him clearly identify the features as rings.

5. Huygens apparently gave some considerable thought to the people he assumed might inhabit such cold distant worlds as Saturn or Titan and wrote: "It is impossible but that their way of living must be very different from ours, having such tedious winters." See "Space: Ears, Rings and Cassini's Gap," *Time* magazine online at http://www.time.com/time/magazine/article/0,9171,952837-2,00.html#ixzz0YsGFKDUV.

6. Quoted in *The Art of Chesley Bonestell* by R. Miller and F. C. Durant, III (2001), London: Paper Tiger, p. 13. For a history of space art, see *Space Art* by R. Miller (1978), Darby, PA: Diane Publishing. Artist Don Davis reports that Bonestell told him the first planet he ever painted was actually Mars, not Saturn.

7. *The Art of Chesley Bonestell* by R. Miller and F. C. Durant, III, p. 44.

8. A. C. Clarke is quoted in *The Art of Chesley Bonestell* by R. Miller and F. C. Durant, III, p. 9.

9. *The Dream of Space Flight: Essays on the Near Edge of Infinity* by W. Wachhorst (2000), New York: Basic Books, pp. 47, 48.

10. *The Art of Chesley Bonestell* by R. Miller and F. C. Durant, III, p. 31. See also biographical details from *A Chesley Bonestell Space Art Chronology* by Melvin Schuetz (1999), Parkland, FL: Universal Publishers/uPUBLISH.com.

11. *The Dream of Space Flight: Essays on the Near Edge of Infinity* by W. Wachhorst, p. 58.

12. *The Art of Chesley Bonestell* by R. Miller and F. C. Durant, III, p. 24.

13. "Re-Thinking Apollo: Envisioning Environmentalism in Space," by H. Henry and A. Taylor (2009), in *Space Culture and Travel: From Apollo to Space Tourism,* Edited by D. Bell and M. Parker. Oxford and Malden, MA: Wiley/Blackwell, p. 196.

14. *The Dream of Space Flight: Essays on the Near Edge of Infinity* by W. Wachhorst, pp. 51, 54, 84. Bonestell in his later years lived on the California coast in Carmel.

15. "Space: Ears, Rings and Cassini's Gap," *Time* magazine online, November 24, 1980, at http://www.time.com/time/magazine/article/0,9171,952837-1,00.html.

16. According to archaeoastronomer Ed Krupp, ancient Egyptian texts indicate that the Great Sphinx of Giza portrays "Horemakhet ('Horus of the Horizon')," who represented "the rising disk of the Sun, fully poised on the eastern horizon." See "The Sphinx Blinks" by E. Krupp (2001), *Sky and Telescope* (March): 86. Krupp serves as the director of the Griffith Observatory, Los Angeles, California.

17. See also *Under the Sea Wind* (1941), re-released in 1996, New York: Penguin Books, and *The Sea Around Us* (1951), re-released in 2003, Oxford and New York: Oxford University Press.

18. *The Edge of the Sea* by R. Carson (1955), re-released in 1998, New York: Houghton Mifflin, pp. xi, 2–3.

19. *The Star Thrower* by L. Eiseley (1994), New York: Harvest/Harcourt Brace, p. 69. Annie Dillard and Cort Conley note that the setting of the essay is Sanibel Island, Florida, though Eiseley's narrator uses the fictitious name Costabel. They write, "Eiseley claimed he picked up the name Costabel by listening to a seashell," see *Modern American Memoirs*, edited by A. Dillard and C. Conley (1995), New York: HarperCollins, p. 416.

20. *The Star Thrower* by L. Eiseley, pp. 71, 91.

21. "The Cassini/Huygens Mission to the Saturnian System," by D. L. Matson, L. J. Spilker, and J. P. LeBreton (2002), *Space Science Reviews* 104, pp. 1–58.

See also *Mission to Saturn: Cassini and the Huygens Probe* by D. M. Harland (2002), Berlin and London: Springer-Verlag.

22. "NASA's New Road to Faster, Cheaper, Better Exploration," by R. A. Kerr (2002), *Science* 298: 1320–22.

23. "Faster, Cheaper, Better: Policy, Strategic Planning, and Human Resource Alignment," an Audit Report by the Office of the NASA Inspector General (2001), Washington, DC: NASA.

24. "Cassini Interplanetary Trajectory Design," by F. Peralta and S. Flanagan (1995), *Control Engineering Practice* 3: 1603–10.

25. "The Cassini Mission Risk Assessment Framework and Application Techniques," by S. Guarro et al. (1995), *Reliability Engineering and System Safety* 49: 293–302. For a sampling of the concern at the time, see "Cassini Mission to Saturn Lifts Off Amid Cloud of Controversy," by M. Clary, in the *Los Angeles Times* online for October 16, 1997, at http://articles.latimes.com/1997/oct/16/news/mn-43330.

26. An overview of the Cassini flybys is available at the NASA/JPL website at http://saturn.jpl.nasa.gov/mission/flybys/, and upcoming "Tour Dates" are listed at http://saturn.jpl.nasa.gov/mission/saturntourdates/.

27. An avid Beatles fan, Porco has posted a picture of the scientific team in London doing the Abbey Road "walk," at http://www.ciclops.org/team/iss_team.php. In 2006, Porco and the Cassini Imaging Team honored Paul McCartney on his sixty-fourth birthday with a video montage of sixty-four scenes of Saturn and its moons. They had artfully set the video to a Beatles soundtrack but the Apple lawyers quickly called so they had to take down that version and put up a silent one. It can be seen, along with a poster montage and the individual images, at http://www.ciclops.org/th_films.php.

28. The benchmark papers of the most scientifically productive instruments are "Cassini Imaging Science: Instrument Characteristics and Anticipated Investigations at Saturn," by C. Porco et al. (2004), *Space Science Reviews* 115: 363–497, "The Cassini Visual and Infrared Mapping Spectrometer Investigation," by R. H. Brown et al. (2004), *Space Science Reviews* 115: 111–68, and "Radar: The Cassini Radar Titan Mapper," by C. Elachi et al. (2004), *Space Science Reviews* 115: 71–110.

29. See the reviews "Saturn: Atmosphere, Ionosphere, and Magnetosphere," by T. Gombosi and A. Ingersoll (2010), *Science* 327: 1476–79, and "An Evolving View of Saturn's Dynamic Rings," by J. N. Cuzzi et al. (2010), *Science* 327: 1470–75.

30. "Cassini Imaging of Jupiter's Satellites, Atmosphere, and Rings," by C. Porco et al. (2009), *Science* 299: 1541–47.

31. "A Test of General Relativity Using Radio Links to the Cassini Spacecraft," by B. Bertotti, L. Iess, and P. Tortora (2003), *Nature* 425: 374–76.

32. "Saturn: Rings," NASA Solar System Exploration at NASA.gov, with explanations for the structure, gaps, and thickness of the rings, at http://solarsystem.nasa.gov/planets/profile.cfm?Object=Saturn&Display=Rings.

33. "Cassini Reveals New Ring Quirks, Shadows During Saturn Equinox," the Cassini Equinox Mission, NASA Jet Propulsion Laboratory website, October

11, 2009, at http://www.nasa.gov/home/hqnews/2009/sep/HQ_09-217_Cassini_Saturn_Rings_Equinox.html.

34. "Giant Propeller in the A Ring," Cassini Equinox Mission, NASA Jet Propulsion Laboratory web site, October 11, 2009, online at http://www.nasa.gov/mission_pages/cassini/multimedia/pia11672.html.

35. "Cassini Radar Views the Surface of Titan," by C. Elachi et al. (2005), *Science* 308: 970–74.

36. "Resolving Rain Over Xanadu," Cassini Equinox Mission, NASA Jet Propulsion Laboratory website, October 12, 2009, at http://saturn.jpl.nasa.gov/news/cassiniscienceleague/science20090910/.

37. *Entering Space: Creating a Spacefaring Civilization* by R. Zubrin (1999), New York: Tarcher/Putnam, pp. 163–66.

38. "NASA reveals first-ever photo of liquid on another world," by T. Patterson (December 18, 2009), CNN.com, online at http://www.cnn.com/2009/TECH/space/12/18/saturn.titan.reflection/index.html.

39. It's too cold for water to be a liquid on Titan's surface, but it can be liquid in combination with other molecules that have lower melting points. Confirmation of liquid methane and ethane on the surface of Titan came in stages. It has been suggested by Voyager data and images from the Hubble Space Telescope in 1995, and when Cassini first arrived at the Saturn system and imaged Titan the evidence was only suggestive; "Imaging of Titan from the Cassini Spacecraft," by C. Porco et al. (2005), *Nature* 434: 159–68. Radar imaging in 2006 confirmed the existence of lakes beyond doubt, "The Lakes of Titan," by E. R. Stofen et al. (2007), *Nature* 445: 61–64. The inventory of organics is discussed in "Titan's Inventory of Surface Organic Materials," by R. Lorenz et al. (2008), *Geophysical Research Letters* 35, L02206.

40. "How to Land on Titan," by S. Lingard and P. Norris (2005), *Ingenia*, Issue 23, article in the online engineering magazine at http://www.ingenia.org.uk/ingenia/issues/issue23/lingard.pdf.

41. Quoted in NASA's *Astrobiology Magazine* online at http://www.astrobio.net/pressrelease/1435/titan-wind-mystery-settled-from-earth.

42. The nature of the plumes and their likely origin in a subsurface ocean, like the evidence for lakes on Titan, has become clarified as Cassini continues its mission. There had been evidence of geological activity on Enceladus even before Cassini. Plumes were confirmed by flybys in 2005, "Enceladus' Water Vapor Plumes," by C. J. Hansen et al. (2006), *Science*311: 1422–25. More details on the largest plume then followed, "The Composition and Structure of the Enceladus Plume," by C. J. Hansen et al. (2011), *Geophysical Research Letters* 38: L11202.

43. *Planetology: Unlocking the Secrets of the Solar System* by T. Jones and E. Stofan (2008), Washington, DC: National Geographic, p. 181.

44. *Alien Ocean: Anthropological Voyages in Microbial Seas* by S. Helmreich (2009), Berkeley and Los Angeles: University of California Press, p. 68.

45. Ibid., p. 255.

46. "Researchers working on the Cassini mission to Saturn and its rings and moons have discovered sodium in the ice grains jetting from the surface of the icy moon Enceladus, using Cassini's Cosmic Dust Analyzer. This is seen as a sign

of salty liquid water beneath the moon's crust. . . ." See "Cassini Sees Salt Spray on Enceladus," by F. Postberg et al. (2009), *Astronomy & Geophysics* 50: 4.09.

47. "The Possible Origin and Persistence of Life on Enceladus and Detection of Biomarkers in the Plume," by C. McKay et al. (2008), *Astrobiology* 8: 909–18. See also "Titan's Astrobiology," by F. Raulin et al., in *Titan from Cassini-Huygens,* edited by R. Brown, J. P. Lebreton, and J. Waite (2009), New York: Springer, p. 216. According to Richard Corfield, British cosmologist James Jeans first speculated that Titan's mass could retain an atmosphere, particularly of gases such as nitrogen or methane. See *Lives of the Planets: A Natural History of the Solar System* by R. Corfield (2007), New York: Basic Books, p. 188.

48. "Microbial life at -13 °C in the brine of an ice-sealed Antarctic lake," *PNAS* (Proceedings of the National Academy of Sciences) by A. E. Murray et al. (2012), online at http://www.pnas.org/content/109/50/20626.

49. *Alien Ocean: Anthropological Voyages in Microbial Seas* by S. Helmreich, p. 256.

50. "Saturn Sublime," by K. S. Robinson, in *Saturn: A New View* by L. Lovett, J. Horvath, and J. Cuzzi (2006), New York: Harry N. Abrams, p. 17.

51. *Flesh and Machines: How Robots Will Change Us* by R. Brooks (2002), New York: Pantheon Books, pp. 11, 236.

52. *The Singularity Is Near: When Humans Transcend Biology* by R. Kurzweil (2005), New York: Viking, pp. 374, 47–48.

53. *Alien Ocean: Anthropological Voyages in Microbial Seas* by S. Helmreich, p. 233.

54. Jo Handelsman's interview with Bruce Gellerman, *Living on Earth* radio program, "Microbes' Big Role," Public Radio International online at http://www.loe.org/shows/shows.htm?programID=07-P13-00013#feature5; also cited in Helmreich, p. 283.

55. "Cancer's Secrets Come Into Sharper Focus," by G. Johnson (August 15, 2011), *New York Times* online at http://www.nytimes.com/2011/08/16/health/16cancer.html?pagewanted=all.

56. *Alien Ocean: Anthropological Voyages in Microbial Seas* by S. Helmreich, pp. 283–84.

57. "Cancer's Secrets Come Into Sharper Focus," online at http://www.nytimes.com/2011/08/16/health/16cancer.html?pagewanted=all.

58. "Enceladus Named Sweetest Spot for Alien Life," by R. Lovett (2011), *Nature* online, May 31, 2011.

6. Stardust—Catching a Comet by the Tail

1. From a cosmic perspective, everything worked out "just so" for biology to eventually occur. A somewhat higher cosmic density would have overcome and reversed the expansion too quickly for generations of stars to live and die and generate enough carbon for organic life-forms. A somewhat lower cosmic density would have led to an expansion too rapid for stars to form at all, leading to a sterile universe. Therefore, the initial conditions of the Big Bang represent a kind of "fine tuning" around the presence of long-lived stars and the interesting consequences that result from them.

2. *The Magic Furnace: The Search for the Origins of Atoms* by M. Chown (2001), Oxford: Oxford University Press.

3. From the NASA Stardust mission page at http://stardust.jpl.nasa.gov/mission/spacecraft.html.

4. Kapton is also an excellent thermal insulator so it's widely used in the aviation and aerospace industries. It was developed by DuPont; see http://www2.dupont.com/Kapton/en_US/.

5. Comet orbits are extreme versions of the nearly circular planetary orbits described by Kepler's laws. The second or "equal areas" law says that the imaginary line connecting a Solar System body to the Sun sweeps out equal areas in equal intervals of time. Comets have highly squashed or eccentric orbits so they travel fastest when they're closest to the Sun. So a comet that travels tens of thousands of astronomical units from the Sun will spend the vast majority of its time near its apogee, or farthest point from the Sun. Since a comet is dark and extremely faint at such a large distance, we only see the tiny fraction of comets that ever venture within the orbits of the planets, and even then they're only visible for a tiny fraction of their orbits. Therefore, the conjecture of a vast repository of perhaps a trillion comets in a spherical entity called the Oort Cloud is a hypothesis consistent with all available data, but very difficult to test directly.

6. See Gary Kronk's "Cometography" website, in particular, http://cometography.com/pcomets/081p.html.

7. Many of these gases don't have a liquid state, so in response to the Sun's energy they "sublime" or pass directly from solid to gaseous form. At the farthest excursion of most comets from the Sun, almost every terrestrial gas or liquid is so cold that it's a solid. For an overall treatise, see the authoritative *Meteorites, Comets, and Planets*, edited by A. M. Davis (2005), Oxford and London: Elsevier Ltd.

8. Information from the Stardust Launch Press Kit, released by NASA in February 1999, and on the mission website, at http://www.nasa.gov/stardust.

9. Minor Solar System bodies like asteroids can be named once they have been observed often enough to have reliable orbits. The first asteroids to be discovered, at the beginning of the nineteenth century, were named after mythological characters: Ceres, Pallas, Juno, and Vesta were the first four. Names must be approved by the International Astronomical Union, but they allow a lot of latitude. As a result, some asteroids have surprising or unusual names (and less than 10 percent of the quarter million logged asteroids have been named). Musical examples include Bach (1814), Chopin (3784), Dvorak (2055), Lennon (4147), McCartney (4148), and Zappafrank (3834). Literary examples include Hugo (2106), Chaucer (2984), Dickens (4370), Clarke (4923), and Asimov (5200). Asteroid Annfrank (5535) was discovered by Karl Reinmuth in 1942 and named after the diarist who died in a concentration camp near the end of World War II.

10. Stardust managed to be low-tech and hi-tech at the same time. Onboard memory was only 128 Mbytes, dozens of times less than most thumb drives and an amount most computer users would scoff at, but when the flight hardware was "frozen" in the early 1990s this was a respectable amount of storage. On the other hand, engineers achieved the amazing feat of sending over a million names of members of the public into space on four microchips in the Stardust payload.

The names were not stored digitally in bits and bytes but were actually etched in writing so small that each name could easily fit in the width of a human hair. In addition to members of the public who submitted their names, one of the chips contains the 58,214 names inscribed on the Vietnam War Memorial in Washington, DC. Two of the chips came back to Earth in the Sample Return Capsule. The other two are drifting off into deep space; their names will only be read by aliens who happen to carry an electron microscope.

11. "Coherent Expanded Aerogels and Jellies" by S. S. Kisler (1931), *Nature* 127: 741.

12. Information from the Stardust Sample Return Press Kit, released by NASA in January 2006, and also on the website http://www.nasa.gov/stardust.

13. "Issues in Planetary Protection: Policy, Protocol, and Implementation" by J. D. Rummel and L. Billings (2004), *Space Policy* 20: 49–54.

14. Comet particles and interstellar dust particles gathered by Stardust are housed at the Astromaterials Acquisition and Curation Office at NASA's Johnson Space Center. The same facility also holds several hundred kilograms of rocks and soil gathered by the Apollo astronauts, cosmic dust collected by high-altitude NASA aircraft, solar wind atoms gathered by the Genesis spacecraft, and meteorites originating from the Asteroid Belt, the Moon, and Mars, collected from the pristine ice of the high Antarctic plateau.

15. Stardust@Home is hosted by the University of California at Berkeley, which also hosted SETI@Home and a number of other citizen science projects; see http://stardustathome.ssl.berkeley.edu/.

16. Galaxy Zoo is a web-based project by which nonexperts classify galaxies according to their shapes and colors, after training against archetypal classifications. To date, over 80,000 users have performed over 30 million classifications, making several important discoveries. Galaxy Zoo is now part of the Zooniverse suite of Citizen Science projects. See http://www.galaxyzoo.org/ and https://www.zooniverse.org/.

17. "Citizen Science: People Power" by E. Hand (2010), *Nature* 466: 685–87.

18. "Dust from Comet Wild 2: Interpreting Particle Size, Shape, Structure, and Composition from Impact Features on the Stardust Aluminum Foils," by A. T. Kearsley et al. (2008), *Meteoritics and Space Science* 43: 41–73.

19. The method was described by amateur astronomer Martin Willes, and it is reproduced in an Astronomy.FM Skylog online at http://www.lcas-astronomy.org/articles/display.php?filename=micrometeorites_on_your_roof&category=miscellaneous.

20. Primary science results from Stardust were reported in "Comet 81P/Wild 2 Under a Microscope" by D. Brownlee et al. (2006), *Science* 314: 1711–16, and in six subsequent papers in a special issue of *Science* magazine.

21. "Stardust: A Mission with Many Scientific Surprises" by D. Brownlee, online at the NASA Stardust website http://stardust.jpl.nasa.gov/news/news116.html.

22. "Cometary Glycine Detected in Samples Returned by Stardust" by J. E. Elsila, D. P. Glavin, and J. P. Dworkin (2009), *Meteoritics and Space Science* 44: 1323–30.

23. "Evidence for Aqueous Activity on Comet 81P/Wild 2 from Sulfide Mineral Assemblages in Stardust Samples and CI Chondrites" by E. L. Berger, T. J.

Zega, L. P. Keller, and D. S. Lauretta (2011), *Geochimica et Cosmochimica Acta* 75: 3501–13.

24. See the overviews in "Whence Comets" by M. F. A'Hearn (2006), *Science* 314: 1708–9, and "NASA Returns Rocks from a Comet" by D. S. Burnett (2006), *Science* 314: 1709–10.

25. The Stardust/NexT mission website is at http://stardustnext.jpl.nasa.gov/index.html.

26. *The Universe of Stars* by H. Shapley (1929). Based on radio talks from the Harvard Observatory, edited by H. Shapley and C. Payne. Cambridge, MA: Harvard Observatory, p. 5.

27. Harlow Shapley, quoted in "The Star Stuff That is Man" by H. G. Garbedian, *New York Times Magazine*, August 11, 1929, Section 5.2.

28. In fission, the nucleus of an atom spontaneously splits into smaller components, with the release of energy. In fusion, the opposite occurs so that atomic particles are added to a light atomic nucleus, with a release of energy. Fusion doesn't occur naturally on Earth, only in stars (and in the early universe).

29. "Riddles in Fundamental Physics" by L. A. Gaumé (2008), *Leonardo* 41, no. 3: 247.

30. "Stardust," Oxford English Dictionary Online, at http://www.oed.com/.

31. "Synthesis of the Elements in Stars: Forty Years of Progress," by G. Wallerstein et al. (1999), *Reviews of Modern Physics* 69: 995–1084.

32. "Riddles in Fundamental Physics" by L. A. Gaumé, p. 247.

33. *Stardust: Supernovae and Life—The Cosmic Connection* by J. Gribbin and M. Gribbin (2000), New Haven and London: Yale University Press, p. 188.

34. "Herschel Detects a Massive Dust Reservoir in Supernova 1987A" by M. Matsuura et al. (2011), *Science Express*, 7 July, DOI 10.1126.

35. *Stardust: Supernovae and Life—The Cosmic Connection* by J. Gribbin and M. Gribbin, p. 195. The strongest evidence for supernova-triggered formation of the Solar System involves unaltered chondrite meteorites with high ratios of short-lived and now extinct radioactive Fe^{60} relative to the stable form of iron, Fe^{56}. This isotope could in principle be made by radiation from the young Sun but other isotopes that would therefore be expected in elevated quantities are not found. The best explanation is that the Fe^{60} was created in a nearby supernova that caused the solar nebula to collapse and impregnated the primordial material with this short-lived species (a half-life of 1.5 million years). See "Short-lived Nuclides in Hibonite Grains from Murchison: Evidence for Solar System Evolution" by K. K. Marhas, J. M. Goswami, and A. M. Davis (2002), *Science* 298: 2182–85, and "An Isotopic View of the Early Solar System" by E. Zinner (2003), *Science* 300: 265–67.

36. Element abundances from *Chemistry*, 9th ed. by R. Chang (2007), New York: McGraw-Hill, p. 52. Helium is rare in organic material due to its disinclination to bond with other atoms.

37. *Stardust: Supernovae and Life—The Cosmic Connection* by J. Gribbin and M. Gribbin, pp. 152, 154.

38. "Recipe for Water: Just add Starlight," ESA Space Science News online at http://www.esa.int/esaSC/SEMW76EODDG_index_0.html.

39. *Great Comets* by R. Burnham (2000), Cambridge: Cambridge University Press, p. 195.

40. *The Tunguska Mystery* by V. Rubtsov and E. Ashpole (2009), Dordrecht and New York: Springer, p. 3. See also *Great Comets* by R. Burnham (2000), Cambridge: Cambridge University Press, p. 203.

41. *The Tunguska Mystery* by V. Rubtsov and E. Ashpole (2009), Dordrecht and New York: Springer, pp. 1, 283.

42. *Impact Jupiter: The Crash of Comet Shoemaker-Levy 9* by D. H. Levy (1995), Cambridge, MA: Basic Books, pp. 169, 189, 190. Congress instructed NASA to detect all potentially Earth-crossing comets and all asteroids larger than about 240 meters or 800 feet across, which is large enough to cause regional havoc if one of them hit the Earth. JPL started the "Sentry" program, a network of small telescopes, to track and identify the threatening objects. There is even a publically accessible "Impact Risks" web page with impact probabilities and a color-coded alert system (think Department of Homeland Security), located at http://usgovinfo.about.com/gi/dynamic/offsite.htm?site=http://neo.jpl.nasa.gov/news/news126.html.

43. For in-depth discussion of this topic, see *Comets and the Origin and Evolution of Life*, 2nd ed., eds. P. J. Thomas, R. D. Ricks, C. F. Chyba and C. P. McKay (2006), Berlin and New York: Springer-Verlag.

44. "We Are Made of Star Dust: Toward a New Periodic Table of Elements" by C. Barlow, online at http://www.thegreatstory.org/Stardustbackground.html.

45. "DNA in Space? Biological Building Blocks Found in Meteorites," by M. D. Lemonick (August 11, 2011), Time.com online at http://www.time.com/time/printout/0,8816,2087758,00.html.

46. "DNA Building Blocks Found in Meteorites," by B. Vastag (August 8, 2011), *Washington Post* online at http://www.washingtonpost.com/national/health-science/dna-building-blocks-found-in-meteorites/2011/08/08/gIQAzNe42I_story.html.

47. "Large Colonial Organisms with Coordinated Growth in Oxygenated Environments 2.1 Gyr Ago" by A. El Albani et al. (2010), *Nature* 466, no. 7302 (July 1): 100–5. Gyr is an abbreviation for one gigayear or one billion years.

48. *Stardust: Supernovae and Life—The Cosmic Connection* by J. Gribbin and M. Gribbin, p. 213.

49. Quoted in ibid., p. 214.

50. Ibid., pp. 213–14.

51. Ibid., p. 214.

52. See http://neo.jpl.nasa.gov/neo/resource.html.

53. *Mining the Sky: Untold Riches from the Asteroids, Comets, and Planets* by J. S. Lewis (1997), New York: Perseus Books. Of course, anyone investing enormous resources in gathering a comet or asteroid to harvest for valuable metals and minerals would have to beware of creating such abundance that they destroyed the market value of their resources. There's a fine line between cornering a market and saturating it.

54. "Core Formation and Metal-Silicate Fractionation of Osmium and Iridium from Gold" by J. M. Brenan and W. F. McDonough (2009), *Nature Geoscience* 2: 798–801.

55. The terms of this international treaty are listed at the United Nations Office of Outer Space Affairs website, at http://www.oosa.unvienna.org/oosa/SpaceLaw/outerspt.html.

7. SOHO—Living with a Restless Star

1. Technically, the photosphere is the region where the Sun becomes completely opaque and so no light reaches us. Since the density varies smoothly, there is a layer of finite thickness where the transition from transparent to opaque occurs. Starting 700,000 kilometers from the center, it only takes 400 kilometers to go from 5 percent of photons escaping to 99.5 percent of photons escaping. The photosphere is really a very slender shell, over which the temperature drops from 7600 to 4500 K. The gas in the photosphere is ionized and the physical mechanism for the opacity is photons scattering off electrons.

2. The physical situation of a cloud is analogous but not identical. Inside a cloud, the density of water vapor droplets is high enough that light bounces off them and can't travel in straight lines, so we can't see inside a cloud and the view from inside in an airplane is opaque and dark. The apparent edge of a cloud is just the region where the water vapor density has dropped to the point where photons no longer scatter off droplets and travel in a straight line. We see it as an edge or surface.

3. "Early Observations of Sunspots?" by D. J. Schove and D. Sarton (1947), *Isis* 37: 69–71. For a review, see "Historical Sunspot Observations: a Review" by J. M. Vaquero (2007), *Advances in Space Research* 40: 929–41, and for a historical catalog, see "A Catalogue of Sunspot Observations from 165 BC to 1684 AD" by A. D. Wittmann and Z. T. Xu (1987), *Astronomy and Astrophysics Supplement Series* 70: 83–94.

4. Galileo and Scheiner knew each other, and were amicable for a while, but ended up in a bitter dispute over who had seen sunspots first. Complicating the argument, Thomas Harriot also observed sunspots in 1610, and Johannes Fabricius was the first to publish his observations, in 1611. For all the gory details, see "Thomas Harriot and the First Telescopic Observation of Sunspots" by J. D. North, in *Thomas Harriot: Renaissance Scientist*, edited by J. W. Shirley (1974). Oxford: Clarendon Press, and "Galileo, Scheiner, and the Interpretation of Sunspots" by W. R. Shea (1970), *Isis* 61: 498–519.

5. This extreme event was called the Carrington Super Flare, observed by amateur astronomers Richard Carrington and Richard Hodgson. See "Timeline: the 1859 Solar Superstorm," July 29, 2008, in *Scientific American* online, at http://www.scientificamerican.com/article.cfm?id=timeline-the-1859-solar-superstorm.

6. "On the Origin of Solar Cycle Periodicity" by A. Grandpierre (2004), *Astrophysics and Space Science* 243: 393–401. The proof that sunspots are magnetic phenomena came from Hale's use of polarimetry. When high-energy particles move in a systematically aligned magnetic field, they emit radiation that's polarized. Hale made images of the Sun in polarized light and showed that the general surface of the Sun was unpolarized, as expected, since the magnetic field lines are tangled and disordered. But sunspot pairs showed high polarization, which was a signature of the ordered magnetic field lines that linked them.

7. "The Sun and Earth's Climate" by J. D. Haigh (2007), *Living Reviews of Solar Physics* 4: 2–59, and "An Influence of Solar Spectral Variations on Radiative Forcing of Climate" by J. D. Haigh, A. R. Winning, R. Toumi, and J. W. Harder (2010), *Nature* 467: 696–99. Even the striking match of the Maunder Minimum and an extended cold snap in Europe has not been proven to be caused by variations in the Sun; see "Abrupt Onset of the Little Ice Age Triggered by Volcanism and Sustained by Sea-Ice/Ocean Feedbacks" by G. H. Miller et al. (2012), *Geophysical Research Letters* 39, L02708.

8. The first physical theory of the solar wind, putting together Carrington and Fitzgerald's speculations and Sydney Chapman's inference of the high temperature of the corona with Ludwig Biermann's observation that comet tails always pointed away from the Sun, was "Dynamics of the Interplanetary Gas and Magnetic Fields" by E. Parker (1958), *Astrophysical Journal* 128: 664.

9. The information at http://spaceweather.com/ is updated every ten minutes. The National Oceanographic and Atmospheric Administration have their own space weather site at http://www.swpc.noaa.gov/today.html.

10. *An Introduction to Space Weather* by M. Moldwin (2008), Cambridge: Cambridge University Press, pp. 28–29.

11. The magnetosphere was discovered in 1958 by the Explorer 1 satellite as part of the International Geophysical Year, and was named the following year; see "Motions in the Magnetosphere of the Earth" by T. Gold (1959), *Journal of Geophysical Research* 64: 1219–24. The magnetosphere "buffers" the Earth from the solar wind at a distance of about 10 Earth radii, deflecting the wind but allowing some radiation and particles to funnel onto the Earth's poles where they power the auroras.

12. "Understanding Solar Behavior and its Influence on Climate" by T. Niroma (2009), *Energy and Environment* 20: 145–59, and *Space Weather, Environment and Societies* by J. Lilensten and J. Bornarel (2006), Dordrecht: Reidel, pp. 117–18.

13. *An Introduction to Space Weather* by M. Moldwin, p. 115.

14. *Space Weather, Environment and Societies* by J. Lilensten and J. Bornarel, pp. 117–18.

15. *An Introduction to Space Weather* by M. Moldwin, p. 113.

16. The NASA website is http://lws.gsfc.nasa.gov/.

17. "The SOHO Mission: an Overview" by V. Domingo, B. Fleck, and I. A. Poland (1995), *Solar Physics* 162: 1–37, and "SOHO: The Solar and Heliospheric Observatory" by V. Domingo, B. Fleck, and I. A. Poland (1995), *Space Science Reviews* 72: 81–84.

18. More precisely, the Lagrange points are the locations where the combined pull of two large gravitational bodies provides the exact amount of centripetal force needed to rotate with them. A third body of comparatively negligible mass can be placed between them and can stay in position with little force applied. L1 is unstable, however. If the satellite moves perpendicular to the line connecting Earth and Sun it will be pulled back into position, but not if it drifts either toward or away from the Sun. With occasional adjustments from small retro rockets, it's easy to keep even a massive spacecraft from drifting away from the Lagrange point.

19. "Ten Years of SOHO" by B. Fleck et al. (2006), *European Space Agency Bulletin*, no. 126, pp. 25-32, and "Four Years of SOHO Discoveries—Some Highlights" by B. Fleck et al. (2000), *European Space Agency Bulletin*, no. 201, pp. 68–86. Also, there's extensive information in the science sections of the NASA and ESA websites for the satellite, at http://sohowww.nascom.nasa.gov/home.html, and at http://www.esa.int/esaSC/120373_index_0_m.html. For the updated Sun-grazing comet haul, see http://sungrazer.nrl.navy.mil/index.php.

20. "ESA, NASA Struggle to Save SOHO" by M. A. Taverna and A. R. Asker (1998), *Aviation Week and Space Technology* 149: 32–33.

21. "The ESA/NASA SOHO Mission Interruption: Using the STAMP Accident Analysis Software for a Software Related Mishap" by C. W. Johnson and C. M. Holloway (2003), *Software: Practice and Experience* 33: 1177–98.

22. See the report online at http://soho.esac.esa.int/whatsnew/SOHO_final_report.html.

23. The pervasiveness of oscillations, vibrations, and harmonics in physical theory even extends to the invisibly small work inside the atom. String theory postulates that all apparently fundamental particles like quarks and electrons are actually composed of infinitesimally small, one-dimensional, mass-energy entities called strings. The diversity of particle properties and interactions are explained by different states of these vibrating and oscillating strings, and interactions between them. See, for example, *Warped Passages: Unraveling the Mysteries of the Universe's Hidden Dimensions* by L. Randall (2005), New York: HarperCollins.

24. The development of solar astronomy from Richard Carrington to George Ellery Hale is entertainingly described in *The Sun Kings: The Unexpected Tragedy of Richard Carrington and the Tale of How Modern Astronomy Began* by S. Clark (2007), Princeton: Princeton University Press.

25. *The Music of the Sun: The Story of Helioseismology* by W. J. Chaplin (2006), London: Oneworld Publications.

26. "Basic Principles of Solar Acoustic Holography" by C. Lindsey and D. C. Brain (2000), *Solar Physics* 192: 261–84.

27. "A Decade of Weather Extremes" by D. Coumou and S. Rahmstorf (2012), *Nature Climate Change* 2: 491–96.

28. "The Mysterious Origins of Solar Flares" by G. D. Holman (2006), in *Scientific American* online, April issue, at http://www.scientificamerican.com/article.cfm?id=the-mysterious-origins-of.

29. They can be visualized elegantly in Google Earth; see http://www.gearthblog.com/satellites.html.

30. *Storms in Space* by J. W. Freeman (2001), Cambridge: Cambridge University Press, pp. 71–72.

31. *The 23rd Cycle: Learning to Live with a Stormy Star* by S. Odenwald (2001), New York: Columbia University Press, pp. 7, 8.

32. *Storms in Space* by J. W. Freeman, p. 73.

33. *Space Weather, Environment and Societies* by J. Lilensten and J. Bornarel (2006), Dordrecht: Reidel, p. 93.

34. *Severe Space Weather Events—Understanding Societal and Economic Impacts: a Workshop Report* by the Space Science Board (2008), Washington, DC: National Academies Press.

35. Research reported at http://science.nasa.gov/science-news/science-at-nasa/2009/03sep_sunspots/.

36. "Multiband Modeling of the Sun as a Variable Star from VIRGO/SOHO Data" by A. F. Lanza (2004), *Astronomy and Astrophysics* 425: 707.

37. "Statistical Investigation and Modeling of Sungrazing Comets Discovered with the Solar and Heliospheric Observatory" by Z. Sekanina (2000), *Astrophysical Journal* 566: 577.

38. *Highlights in Space* by the Office of Outer Space Affairs, United Nations Office at Vienna (2004), New York: United Nations, p. 142.

39. *Sun, Earth and Sky* by K. Lang (2006), 2nd ed., Singapore: Springer, pp. 174, 176.

40. *Varese: A Looking Glass Diary* by L. Varese (1972), New York: W. W. Norton, pp. 228–29, 101.

41. Stephen P. McGreevy's ground-based ELF-VLF recordings, at http://www-pw.physics.uiowa.edu/mcgreevy/.

42. *Skywatchers, Shamans & Kings: Astronomy and the Archaeology of Power* by E. C. Krupp (1997), New York: John Wiley & Sons, p. 128.

43. *Mysteries and Discoveries of Archaeoastronomy: From Giza to Easter Island* by G. Magli (2009), New York: Springer Praxis, pp. 62, 63.

44. *The Cambridge Concise History of Astronomy* by M. Hoskins and C. Ruggles (1999), Cambridge: Cambridge University Press, p. 6.

45. "If the Stones Could Speak: Searching for the Meaning of Stonehenge," by C. Alexander, from National Geographic.org (June 2008), at http://ngm.nationalgeographic.com/2008/06/stonehenge/alexander-text/1.

46. *Mysteries and Discoveries of Archaeoastronomy: From Giza to Easter Island* by G. Magli, p. 32.

47. "If the Stones Could Speak: Searching for the Meaning of Stonehenge," by C. Alexander, at http://ngm.nationalgeographic.com/2008/06/stonehenge/alexander-text/1.

48. *Skywatchers, Shamans & Kings: Astronomy and the Archaeology of Power* by E. C. Krupp, p. 134. Krupp may be referring to the white quartz wall that defines the "reconstructed" exterior of Newgrange. Geraldine and Matthew Stout point out that the wall is a fabrication from stone material found on site but may not be indicative of the original megalithic structure.

49. *The Cambridge Concise History of Astronomy* by M. Hoskins and C. Ruggles, p. 2.

50. *Mysteries and Discoveries of Archaeoastronomy: From Giza to Easter Island* by G. Magli, p. 35.

51. *The Stars and the Stones: Ancient Art and Astronomy in Ireland* by M. Brennan (1983), London: Thames and Hudson, 1983, pp. 7, 10.

52. "Chankillo: A 2300-Year-Old Solar Observatory in Coastal Peru," by I. Ghezzi and C. Ruggles (2007), *Science* 315: 1239.

53. "Chankillo: A 2300-Year-Old Solar Observatory in Coastal Peru," by I. Ghezzi and C. Ruggles, p. 1241.

54. "The Sphinx Blinks" by E. C. Krupp (2001), *Sky and Telescope* (March): 86. The trilithons that create the interior horseshoe in Stonehenge were constructed sometime between 2550 and 1600 BC.

55. *Lexia to Perplexia* by Talan Memmott (2000) is a web-based text consisting of ten linked web pages or chapters and is online at http://www.altx.com/ebr/ebr11/11mem/plex/appendix-1.html. The quoted material is from the chapter titled "Metastrophe: Temporary miniFestos."

56. Ibid.

57. *Storms in Space* by J. W. Freeman (2001), Cambridge: Cambridge University Press, p. 53.

58. *Touch the Sun: a NASA Braille Book* by N. Grice (2005), Washington, DC: Joseph Henry Press.

59. "Solar Dynamics Observatory—Exploring the Sun in High Definition," a NASA Factsheet, Goddard Space Flight Center, FS-2008-04-102-GSFC.

60. One of the reasons astronomers prefer to search for planets around Sun-like stars is the fact that more massive stars are more active and could present a larger threat to life on any planets orbiting them (in addition to having shorter lifetimes and so allowing less time for complex life to develop). Low-mass red dwarfs are abundant and might account for ten times as many habitable planets, but recent research suggests that they also have flares that might be hazardous to life; see "M Dwarf Flares from Time-Resolved SDSS Spectra" by E. J. Hilton, A. A. West, S. L. Hawley, and A. F. Kowalski (2010), *Astronomical Journal*140: 1402–13.

8. Hipparcos—Mapping the Milky Way

1. *Albert Einstein: Maker of Universes* by H. G. Garbedian (1939), New York and London: Funk and Wagnalls, p. 77.

2. "How Fast Are You Moving When You're Sitting Still?" by A. Fraknoi (2007), *Universe in the Classroom*, ol. 71, Astronomical Society of the Pacific. These motions within motions seem epicyclic, and it might be wondered whether they continue onto ever larger scales with no end. In fact, on scales larger than about a hundred million light-years, the systematic motions do converge. The Milky Way and its few dozen galactic neighbors in the Local Group, plus the Virgo Cluster, have a net motion within a vast entity called the Local Supercluster, which contains several hundred thousand galaxies. The largest reference frame for motions is the relic radiation from the Big Bang itself, known as the Cosmic Microwave Background (CMB). Since this represents the universe as a whole, it acts as a benchmark, as the universe can't be going anywhere! See http://www.astrosociety.org/education/publications/tnl/71/howfast.html.

3. *The Making of History's Greatest Star Map* by M. Perryman (2010), New York: Springer, p. 37.

4. *On the Origin of Stories: Evolution, Cognition, and Fiction* by B. Boyd (2009), Cambridge, MA, and London: Belknap Press of Harvard University Press, p. 209.

5. "Messages from the Stone Age," by K. Ravilious (2010), *New Scientist*, February 20–26, p. 32.

6. Quoted in ibid., p. 33.

7. Clive Ruggles and Michel Cotte, eds. (2010), "Heritage Sites of Astronomy and Archaeoastronomy in the Context of the UNESCO World Heritage Conven-

tion: A Thematic Study," available at the International Council on Monuments and Sites website at http://openarchive.icomos.org/267/.

8. http://seedmagazine.com/content/article/symbols_from_the_sky/.

9. See "Archeociel: Chantal Jeques Wolkiewiez" at http://www.archeociel.com/Accueil_eng.htm#Abstract.

10. *Star Tales* by I. Ridpath (1988), New York: Universe Books, p. 2.

11. "Star Search," by Harald Meller (2004), *National Geographic* 205, no. 1 (January): 76–87. See also "Solar Circle," by Ulrich Bolser (2006), *Archaeology* 59, no. 4 (July/August): 30–35.

12. Quoted in *The History and Practice of Ancient Astronomy* by J. Evans (1988), New York and Oxford: Oxford University Press, p. 3.

13. Ibid. The accuracy of various astronomical details in Homer's *Odyssey* has allowed astronomers to pinpoint the date when Odysseus defeats the suitors and reclaims his wife and lands. In 2008, researchers used details of the total solar eclipse of April 1178 BC, briefly described in the text, to date events in the narrative and to suggest that portions of the text may be based on fact. "Astronomers hit a homer with 'Odyssey'," by Thomas H. Maugh II (2008), *Los Angeles Times*, June 24, online at http://articles.latimes.com/2008/jun/24/science/sci-odyssey24. Holly Henry wishes to thank California State University colleague Sunny Hyon for pointing me to this article and multiple other news reports that proved useful in writing this book.

14. Quoted in ibid., p. 4.

15. *Star Maps: History, Artistry, and Cartography* by N. Kanas (2007), New York: Springer/Praxis, p. 50.

16. "Astronomy in Antiquity," by M. Hoskin, in *The Cambridge Concise History of Astronomy*, edited by Michael Hoskin (1999), Cambridge: Cambridge University Press, p. 39.

17. *Hipparcos, The New Reduction of the Raw Data* by F. Van Leeuwen (2007), New York: Springer, p. 6.

18. "The Epoch of the Constellations on the Farnese Atlas and their Origin in Hipparchus's Lost Catalog," by B. Shaeffer (2005), *Journal for the History of Astronomy* 36: 169, 173.

19. Ibid., pp. 178, 174.

20. Stars near the Sun share the same general motion around the center of the galaxy, and their speeds relative to the Sun are quite modest. Hipparcos measured star distances directly by trigonometry, using the span of the Earth's orbit to form a triangle where there was enough information to calculate the distance. This is the parallax effect. Hipparcos also made enough observations of some nearby stars to see them change their position on the sky, called by astronomers a proper motion, leading to a measurement of their transverse speed.

21. *The Making of History's Greatest Star Map* by M. Perryman, p. 40.

22. The Human Genome Project was a thirteen-year effort to identify all of the 20,000–25,000 human genes and map the sequence of 3 billion base pairs that make up human DNA. Sponsored by the U.S. Department of Energy and the National Institutes of Health, it began in 1989 and a complete version of the human genome was published in 2003, although many research investigations continue. A parallel research effort by the private Celera Corporation, headed by

Craig Venter, began in 1998 and provided strong competition for the federally funded project. For an overview of the science, see *Drawing the Map of Life: Inside the Human Genome Project* by V. K. McIlheny (2010), New York: Basic Books, and for an entertaining account of the competing project, see *The Genome War: How Craig Venter Tried to Capture the Code of Life and Save the World* by J. Shreeve (2005), New York: Ballantine Books.

23. "When E.T. Phones the Pope," by M. Kauffman (2009), *Washington Post*, November 8, 2009. The Vatican staffs a working observatory in southern Arizona and participates in an astronomer exchange program with the University of Arizona, Kitt Peak Observatory, and the U.S. National Observatory.

24. *Frontiers of Astrobiology* edited by C. Impey, J. Lunine, and J. Funes, S. J. (2012), Cambridge: Cambridge University Press.

25. Quoted in "Vatican Hosts Astrobiology Experts," by D. Ariel (2009), *Denton Record-Chronicle*, November 11, 2009, online at http://www.dentonrc.com/.

26. Ibid.

27. "Probability Distribution of Terrestrial Planets in Habitable Zones around Host Stars," by J. Guo et al. (2009), *Astrophysics and Space Science* 323: 367–73.

28. The trigonometric measurements of distance typically involve forming a skinny triangle where the base is the diameter of the Earth-Sun orbit, and the small apex angle is measured by watching the shift of a nearby star on the sky over a year as seen against a backdrop of more distant stars. Armed with a distance, similar trigonometry will allow the angular size of an object to be converted into its physical size. Also, distance plus the inverse square law of light propagation can be used to convert its apparent brightness into intrinsic or absolute brightness. Knowing the amount of the radiation emitted by the star, its mass can be estimated. Distance is the key to all these calculations.

29. For reference, there are about 6,000 stars visible to the naked eye from a location with no artificial light and no Moon visible. This number is an average; from person to person it can vary by 30 percent or more.

30. *Ptolemy's Almagest* translated by G. J. Toomer (1998), Princeton: Princeton University Press.

31. *Parallax: The Race to Measure the Cosmos* by A. W. Hirschfeld (2002), New York: Henry Holt.

32. When launching satellites and space probes, it's worth getting an assist from the rotation of the Earth, which is over 1,000 mph at the equator. Thus, NASA does its launches from Cape Canaveral in Florida, close to the most southerly point in the continental United States. The European Space Agency does even better by using a remote facility just 300 miles north of the equator.

33. The overarching and comprehensive summary of Hipparcos's science is a book by the project scientist, *Astronomical Applications of Astrometry: Ten Years of Exploitation of the Hipparcos Satellite Data* by M. Perryman (2009), Cambridge: Cambridge University Press.

34. See the materials online at http://www.rssd.esa.int/index.php?project=HIPPARCOS&page=star_globe.

35. "The Hipparcos Catalogue," by M. Perryman et al. (1997), *Astronomy and Astrophysics* 323: L49–L52.

36. "The Tycho Catalogue," by E. Høg et al. (1997), *Astronomy and Astrophysics* 323: L57–L60.

37. "The Tycho-2 Catalogue of the 2.5 Million Brightest Stars," by E. Høg et al. (2000), *Astronomy and Astrophysics* 355: L27–L30.

38. *Hipparcos, The New Reduction of the Raw Data* by F. Van Leeuwen (2007), New York: Springer.

39. For Michael Perryman's bio, see http://www.esa.int/esaMI/Space_Year_2007/SEMGVSMPQ5F_0.html.

40. *The Making of History's Greatest Star Map* by M. Perryman (2010), Berlin and Heidelberg: Springer.

41. *Measuring the Universe: The Extragalactic Distance Ladder* by S. Webb (1999), New York: Springer.

42. Parallax does assume that space is Euclidean, or that light travels in straight lines so that trigonometry in space is as reliable as trigonometry on a piece of paper. General relativity says that space can be curved but the effect is only significant when gravity is strong. Throughout the volume surveyed by Hipparcos, the assumption on Euclidean geometry is very secure.

43. "The Pulsation Mode of the Cepheid Polaris," by D. G. Turner et al. (2012), *The Astrophysical Journal*, arXiv:1211.6103.

44. "The Hyades: Distance, Structure, Dynamics, and Age," by M. Perryman et al. (1998), *Astronomy and Astrophysics* 331: 81–120, and "A Distance of 133–137 pc to the Pleiades Star Cluster," by X. Pan et al. (2004), *Nature* 427: 326–28.

45. "Galactic Kinematics of Cepheids from Hipparcos Proper Motions," by M. Feast and P. Whitelock (1997), *Monthly Notices of the Royal Astronomical Society* 291: 683–93.

46. "The Hipparcos Catalogue as a Realisation of the Extragalactic Reference Frame," by J. Kovalesky et al. (1997), *Astronomy and Astrophysics* 323: 620–33.

47. "Hipparcos Variable Star Detection and Classification Efficiency," by P. Dubath et al. (2011), preprint number arXiv 1107.3638.

48. "Double Star Data in the Hipparcos Catalogue," by L. Lindegren et al. (1997), *Astronomy and Astrophysics* 323: L53–L56.

49. "Hipparcos Sub-Dwarf Parallaxes: Metal-Rich Clusters and the Thick Disk," by I. N. Reid (1998), *Astronomical Journal* 116: 204–28.

50. "Debris Streams in the Solar Neighborhood as Relics from the Formation of the Milky Way," by A. Helmi et al. (1999), *Nature* 402: 53–55.

51. "Ice Age Epochs and the Sun's Path Through the Galaxy," by D. R. Gies and J. W. Helsel (2005), *Astrophysical Journal* 636: 844–48.

52. "HD 209458 Planetary Transits from Hipparcos Photometry," by N. Robichon and F. Arenou (2000), *Astronomy and Astrophysics*- 355: 295–98.

53. "Determination of the PPN Parameter Gamma with the Hipparcos Data" by M. Froeschle, F. Mignard, and F. Arenou (1997), in *Proceedings of the ESA Symposium "Hipparcos—Venice '97,"* Venice, Italy, ESA SP-402, Noordwijk, Netherlands: European Space Agency.

54. "The Gaia Mission: Science Organization and Present Status," by L. Lindegren et al. (2008), in *A Giant Step: from Milli- to Micro-Arcsecond Astrometry*, IAU Symposium 248, edited by W. Lin, I. Platais, and M. Perryman, Berlin: Dordrecht.

55. "The Three Dimensional Universe with Gaia," Proceedings of a Symposium at the Paris Observatory, Meudon, ESA SP-576, Noordwijk, Netherlands: European Space Agency.

56. "Gaia Overview," ESA, online at http://sci.esa.int/science-e/www/area/index.cfm?fareaid=26.

57. Gaia's education potential is discussed at http://www.esa.int/export/esaSC/120377_index_0_m.html.

9. Spitzer—Unveiling the Cool Cosmos

1. Demonstrating that space is not a perfect vacuum took a very long time because the effects of gas and dust between stars are both subtle and profound. Before the invention of the telescope, some had speculated that the ragged, dark regions of the Milky Way were obscured in some way, but they had no evidence. In 1847, the celebrated German astronomer Friedrich Georg Wilhelm von Struve demonstrated that the interstellar medium dimmed or extinguished the light of distant stars by about a factor of two for every 2,000 light-years the starlight traveled. He was part of a remarkable dynasty of astronomers spanning five generations. In the 1930s, the American astronomer Robert Trumpler showed that starlight was dimmed and reddened and proposed a mechanism where the effect was caused by microscopic dust grains thinly dispersed in the space between stars.

2. "Cosmic census finds crowd of planets in our galaxy," by Seth Borenstein (February 19, 2011), MSNBC.com online at http://www.msnbc.msn.com/id/41686017/ns/technology_and_science-space/t/cosmic-census-finds-crowd-planets-our-galaxy/.

3. The planet was discovered with a small ground-based telescope in 2008; see "WASP-12b: The Hottest Transiting Exoplanet Yet Discovered," by L. Hebb et al. (2009), *The Astrophysical Journal* 693: 1920–28, and the characterization that revealed its carbon composition was from Spitzer two years later, "A High C/O Ratio and Weak Temperature Inversion in the Atmosphere of Exoplanet WASP-12b" by N. Madhusudhan et al. (2010), *Nature* 469: 64–67.

4. "NASA's Spitzer Reveals First Carbon-Rich World," online at http://www.spitzer.caltech.edu/news/1231-ssc2010-10-NASA-s-Spitzer-Reveals-First-Carbon-Rich-Planet.

5. "Spitzer View on the Evolution of Star-Forming Galaxies from z = 0 to z = 3," by P. Pérez-Gonzáles et al. (2007), *Astrophysical Journal* 640: 92–102.

6. *William Herschel: His Life and Works* by E. Holden (1881), New York: Scribner's and Sons.

7. "Low-Level Laser Therapy Facilitates Superficial Wound Healing in Humans: A Triple-Blind, Sham-Controlled Study," by J. Hopkins et al. (2004), *Journal of Athletic Training* 39, no. 3: 223–29. The National Center for Biotechnology Information functions under the aegis of the U.S. National Institutes of Health.

8. "Light Emitting Diodes Bring Relief to Young Cancer Patients" (November 5, 2003), NASA Press Release available online at http://www.nasa.gov/centers/marshall/news/news/releases/2003/03-199.html.

9. Atoms and molecules are in constant microscopic motion and temperature is a measure of that motion. Unlike the familiar Celsius and Fahrenheit scales,

physicists and astronomers use a temperature scale where zero corresponds to no microscopic motion, or absolute cold. This is the Kelvin scale; zero on the Kelvin scale is -273°C while the freezing point of water is 273 K. All objects in the universe emit thermal radiation as a result of their temperature. Thermal radiation has a smooth spectrum with a peak wavelength inversely proportional to the temperature. For objects at thousands of Kelvin, like the surface of the Sun, that peak is at visible wavelengths. For objects at hundreds of Kelvin, like the Earth and everything on it, that peak is at invisible infrared wavelengths.

10. "History of Infrared Telescopes and Astronomy," by G. Rieke (2009), *Experimental Astronomy* 25: 125–41.

11. "NASA's New Airborne Observatory," by L. Keller and J. Wolf (2010), *Sky and Telescope* (October): 22–28.

12. For summaries and overviews, see "The Infrared Astronomical Satellite (IRAS) Mission," by G. Neugebauer et al. (1984), *Astrophysical Journal* 278: L1–L6, and "The Infrared Space Observatory (ISO) Mission," by M. Kessler et al. (1986), *Astronomy and Astrophysics* 315: L27–L31.

13. "The Spitzer Space Telescope Mission," by M. Werner et al. (2004), *Astrophysical Journal Supplement* 154: 1–9.

14. "Variable Extinction at the Galactic Center," by M. Lebofsky (1979), *Astronomical Journal* 84: 324–28.

15. "A Spitzer/IRAC Survey of the Orion Molecular Clouds," by S. Megeath et al. (2005), in *Massive Star Birth, IAU Symposium 227*, ed. R. Cesaroni, E. Churchwell, M. Fells, and C. Walmesley, Cambridge: Cambridge University Press, pp. 1–6.

16. "The $0.4 < z < 1.3$ Star Formation History of the Universe as Viewed in the Infrared," by B. Magnelli et al. (2009), *Astronomy and Astrophysics* 496: 57–75.

17. "New Measurements of Cosmic Infrared Background Fluctuations from Early Epochs," by A. Kashlinsky et al. (2007), *Astrophysical Journal* 654: L5–L8.

18. Quoted in "NASA Telescope Pick's up Glow of Universe's First Objects," from the NASA Spitzer mission news web page at http://www.nasa.gov/centers/goddard/news/topstory/2006/spitzer_firststars.html.

19. "The Spitzer/GLIMPSE Surveys: A New View of the Milky Way," by E. Churchwell et al. (2009), *Publications of the Astronomical Society of the Pacific* 121: 213–30.

20. "Spitzer Detection of Polycyclic Aromatic Hydrocarbons and Silicate Dust Features in the Mid-Infrared Spectra of $z = 2$ Ultraluminous Infrared Galaxies," by L. Yan et al. (2005), *Astrophysical Journal* 628: 604–10.

21. "How the Spitzer Space Telescope Unveils the Unseen Cosmos," by M. Werner (2009), *Astronomy* 37, no. 3, March Issue, p. 44.

22. Organized by the Citizen Science Association, Zooniverse.org started as "Galaxy Zoo" and is dedicated to increasing public participation in scientific research and is a partnership of the Adler Planetarium, Johns Hopkins University, the University of Minnesota, as well as partners from Britain including the University of Oxford, the National Maritime Museum, the Royal Observatory in Greenwich, and the University of Nottingham. See their website at http://www.zooniverse.org/.

23. Online at http://www.milkywayproject.org/. As of December 2012, more than 3 million gas bubbles as sites of possible star formation have been identified by Citizen Scientists.

24. "The Spitzer Warm Mission Science Prospects," by J. Stauffer et al. (2007), *American Institute of Physics Conference Proceedings* 243: 43–66, and "NASA's Spitzer Telescope Warms Up to a New Career," online at http://www.nasa.gov/mission_pages/spitzer/news/spitzer-20090506.html.

25. See the NASA press release titled "NASA's Kepler Mission Discovers a World Orbiting Two Stars" online at http://www.nasa.gov/mission_pages/kepler/news/kepler-16b.html.

26. "Photosynthesis: Likelihood of Occurrence and Possibility of Detection on Earth-like Planets," by R. Wolstencroft and J. Raven (2002), *Icarus* 157: 535–48.

27. "Revealing the Dawn of Photosynthesis," by R. Hooper (2006), *New Scientist* (August 19).

28. "Star Light, Star Bright, Any Oxygen Tonight?" by L. Mullen (March 17, 2003), *Astrobiology Magazine* online at http://www.astrobio.net/exclusive/404/star-light-star-brightany-oxygen-tonight.

29. "Universality in Intermediary Metabolism," by E. Smith and H. Morowitz (2004), *Proceedings of the National Academy of Sciences* 101: 13168–71.

30. "The Day Earth Came to Life," by J. Trefil and W. O'Brien-Trefil (2009), *Astronomy* (September): 29.

31. *The Story of Light* by B. Bova (2001), Naperville, IL: Sourcebooks, pp. 37, 36.

32. *Animal Eyes* by M. Land and D.-E. Nilsson (2002), Oxford: Oxford University Press, p. 1.

33. *In the Blink of an Eye: How Vision Sparked the Big Bang of Evolution* by A. Parker (2003), New York: Basic Books, pp. 24–25. The Light Switch theory is not universally accepted and there are a number of quite different ideas for what caused the surge of evolution that included the development of complex eyes. With the fossil record half a billion years ago so sparse, it may never be possible to know exactly how the evolution played out and why.

34. *Animal Eyes* by M. Land and D.-E. Nilsson, p. 12.

35. *Light* by M. Sobel (1987), Chicago and London: University of Chicago Press, p. 47.

36. "A Bees-eye View: How Insects See Flowers Very Differently to Us," by M. Hanlon (August 8, 2007), *Daily Mail* online at http://www.dailymail.co.uk/sciencetech/article-473897/A-bees-eye-view-How-insects-flowers-differently-us.html#ixzz19NWyoPIn.

37. *Animal Eyes* by M. Land and D.-E. Nilsson, pp. 12–13, 15.

38. "So Much More Than Plasma and Poison," by N. Angier (June 6, 2011), New York Times.com online at http://www.nytimes.com/2011/06/07/science/07jellyfish.html?pagewanted=all. See also "Box Jellyfish Use Terrestrial Visual Cues for Navigation," by A. Garm, M. Oskarssoon, and D.-E. Nilsson (2011), *Current Biology* 21, .o. 9, pp. 798–803.

39. *The Deep*, edited by C. Nouvain (2007), Chicago and London: University of Chicago Press, p. 18.

40. "The Nocturnal Ballet of Deep-Sea Creatures," by M. Youngbluth, in *The Deep*, edited by C. Nouvain (2007), Chicago and London: University of Chicago Press, p. 71.

41. "Spookfish has Mirrors for Eyes," by J. Morgan, BBC News online at http://news.bbc.co.uk/2/hi/7815540.stm.

42. Countershading is a form of camouflage, common in land and sea animals. It's designed to reduce the visual impact of natural shadows and so reduce visual cues a predator might get about the size and shape of their prey. In marine animals countershading acts to match the background light, if seen from below, and dark if seen from above.

43. *The Silent Deep: The Discovery, Ecology and Conservation of the Deep Sea* by T. Koslow (2007), Chicago: University of Chicago Press, p. 52.

44. "Living Lights in the Sea." by E. Widder, in *The Deep*, edited by C. Nouvain (2007), Chicago and London: University of Chicago Press, pp. 85–86. A deep-sea explorer, Widder is founder and director the Ocean Research and Conservation Association (ORCA), which focuses on ocean water quality and ecosystem health.

45. *The Silent Deep: The Discovery, Ecology and Conservation of the Deep Sea* by T. Koslow, p. 58.

46. *Aglow in the Dark: The Revolutionary Science of Bioluminescence* by V. Pieribone and D. Gruber (2006), Cambridge, MA: Belknap Press of Harvard University Press, p. 143.

47. The principle is a law of physics called conservation of angular momentum. In a new solar system the initial gas cloud shrinks by a large factor and the rotation speed is correspondingly amplified. Planets all orbit stars in the same direction and spin in the same sense as their orbits due to this principle.

48. "Frequency of Debris Disks Around Solar-Type Stars: First Results from a MIPS Survey," by G. Bryden et al. (2006), *Astrophysical Journal* 636: 1098–1128.

49. "The Formation and Evolution of Planetary Systems: First Results from a Spitzer Science Legacy Program," by M. Meyer et al. (2004), *Astrophysical Journal Supplement* 154, pp. 422–44.

50. "Fullerenes in Interstellar and Circumstellar Environments," by J. Cami, J. Bernard-Salas, E. Peeters, and S. Malek (2011), in *The Molecular Universe, IAU Symposium 280*, Dordrecht: Reidel, p. 23.

51. "Spitzer IRS Spectroscopy of IRAS-Discovered Debris Disks," by C. Chen et al. (2006), *Astrophysical Journal Supplement* 166: 251–377.

52. The number of exoplanets is a rapidly moving target, with a doubling time of less than two years. Most of the more than 700 confirmed exoplanets have been found by the Doppler spectroscopic method, but as of early 2012 over 2200 candidate exoplanets had been found by the transit eclipse method using NASA's Kepler satellite; see "Planetary Candidates Observed by Kepler. III. Analysis of the First 16 Months of Data" by N. Batalha et al. (2012), *Astrophysical Journal Supplements*, in press, astro-ph/1202.5852.

53. *Exoplanets* edited by S. Seager (2011), Tucson: University of Arizona Press.

54. The Planet Hunters website is online at http://www.planethunters.org/. The Yale University news release on citizen scientists finding planets is online at http://www.astro.yale.edu/news/20110922-2-probable-planets-found-people-like-you.

55. "Spitzer's Cold Look at Space," by M. Werner (2009), *American Scientist* 97: 458–68.

56. "A Map of the Day-Night Contrast of the Exoplanet HD 189733b," by H. Knutson et al. (2007), *Nature* 477: 183–86.

57. "The Phase-Dependent Infrared Brightness of the Extrasolar Planet Upsilon Andromeda b," by J. Harrington et al. (2006), *Science* 314: 623–26.

58. Though on Earth these marine worms are on average 1.5 inches in height, *Spirobranchus giganteus* populate tropical coral reefs across the globe. Holly Henry wishes to thank California State University colleague Nancy Best for identifying in Cameron's film a modified version of the Christmas Tree Worms she has encountered while scuba diving.

59. This exoplanet, discovered by Xavier Dumusque and colleagues with the HARPS spectrograph at the European Southern Observatory in Chile, is only 4.4 light-years from Earth and does not orbit Alpha Centauri B in the habitable zone. But a super-Earth (HD 40307g) reported in November 2012 orbits in its host star's "Goldilock's zone" and is located 42 light-years from Earth.

60. "A Spitzer Search for Planetary Mass Brown Dwarfs with Circumstellar Disks: Candidate Selection," by P. Harvey, D. Jaffe, K. Allers, and M. Liu (2010), *Astrophysical Journal* 720: 1374–79.

61. "The First Hundred Brown Dwarfs Discovered by the Wide-Field Infrared Survey Explorer (WISE)" by D. Kirkpatrick et al. (2011), *Astrophysical Journal Supplement* 197: 19–35.

62. "The MEarth Project: Searching for Transiting Habitable Super-Earths Around Nearby M Dwarfs," by J. Irwin, D. Charbonneau, P. Nutzman, and E. Falco (2008), in *Proceedings of the International Astronomical Union* 4: 37–43.

63. "A Super-Earth Transiting a Nearby Low-Mass Star," by D. Charbonneau et al. (2009), *Nature* 462: 891–94.

64. "A Ground-Based Transmission Spectrum of the Super-Earth Exoplanet GJ 1214b," by J. Bean, E. Miller-Ricci, and D. Homeier (2010), *Nature* 468: 669–72.

65. *Transiting Exoplanets* by C. Haswell (2010), Cambridge: Cambridge University Press and the Open University, p. 217.

10. Chandra—Exploring the Violent Cosmos

1. *An Acre of Glass: A History and Forecast of the Telescope* by J. B. Zirker (2005), Baltimore, MD: Johns Hopkins University Press.

2. "How Many Stars?" by F. Cain (2010), from http://www.universetoday.com/24328/how-many-stars/.

3. Biographical material on Röntgen and a summary of his discovery is online at the Official Web Site of the Nobel Prize, at http://nobelprize.org/nobel_prizes/physics/laureates/1901/rontgen-bio.html.

4. "Röntgen's Ghosts: Photography, X-Rays, and the Victorian Imagination," by A. W. Grove (1997), *Literature and Medicine* 16, no. 2 (Fall): 142.

5. "'The New Light': X Rays and Medical Futurism," by N. Knight, in *Imagining Tomorrow: History, Technology, and the American Future*, edited by J. J. Corn (1986), Cambridge, MA and London: MIT Press, pp. 13, 14.

6. *Inventing Modern: Growing Up with X-Rays, Skyscrapers, and Tailfins* by J. H. Lienhard (2003), Oxford: Oxford University Press, p. 44.

7. "X-ray Mania: The X Ray in Advertising, Circa 1895" by E. S. Gerson (2004), *RadioGraphics: The Journal of Continuing Medical Education in Radiology* 24: 544–51.

8. "The X-ray Shoe Fitter—An Early Application of Roentgen's 'New Kind of Ray,'" by D. Lapp (2004), *The Physics Teacher* 42: 355.

9. "X-Ray Mania: The X Ray in Advertising, Circa 1895," by E. S Gerson, pp. 546–47.

10. *Inventing Modern: Growing Up with X-Rays, Skyscrapers, and Tailfins*, p. 16.

11. "'The New Light': X Rays and Medical Futurism," by N. Knight, p. 22.

12. "Röntgen's Ghosts: Photography, X-Rays, and the Victorian Imagination," p. 164.

13. Siegel and Shuster took artistic and scientific license with Superman's superpowers, not least with his X-ray vision. In the comic, his eyes emit X-rays, which he uses for more piercing vision. But of course the X-rays that can traverse solid matter had to be registered in some medium, so it's not an analogous situation to light, which reflects off objects and is detected by our eyes.

14. *Naked to the Bone: Medical Imaging in the Twentieth Century*, by B. Holztmann Kevles (1998), New York: Perseus Publishing, p. 31.

15. "X Rays and the Quest for Invisible Reality in the Art of Kupka, Duchamp, and the Cubists," by L. D. Henderson (1988), *Art Journal* 47, no. 4: 323–40.

16. "The Image and Imagination of the Fourth Dimension in Twentieth-Century Art and Culture," by L. D. Henderson (2009), *Configurations* 17, no. 1 (Winter): 133, 146. See also *The Fourth Dimension and Non-Euclidean Geometry in Modern Art*, by L. D. Henderson (2010), new edition, Cambridge, MA: The MIT Press.

17. "Merging Art and Science" by A. I. Miller (2011), in *Art and Science: Merging Art and Science to Make a Revolutionary New Art Movement*, London: GV Art, p. 3.

18. "Subrahmanyan Chandrasekhar. 19 October 1910–21 August 1995," by R. Tayler (1996), *Biographical Memoirs of the Royal Society* 42: 80–94. More personal reminiscences of Chandra are included in "Leaves from an Unwritten Diary: S. Chandrasekhar, Reminiscences and Reflection," by S. Vishveshwara (2000), *Current Science* 8: 1025–33.

19. As we've seen already with the Mars Exploration Rovers and the Spitzer Space Telescope, NASA likes to involve the public in the naming of its major facilities. In the competition to name their X-ray "Great Observatory," NASA received over 6,000 entries from fifty countries.

20. From the Chandra Observatory history pages, at http://chandra.harvard.edu/xray_astro/history.html.

21. "X-rays from the Sun," by C. Keller (1995), *Cellular and Molecular Life Sciences* 51: 710–20.

22. "Evidence for X-rays from Sources Outside the Solar System," by R. Giacconi, H. Gursky, F. Paolini, and B. Rossi (1962), *Physical Review Letters* 9: 439–43. In 2002, Giacconi received the Nobel Prize in Physics for this and other

discoveries. For a more personal perspective from one of the pioneers of X-ray astronomy, see "Forty Years on from Aerobee 155: a Personal Perspective," by K. Pounds (2002), in *X-ray Astronomy in the New Millennium*, edited by R. Blandford, A. Fabian, and K. Pounds, *Royal Society of London Philosophical Transactions A* 360: 1905.

23. "An Education in Astronomy," by R. Giacconi (2005), *Annual Review of Astronomy and Astrophysics* 43: 1–30.

24. *The Violent Universe: Joyrides Through the X-Ray Cosmos* by K. Weaver (2005), Baltimore, MD: Johns Hopkins University Press.

25. *The Universe in X-Rays* edited by J. Trümper and G. Hasinger (2008), New York: Springer.

26. "The Development and Scientific Impact of the Chandra X-Ray Observatory," by D. Schwartz (2004), *International Journal of Modern Physics D* 13: 1239–47.

27. The two types of radiation are called thermal and non-thermal. Any gas, liquid, solid, or plasma with a single temperature will emit a smooth spectrum with peak intensity at a wavelength that inversely scales with the temperature—hotter objects have thermal radiation peaking at shorter wavelengths. Non-thermal radiation has no characteristic peak; it extends over a very broad wavelength range, factors of hundreds or thousands.

28. *Black Holes and Time Warps: Einstein's Outrageous Legacy* by K. Thorne and S. Hawking (1995), New York: W. W. Norton.

29. *Accretion Power in Astrophysics* by J. Frank, A. King, and D. Raine (2002), Cambridge: Cambridge University Press.

30. "New Evidence for Black Hole Event Horizons from Chandra," by M. Garcia et al. (2001), *Astrophysical Journal* 553: L47–L50.

31. Isolated black holes do accrete mass, but at a generally much lower rate than in binary systems. However, some small fraction of the isolated black holes should be detectable in X-rays; see "X-Rays from Isolated Black Holes in the Milky Way," by E. Algol and M. Kamionkowski (2002), *Monthly Notices of the Royal Astronomical Society* 334: 553–62.

32. "The Mass of the Black Hole in the X-Ray Binary M33 X-7 and the Evolutionary Status of M33 X-7 and IC 10 X-1," by M. Abubekerov et al. (2009), *Astronomy Reports* 53: 232–42.

33. The emission of gravitational radiation from an intense and changing gravity field is a key prediction of Einstein's theory of general relativity. Gravitational radiation has been observed indirectly in pulsars where the pulsar very gradually "spins down" as it radiates gravity waves. Gravity waves from a binary black hole will be much stronger and potentially provide a dramatic confirmation of general relativity.

34. In the complete picture of heavy element creation in the universe, there are two processes that generate nuclei heavier than iron. The s-process or "slow" process operates in the atmosphere of giant stars, when neutrons are steadily added to nuclei to increase their atomic number. The r-process or "rapid" process is the dramatic creation of heavy elements in the billion-degree blast wave that results from a supernova explosion.

35. "Chandra High Resolution X-Ray Spectrum of the Supernova Remnant 1E 0102.2-7219," by K. Flanagan et al. (2004), *Astrophysical Journal* 605: 230–46.

36. See "Expansion Velocity of Ejecta in Tycho's Supernova Remnant Measured by Doppler-Broadened X-Ray Line Emission," by A. Hayato et al. (2010), *Astrophysical Journal* 725: 894–903, and "Discovery of Spatial and Spectral Structure in the X-Ray Emission from the Crab Nebula," by M. Weisskopf et al. (2000), *Astrophysical Journal* 536: L81–L84.

37. "New Evidence Links Stellar Remains to Oldest Recorded Supernova," from ESA News online at http://www.esa.int/esaCP/SEMGE58LURE_index_0.html.

38. "Thermal Radiation from Neutron Stars: Chandra Results," by G. Pavlov, V. Zavlin, and D. Sanwal (2002), in *Seminar on Neutron Stars, Pulsars, and Supernova Remnants*, MPE Report 278, edited by W. Becker, H. Lesch, and J. Trumper, Garching, Germany: Max Planck Institute for Astrophysics.

39. "Progress in X-Ray Astronomy" by R. Giacconi (1973), *Physics Today* 26: 38–47.

40. "John Wheeler, Physicist Who Coined the Term 'Black Hole,' is Dead at 96," by D. Overbye, *New York Times*, April 14, 2008 online at http://www.nytimes.com/2008/04/14/science/14wheeler.html?pagewanted=1&_r=1.

41. *Gravity's Fatal Attraction: Black Holes in the Universe*, by M. Begelman and M. Rees (2010), 2nd ed., Cambridge: Cambridge University Press, p. 13.

42. *Black Holes and Time Warps: Einstein's Outrageous Legacy* by K. Thorne and S. Hawking (1995), New York: W. W. Norton, p. 23.

43. Excerpted from the liner notes for the Rush Remasters series, *A Farewell to Kings*, by Rush, published by Anthem and Mercury Records, copyrighted in 1977. Ellipses in brackets indicate deleted text to distinguish from ellipses in the original.

44. "On the Orbital and Physical Parameters of the HDE 226868/Cygnus X-1 Binary System," by L. Iorio (2008), *Astrophysics and Space Science* 315: 335–40.

45. The bet was a year's subscription to *Penthouse Magazine* for Thorne against four years of the British satirical magazine *Private Eye* for Hawking. Hawking was no doubt happy to lose the bet, since losing it meant the existence of black holes had been verified, which was a validation of much of his life's work.

46. *Revealing the Universe: The Making of the Chandra X-ray Observatory*, by W. Tucker and K. Tucker (2001), Cambridge: Harvard University Press, p. 44.

47. *Night Train* by M. Amis (1988), New York: Harmony Books.

48. Online at http://public.web.cern.ch/public/en/lhc/safety-en.html.

49. From "Tomorrow is Yesterday," written by D. C. Fontana, directed by Michael O'Herlihy, *Star Trek*, The Original Series, Season 1, Episode 19, Airdate 26 January 1967.

50. "How Black is Cygnus X-1?" *Science News* (1975), 107, p. 150.

51. "How an Amateur Astronomer Captured a Black Hole Scoop," by S. G. Cullen (2010), *Astronomy* 38 (June): 56–57.

52. *Touch the Invisible Sky: A Multi-Wavelength Braille Book Featuring Tactile NASA Images*, by N. Grice, S. Steel and D. Daou (2007), Puerto Rico and Columbia: Ozone Publishing.

53. Earlier books in the series were *Touch the Universe: a NASA Braille Book of Astronomy* by N. Grice (2002), Washington, DC: Joseph Henry Press, and *Touch the Sun: a NASA Braille Book* by N. Grice (2005), Washington, DC: Joseph Henry Press.

54. *Black Holes and Time Warps: Einstein's Outrageous Legacy* by K. Thorne and S. Hawking (1995), New York: W. W. Norton, p. 524.

55. The logical process was as follows. The density of stars is very high in the center of the Milky Way, many thousands of times higher than in the solar neighborhood. However, there is a limit to how many stars can be placed in a finite volume—as more stars are added, their gravity-induced velocities increase, "puffing" up the distribution and so setting a natural limit in Newtonian gravity to the density of any star cluster. The mass density inferred by stellar motions of several million mph within a few light weeks of the center of the galaxy is a million times higher than any plausible star cluster, so the cluster hypothesis fails spectacularly. A black hole becomes the only viable explanation.

56. *The Galactic Supermassive Black Hole* by F. Melia (2007), Princeton: Princeton University Press.

57. "Rapid X-ray Flaring from the Direction of the Supermassive Black Hole in the Galactic Center," by F. Baganoff et al. (2001), *Nature* 413: 45–48.

58. "Variable Positron Annihilation Radiation from the Galactic Center Region," by G. Riegler et al. (1981), *Astrophysical Journal* 248: L13–L16.

59. "Revelations in our Own Backyard: Chandra's Unique Galactic Center Discoveries," by S. Markoff (2010), *Proceedings of the National Academies of Science* 107: 7196–7201.

60. For a summary of the properties and the manifestations in X-rays, see the Chandra Field Guide, online at http://chandra.harvard.edu/xray_sources/quasars.html.

61. "The Relation Between Black Hole Mass, Bulge Mass, and Near-Infrared Luminosity," by A. Marconi and L. Hunt (2003), *Astrophysical Journal* 598: L21–L24.

62. "In-Depth Chandra Study of the AGN Feedback in Virgo Elliptical Galaxy M84," V. Finoguenov et al. (2008), *Astrophysical Journal* 686: 911–17.

63. "Discovery of Binary Active Galactic Nucleus in the Ultraluminous Infrared Galaxy NGC 6240 using Chandra," by S. Komossa et al. (2003), *Astrophysical Journal* 582: L15–L19.

64. "Deepest X-Rays Ever Reveal Universe Teeming with Black Holes," a Chandra X-Ray Observatory press release online at http://chandra.harvard.edu/press/01_releases/press_031301.html.

65. "Radiation Pressure, Absorption, and AGN Feedback in the Chandra Deep Fields," by S. Raimundo et al. (2010), *Monthly Notices of the Royal Astronomical Society* 408: 1714–20.

66. "AGN Feedback Cause Downsizing," E. Scannapieco, J. Silk, and R. Bouwens (2005), *Astrophysical Journal* 635: L13–L16.

67. The two dominant components of the universe remain enigmatic; however, dark energy is even more poorly understood than dark matter. The evidence for mass in the outer parts of galaxies accumulated in the 1970s as the disks of spiral galaxies were mapped and shown to have velocities too high to be explained by the visible matter. This argument was extended to elliptical galaxies until it became clear that all galaxies have a dominant mass component that doesn't shine or interact with radiation. Gravitational lensing data have affirmed the existence of dark matter on all scales in the universe. Dark energy was hypothesized after supernovae in distant galaxies were observed to be unaccountably faint, implying the galaxies were farther away than expected, which in turn implied that the universe has been accelerating in the past 5 billion years. Dark energy is therefore something that repels matter and its physical nature is not understood at all.

68. "A Direct Empirical Proof of the Existence of Dark Matter," by D. Clowe et al. (2006), *Astrophysical Journal* 648: L108–L113.

69. "Mysterious Dark Energy Confirmed by New Method," by J. Bryner (2008), from Space.com on December 16, 2008, online at http://www.space.com/6230-mysterious-dark-energy-confirmed-method.html, and "Arrested Development of the Universe," by P. Edmonds (2008), in the Chandra Chronicles online at http://chandra.harvard.edu/chronicle/0408/darkenergy/.

11. HST—The Universe in Sharp Focus

1. Hubble's preeminence stems in part from a versatile suite of instruments that have between them more than 400 modes of observation. The observatory has been serviced five times by the now-defunct Space Shuttle: in 1993, 1997, 1999, 2002, and 2009. It will be allowed to die a natural death, which will be caused in all likelihood by the sequential loss of gyroscopes due to natural aging that will render the HST unable to lock on any celestial target. For more details, see http://hubblesite.org/the_telescope/team_hubble/servicing_missions.php.

2. The scientific return of Hubble is high because astronomers clamor for time on the facility. Each annual cycle of observing, a call is put out for proposals from the international community, and several thousand carefully crafted proposals are received, only about 15 percent of which can get time on the telescope. Reading and evaluating the proposals and choosing the best involves several hundred astronomers, who read their subset of the proposals according to scientific subfield, then gather at the Space Telescope Science Institute in Baltimore to carry out the selection process. The peer review mechanism includes elaborate mechanisms to avoid conflict of interest, since research grants come attached with the observing time and astronomers are often evaluating proposals by their peers and rivals! They do this without remuneration as a volunteer activity. Each cycle, the director has a small amount of discretionary time.

3. In the list here, it is actually number 50, but counting the Large Binocular Telescope's two mirrors, it is number 51, and a dozen or more large telescopes are planned: http://astro.nineplanets.org/bigeyes.html.

4. This influential document was never published in the peer-reviewed or public literature, but it has been reprinted in *Exploring the Unknown: Selected Docu-*

ments in the History of the U.S. Civil Space Program, Volume 5, Exploring the Cosmos, edited by J. Lodgson, A. Snyder, R. Launius, S Garber, and R. Newport (2001), NASA SP-2001-4407, Washington, DC: National Aeronautics and Space Administration.

5. For more information, see http://science1.nasa.gov/missions/oao/.

6. A chronology can be found at NASA's History Division site, http://history.nasa.gov/hubble/chron.html. The story is told in more detail by Robert Zimmerman in *The Universe in a Mirror: The Saga of the Hubble Space Telescope and the Visionaries Who Built It* (2008), Princeton and Oxford: Princeton University Press.

7. The mirror was too flat at the edges by 2.2 microns, or 2.2 millionths of a meter. However, the overall accuracy of the surface was 200 times better, with variations of only 0.001 microns or one billionth of a meter.

8. From the May 17, 1990 episode of *Late Night with David Letterman* were the following top excuses: (10) The guy at Sears promised it would work fine. (9) Some kids on Earth must be fooling around with a garage door opener. (8) There's a little doo-hickey rubbing against the part that looks kind of like a cowboy hat. (7) See if you can think straight after twelve days of drinking Tang. (6) Some bum with a squeegee smeared the lens at a red light. (5) The blueprints were drawn up by that "Hey, Vern!" guy, Earnest. (4) Those damn raccoons! (3) Shouldn't have used G.E. components. (2) Ran out of quarters. (1) Race of super-evolved galactic beings/jokesters is screwing with us again. Letterman also riffed: " . . . apparently, they can't get this telescope to focus . . . for that much money you might as well get the auto-focus."

9. "The Hubble Space Telescope Optical Systems Failure Report" by L. Allen et al. (1990), NASA-TM-103443, Washington, DC: National Aeronautics and Space Administration.

10. "Engineering the COSTAR," by J. Crocker (1993), *Optics and Photonics News* 4: 11. COSTAR was removed from the telescope during the fifth servicing mission in 2009 and is now on display in the Smithsonian Air and Space Museum in Washington, DC.

11. With six gyroscopes, Hubble has a lot of redundancy on this critical component. Perplexingly, the gyros have been failing throughout the mission at rates never seen in ground testing. Space conditions are a little different due to cosmic rays and other "hard" forms of incoming energy. Three gyros provide stabilization in three dimensions as required for fine pointing and motion. Hubble has had to make do with two gyros on occasion and uses a non-axial instrument, the Fine Guidance Sensors, as the third pointing device. In 1999, just ahead of the third servicing mission, Hubble lost so many gryos it had to be shut down for a few weeks. If it is ever reduced to a single gyro, it will not be able to conduct routine science operations.

12. *Hubble: Imaging Space and Time* by D. Devorkin and R. Smith (2008), Washington, DC: National Geographic, p. 10.

13. *The Universe in a Mirror: The Saga of the Hubble Space Telescope and the Visionaries Who Built It* by R. Zimmerman, p. 197.

14. *Hubble: Imaging Space and Time* by D. Devorkin and R. Smith, p. 10.

15. By the early 1930s, the Mount Wilson Observatory opened to visitors for evening observations of the stars and attracted as many as 4,000 visitors over a

single weekend. Famous scientists, actors, and literati visited the Observatory, including British novelist Aldous Huxley, British actor George Arliss, as well as Albert Einstein. See *Edwin Hubble: Mariner of the Nebulae* by G. Christianson (1995), New York: Farrar, Straus and Giroux, pp. 212, 213.

16. "Nominee Backs a Review of NASA's Hubble Decision," by G. Gugliotta (2005), *Washington Post*, online at http://www.washingtonpost.com/wp-dyn /articles/A47810-2005Apr12.html, and "NASA Gives Green Light to Hubble Rescue," by A. Boyle (2006), Space news from MSNBC, online at http://www .msnbc.msn.com/id/15489217/#.Tj2yCWEmCuI.

17. "Hubble Left in Good Repair," by E. Berger (2009), *Houston Chronicle* on May 19, 2009, p. 1.

18. "Scientific Impact of Large telescopes," by C. Benn and S. Sanchez (2001), *Publications of the Astronomical Society of the Pacific* 113: 385–96.

19. Quoted in *Hubble: 15 Years of Discovery* by L. Christensen and R. Fosbury (2006), New York: Springer Science, p. 5. Robert Zimmerman notes: "As of early 2006, more than 6,000 scientific papers had been published based on Hubble data, totaling about 35 percent of the entire scientific output from all NASA research projects and more than three times the output of its nearest competitors, the Voyager and Viking probes . . . . This amazing unending stream of data has not only helped revolutionize the field of astronomy, it has changed our very perception of the universe." *The Universe in a Mirror: The Saga of the Hubble Space Telescope and the Visionaries Who Built It* by R. Zimmerman, pp. 165–66.

20. *Impact Jupiter: The Crash of Comet Shoemaker-Levy 9* by D. Levy (1995), Cambridge, MA: Basic Books.

21. "Spacescapes: Romantic Aesthetics and the Hubble Space Telescope Images," by E. Kessler (2006), PhD diss., University of Chicago, 2006, p. 11. Kessler's book deriving from her PhD thesis, *Picturing the Cosmos: Hubble Space Telescope Images and the Astronomical Sublime*, Minneapolis: University of Minnesota Press (2012), was not available from the publisher at the time of this writing.

22. Ibid., p. 16.

23. The rods are preferentially at the edge of the visual field, which explains why amateur astronomers are experts at "averted vision," where you constantly move your head slightly to put the image off to the side of your eye where it will hit the more sensitive rods. For more information, see "Color in Astronomical Images," by J. Lodriguss, online at http://www.astropix.com/HTML/I_ASTROP/ COLOR.HTM.

24. "Spacescapes: Romantic Aesthetics and the Hubble Space Telescope Images," by E. Kessler, p. 107.

25. The instruments actually house dozens of filters, some admitting a wide range of wavelengths and others a narrow range, spanning the entire spectrum from ultraviolet to infrared light. These include the classical red, green, and blue filters, which are often used to convey color in a TV or digital camera.

26. The generation of "true color" images is a complicated procedure that was mastered early on by the staff at the Space Science Telescope Institute. Even a pair of images through red and blue filters can be used to approximate true color,

but best results come from three or more different filters, typically "red, green, and blue." For the infrared cameras, a similar strategy is used even though they are not detecting visible light. The results are mapped from longer wavelengths into the visible spectrum in a consistent but essentially arbitrary way, so are said to represent "false color."

27. "Spacescapes: Romantic Aesthetics and the Hubble Space Telescope Images," by E. Kessler, p. 41.

28. Ibid., p. 42.

29. "The Public Impact of Hubble Space Telescope," by C. Christian and A. Kinney, Space Telescope Science Institute, February 22, 1999, online at http://www.stsci.edu/~carolc/publications/public_impact.PDF.

30. Quoted in "Spacescapes: Romantic Aesthetics and the Hubble Space Telescope Images," by E. Kessler (2006), PhD diss., University of Chicago, p. 196,

31. Ibid., p. 178.

32. See the Hubble Heritage homepage online at http://heritage.stsci.edu/gallery/gallery.html.

33. Quoted in "Spacescapes: Romantic Aesthetics and the Hubble Space Telescope Images," by E. Kessler, pp. 153–54.

34. Information on the Red Spider Nebula is available in *Hubble: 15 Years of Discovery* by L. Christensen and B. Fosbury (2006), New York: Springer Science, p. 112.

35. "The Hubble Space Telescope Medium Deep Survey with the Wide Field Planetary Camera. I. Methodology and Results on the Field near 3C 273," by R. Griffiths et al. (1994), *Astrophysical Journal* 437: 67–82.

36. "The Hubble Space Telescope Quasar Absorption line Key Project. I. First Observational Results, Including Lyman-Alpha and Lyman Limit Systems," by J. Bahcall et al. (1993), *Astrophysical Journal Supplements* 87: 1–43.

37. For explanation, see http://outreach.atnf.csiro.au/education/senior/astrophysics/variable_cepheids.html.

38. "The HST Key Project on the Extragalactic Distance Scale. XXVIII. Combining the Constraints on the Hubble Constant," by J. Mould et al. (2000), *Astrophysical Journal* 539: 786–94.

39. "A Common Explosion Mechanism for Type 1a Supernovae," by P. Mazzali et al. (2007), *Science* 315: 825–28.

40. "Observational Evidence from Supernovae for an Accelerating Universe and a Cosmological Constant," by A. Reiss et al. (1998), *Astronomical Journal* 116: 1009–38, and "Measurements of Omega and Lambda from 42 High-Redshift Supernovae," by S. Perlmutter et al. (1999), *Astrophysical Journal* 517: 565–86.

41. "Supernovae, Dark Energy, and the Accelerating Universe," by S. Perlmutter (2003), *Physics Today* (March): 53–60.

42. "The Distribution of Dark Matter in the Coma Cluster," by D. Merritt (1987), *Astrophysical Journal* 313: 121–35.

43. "A Systematic Search for Gravitationally Lensed Arcs in the Hubble Space Telescope WFPC2 Archive," by D. Sand, T. Treu, R. Ellis, and G. Smith (2005), *Astrophysical Journal* 627: 32–52.

44. "Supermassive Black Holes in Galactic Nuclei: Past, Present and Future Research," by L. Ferrarese and H. Ford (2005), *Space Science Reviews* 116: 523–624.

45. "A Fundamental Relationship between Supermassive Black Holes and their Host Galaxies," by L. Ferrarese and D. Merritt (2000), *Astrophysical Journal* 539: L9–L13.

46. "The Hubble Deep Field: Observations, Data Reduction, and Galaxy Photometry," by R. Williams et al. (1996), *Astronomical Journal* 112: 1335–89.

47. "The Hubble Ultra Deep Field," by S. Beckwith et al. (2006), *Astronomical Journal* 132: 1729–55.

48. "Optical Images of an Exoplanet 25 Light Years from Earth," by P. Kalas et al. (2006), *Science* 322: 1345–48.

49. "When Exoplanets Transit Their Parent Stars," by D. Charbonneau, T. Brown, A. Burrows, and G. Laughlin (2006), in *Protostars and Planets V*, edited by B. Reipurth, D. Jewitt, and K. Keil, Tucson: University of Arizona Press, pp. 701–16.

50. "Water Vapor in the Atmosphere of a Transiting Extrasolar Planet," by G. Tinetti et al. (2007), *Nature* 448: 163–68.

51. "Planets in the Galactic Bulge: Results of the SWEEPS Survey," by K. Sahu et al. (2008), in *Extreme Solar Systems*, ASP Conference Series vol. 398, edited by D. Fischer, F. Razio, S. Thorsett, and A. Wolszczan, pp. 93–98.

52. "18 Years of Science with the Hubble Space Telescope," by J. Dalcanton (2009), *Nature* 457: 46.

53. *Chasing Hubble's Shadows: The Search for Galaxies at the Edge of Time* by J. Kanipe (2006), New York: Hill and Wang, pp. 160, 5–6.

54. Ibid., p. 7.

55. Larger telescopes have better angular resolution and so can see smaller details on celestial objects and resolve individual objects in crowded regions. From the ground, this only works down to a resolution limit set not by the optics but by the blurring of images due to turbulence in the Earth's atmosphere. The better resolution of the 100-inch telescope compared to previous telescopes let Hubble pick out individual stars in the Andromeda nebula. Because of its location on orbit, the Hubble Space Telescope provides an angular resolution about ten times better than can be achieved from the ground, 0.05 seconds of arc as opposed to 0.5 seconds of arc. This gives HST extreme acuity in measuring the brightness of stars, especially when astronomers are looking for Cepheid variables to accurately measure distances to spiral nebulae. With the HST, this can be done out to a distance of 60 million light-years. Also, astronomers and cosmologists can use the known intrinsic brightness of supernovae that occur in binary star systems to calculate even larger distances. The Hubble Telescope's observations of supernovae in the distant universe have extended the distance measurements to about 9–10 billion light-years, easily sufficient to detect the cosmic acceleration that began about 5 billion years ago.

56. Hubble has accumulated disproportional credit for the discovery of cosmic expansion in the decades since his death. In fact, Alexander Friedmann not only published the first non-static solutions to Einstein's equations of general

relativity, in 1922, but he used Vesto Slipher's redshifts to infer expansion and gave an approximate value for the current expansion rate. Hubble never credited Slipher for the use of his redshifts. Aspects of the attribution of the concept of the expanding universe are discussed in "Lemaître's Hubble Relationship" by M. Way and H. Nussbaumer (2011), *Physics Today* 64: 8.

57. *Edwin Hubble: Mariner of the Nebulae* by G. Christianson (1995), New York: Farrar, Straus and Giroux, pp. 182, 183.

58. *The Realm of the Nebulae* by E. Hubble (1958), New York: Dover Publications, p. 202.

59. *Up Till Now: The Autobiography* by W. Shatner and D. Fisher (2008), New York: Thomas Dunne Books/St. Martin's Griffin, p. 150. The interconnections between NASA and *Star Trek* are numerous. Jon Wagner and Jan Lundeen note that the 1992 *Star Trek* exhibit at the Smithsonian National Air and Space Museum, which they refer to as the "Showcase of our Nation's Collective Memory," attracted more visitors than NASA's historic spacecraft on display. See *Deep Space and Sacred Time:* Star Trek *in the American Mythos* by J. Wagner and J. Lundeen (1998), Westport, CT: Praeger, p. 1. *Star Trek* actors are often asked to narrate educational science films as they've become so deeply associated with astronomy and the space sciences. For instance, in 2011, when the Space Shuttle fleet was finally retired, NASA aired an 80-minute NASA documentary titled *Space Shuttle* narrated by William Shatner.

60. *Star Trek: An Annotated Guide to Resources* by S. Gibberman (1991), Jefferson, NC and London: McFarland and Company, p. 147. Nichols continues to promote interest in space exploration in volunteering for events sponsored by organizations such as the Traveling Space Museum, which specializes in space education for school age children.

61. *NASA/TREK: Popular Science and Sex in America* by C. Penley (1997), New York and London: Verso, p. 19. Jemison is not only the first African American woman in space, but apparently the only astronaut to also appear in an episode of *Star Trek: The Next Generation.*

62. *The Nature of the Universe* by F. Hoyle (1950), New York: Harper and Brothers, p. 10.

63. *In the Shadow of the Moon* directed by D. Sington (2007), including Harrison Schmitt, Alan Bean, Michael Collins, Velocity/THINKFilm.

12. WMAP—Mapping the Infant Universe

1. *Anaximander and the Origins of Greek Cosmology* by C. Kahn (1994), Indianapolis, IN: Hackett Publishing.

2. *Music of the Spheres: Music, Science, and the Natural Order of the Universe* by J. James (1995), New York: Springer.

3. *Conceptions of Cosmos: From Myths to the Accelerating Universe* by H. Kragh (2007), Oxford: Oxford University Press, p. 149.

4. Lemaître actually made this point in a paper in 1927, two years before Hubble and Humason announced their findings that galaxies are receding at a rate that increases with increasing distance, which is the sign of uniform expansion in three dimensions. Lemaître's prescience was extraordinary. He not

only anticipated the physical explanation of the recession of galaxies, and the existence of radiation left over from the Big Bang, but he postulated the origin as a space-time singularity where relativity and quantum theory must have both held sway, and he speculated about vacuum energy that might have the behavior of Einstein's cosmological constant. These are concepts that only fully emerged in the 1990s and are now considered standard ingredients of big bang cosmology.

5. "The Beginning of the World from the Point of View of Quantum Theory," by G. Lemaître (1931), *Nature* 127: 706.

6. *The Day Without Yesterday: Lemaitre, Einstein, and the Birth of Modern Cosmology* by J. Farrell (2005), New York: Thunder Mouth Press, p. 106.

7. "The Beginning of the World from the Point of View of Quantum Theory," by G. Lemaître, p. 706.

8. *The Primeval Atom: An Essay on Cosmogony* by G. Lemaître (1950), translated by Betty and Serge Korff, Toronto and New York: D. Van Nostrand, p. 133.

9. Quoted in *The Day Without Yesterday: Lemaitre, Einstein, and the Birth of Modern Cosmology* by J. Farrell (2005), New York: Thunder Mouth Press, p. 99.

10. Quoted in *The Day We Found the Universe* by M. Bartusiak (2010), New York: Vintage/Random House, p. 257.

11. *Finding the Big Bang* by J. Peebles, L. Page, Jr., and R. B. Partridge (2009), Cambridge: Cambridge University Press, p. 18.

12. Quoted in *The Day Without Yesterday: Lemaitre, Einstein, and the Birth of Modern Cosmology*, p. 163.

13. Quoted in *Conceptions of Cosmos: From Myths to the Accelerating Universe* by H. Kragh (2007), Oxford: Oxford University Press, p. 232.

14. "The Origin of Chemical Elements," R. Alpher, H. Bethe, and G. Gamow (1984), *Physical Review* 73: 803–4.

15. "The Evolution of the Universe," by G. Gamow (1948), *Nature* 162: 680–82.

16. From the BBC Archives, at http://www.bbc.co.uk/science/space/universe/scientists/fred_hoyle.

17. "Molecular Lines from the Lowest States of Diatomic Molecules Composed of Atoms Probably Present in Interstellar Space," by A. McKellar (1941), *Publications of the Dominion Astrophysical Observatory* 7: 251–72.

18. *3K: The Cosmic Microwave Background Radiation* by B. Partridge (1995), Cambridge: Cambridge University Press.

19. Originally recounted on a story on NPR's *All Things Considered* called "The Big Bang's Echo" by R. Shoenstein, broadcast on May 17, 2005, and then documented on the American Institute of Physics history pages, at http://www.aps.org/programs/outreach/history/historicsites/penziaswilson.cfm.

20. The universe from about ten thousand years after the Big Bang to about 100–200 million years after the Big Bang is approximately like an ideal gas, where temperature, density, and pressure are linearly related and described by the classical physics of Boyle's law. Earlier than ten thousand years, the expansion is governed by radiation, not matter. After a few hundred million years, the steady action of gravity creates slight concentrations of matter that collapse to

form stars and galaxies, and those processes are nonlinear (i.e., the collapse is not proportional to density) and difficult to model precisely.

21. That radiant energy was of course much higher earlier in the universe. At the time the microwaves were "liberated," the universe was a thousand times smaller so the energy density was a billion times higher and any region emitted a dull red glow of 10 kW.

22. An entertaining expression of this obscure fact can be found in a video created by NASA/JPL for the Spitzer Space Telescope, http://www.spitzer.caltech.edu/video-audio/1387-irastro024-Big-Bang-Musical.

23. Normally a physical measurement is checked or calibrated against a carefully set up, externally defined scale. For example, a thermometer is calibrated against a substance of known freezing or melting point, and a car odometer is calibrated against a precisely determined distance. But the radiation from the universe is so close to thermal that no manmade object is better!

24. "Four-Year COBE DMR Microwave Background Observations: Maps and Basic Results," by C. Bennett et al. (1996), *Astrophysical Journal* 464: L1–L4.

25. For perspective, see, for example, http://www.nytimes.com/2006/10/08/weekinreview/08johnson.html.

26. A more esoteric problem with the standard Big Bang is the absence of "relics." Relics are leftover space-time anomalies like single magnetic poles, or monopoles, which should have been produced in abundance in the early universe, and so should still be occasionally seen even though the universe is now much larger. Monopoles or other relics have never been seen. Inflation explains this by stretching out space by such a large factor that monopoles should be extremely rare.

27. *The Inflationary Universe* by A. Guth (1998), New York: Basic Books.

28. The slight increase in temperature in the direction of motion, and the slight decrease of temperature in the opposite direction, is analogous to the Doppler effect. The microwaves are slightly compressed looking in the direction of our forward motion, so their wavelength is slightly reduced, which raises their energy or temperature. The converse happens looking away from our direction of motion. Formally, the pattern of the temperature difference across the sky is a dipole.

29. "The Microwave Anisotropy Probe (MAP) Mission," by C. Bennett et al. (2003), *Astrophysical Journal* 583: 1–23. See also the extensive and well-maintained web pages intended for public audiences at http://lambda.gsfc.nasa.gov/product/map/current/.

30. See the website of the European Space Agency, at http://www.esa.int/SPECIALS/Planck/index.html. Planck is refining measurements of the age of the universe and cosmological parameters, but only by small amounts.

31. *Music of the Big Bang: The Cosmic Microwave Background and the New Cosmology* by A. Balbi (2008), Berlin and Heidelberg: Springer/Verlag.

32. "The Cosmic Symphony," by W. Hu and M. White (2004), *Scientific American* (February): 44–53.

33. Radio and microwaves, like those that comprise the cosmic microwave background, ultraviolet radiation, visible light, and X-rays are all forms of electromagnetic radiation, and are subdivided into these categories based on frequency. And all travel at the speed of light.

34. *The Cambridge Companion to Electronic Music* by N. Collins and J. d'Escrivan (2007), Cambridge: Cambridge University Press, p. 58. For a short biography of Mathews, see "Max Mathews, Pioneer in Making Computer Music, Dies at 84," by W. Grimes (April 23, 2011), New York Times.com online at http://www.nytimes.com/2011/04/24/arts/music/max-mathews-father-of-computer-music-dies-at-84.html.

35. *A History of Rock Music 1951–2000* by P. Scaruffi (2003), Lincoln, NE: iUniverse, p. 74.

36. Ibid., p. 92.

37. Ibid., p. 94.

38. Holly Henry is grateful to astronomer and California State University colleague Leo Connolly for pointing out the common use of NGC 891 as a ready analog for the Milky Way and for background on the history of space rock as a genre.

39. *The Cambridge Companion to Electronic Music* by N. Collins and J. d'Escrivan (2007), Cambridge: Cambridge University Press, pp. 162–63.

40. Liner notes to *Apollo: Atmospheres & Soundtracks* by B. Eno, with D. Lanois and R. Eno (1983), EG Music/BMI.

41. *The Songlines* by Bruce Chatwin (1986), New York: Penguin.

42. See the project description on Bjork's websites online at http://www.bjork.com and at http://bjork.fr/Biophilia,1542.

43. *The Music of Pythagoras: How an Ancient Brotherhood Cracked the Code of the Universe and Lit a Path from Antiquity to Outer Space* by K. Ferguson (2008), New York: Walker and Company.

44. *Music of the Big Bang: The Cosmic Microwave Background and the New Cosmology* by A. Balbi (2008), Berlin and Heidelberg: Springer/Verlag, p. 81.

45. To hear recordings of Earth's radio emissions, see "Sciencecasts: The Sound of Earthsong," online at http://www.youtube.com/watch?v=MkTL2Ug6llE, or radio emissions by Saturn; see "Cassini: Unlocking Saturn's Secrets" online at http://www.nasa.gov/mission_pages/cassini/multimedia/pia07966.html.

46. *The Sun's Heartbeat: And Other Stories from the Heart of the Star that Powers our Planet* by B. Berman (2011), New York: Little, Brown.

47. "Sunspot Breakthrough," by T. Phillips (August 24, 2011), NASA Press Release online at http://www.nasa.gov/mission_pages/sunearth/news/sunspot-breakthru.html.

48. *The Mysterious Universe* by J. Jeans (1937), Cambridge: Cambridge University Press, p. 69.

49. *Music of the Big Bang: The Cosmic Microwave Background and the New Cosmology* by A. Balbi, p. 126.

50. "Detection of the Baryon Acoustic Peak in the Large-Scale Correlation Function of SDSS Luminous Red Galaxies" by D. L. Eisenstein et al. (2005), *Astrophysical Journal* 633: 560–74.

51. *The Wraparound Universe.* by J.-P. Luminet, translated by E. Novak (2008), Wellesley, MA: A. K. Peters, pp. 289, 239.

52. "Cries from the Infant Universe," interview of Mark Whittle by Richard Drumm, June 27, 2009, online at http://cosmoquest.org/blog/365daysofastronomy/2009/12/03/december-3rd-when-the-universe-was-young/.

53. To hear Whittle's recreation of the Big Bang, see "Cries from the Infant Universe" listed in the previous endnote. John G. Cramer, University of Washington, has produced a similar sound profile of the CMB peaks, "The Sound of the Big Bang," online at http://faculty.washington.edu/jcramer/BBSound.html.

54. "When the Universe Was Young," interview of Mark Whittle by Richard Drumm, December 3, 2009, online at http://cosmoquest.org/blog/365daysofastronomy/2009/12/03/december-3rd-when-the-universe-was-young/.

55. *Music of the Big Bang: The Cosmic Microwave Background and the New Cosmology* by A. Balbi, pp. 54–55.

56. "The Beginning of the World from the Point of View of Quantum Theory," by G. Lemaître (1931), *Nature* 127: 706.

57. Indeed, it is interesting that the medium used to transmit a "message in a bottle" when the Voyager spacecraft was tossed into the interstellar ocean was an old-fashioned phonograph record. The scope of humanity's achievements was conveyed in the wiggles of a spiral groove on a gold-plated record. It was a presumption that intelligent aliens that might discover the artifact in the future would be able to decode it. While the choice of technology was criticized—it was mainstream in 1977 but is now almost obsolete on Earth—the team designing the record pushed back, noting that the analog technology of a phonograph record would last a billion years or more in space, while the longevity of our current storage devices like CD's and DVD's is very unclear.

58. "Three Year Wilkinson Microwave Anisotropy Probe (WMAP) Observations: Implications for Cosmology," by D. Spergel et al. (2007), *Astrophysical Journal Supplements* 170: 377–408. For the less technical summary, see "With Its Ingredients MAPped, Universe's Recipe Beckons," by C. Seife (3002), *Science* 300: 730–31.

59. "Wilkinson Microwave Anisotropy Probe Data and the Curvature of Space," by J.-P. Uzan, U. Kirchner, and G. Ellis (2003), *Monthly Notices of the Royal Astronomical Society* 344: L65–L68.

60. "Seven Year Wilkinson Microwave Anisotropy Probe (WMAP) Observations: Cosmological Interpretations," by E. Komatsu et al. (2011), *Astrophysical Journal Supplement* 192: 18–45. The WMAP technical listing is at http://lambda.gsfc.nasa.gov/product/map/current/map_bibliography.cfm.

61. "Three Year Wilkinson Microwave Anisotropy Probe (WMAP) Observations: Polarization Analysis," by L. Page et al. (2007), *Astrophysical Journal Supplement* 170: 335–76.

62. "Seven Year Wilkinson Microwave Anisotropy Probe (WMAP) Observations: Power Spectra and WMAP-Derived Parameters," by D. Larson et al. (2011), *Astrophysical Journal Supplement* 192: 16–35.

63. Inflation posits that quantum fluctuations were exponentially expanded to become patches large enough to be the seed for later galaxies. The random nature of quantum fluctuations opens the possibility that in a space-time context preceding the Big Bang, other fluctuations could have given rise to universes separate from ours, with randomly different properties and even different laws of physics operating. The multiverse is a clever theoretical construct, but will be impossible to test unless these other universes leave imprints in the one universe we can actually study.

13. Conclusion—New Horizons, New Worlds

1. The word atom comes from the ancient Greek adjective *atomos*, meaning uncuttable.

2. Atomism emerged from an earlier, and even more esoteric, debate between Parmenides and Heraclitus. Parmenides argued that there was an underlying unitary mass and that all change was an illusion; Heraclitus by contrast believed in the primacy of change and was famous for saying that you can't step into the same river twice.

3. *Democritus (The Great Philosophers)* by P. Cartledge (1997), London: Routledge.

4. The crystalline spheres were an invention of Anaximander in the sixth century BC, and it's not clear if they were a formal mathematical device or intended to be physically real; see G.E.R. Lloyd (1978), "Saving the Phenomena," *Classical Quarterly* 28: 202–22.

5. *Matter, Space, and Motion* by R. Sorabji (1988), Ithaca, NY: Cornell University Press.

6. The best examples come from thermodynamics. Imagine a brick on a table in front of you. Atoms in the brick are fixed within a material of the brick, but each one has a random vibrational motion associated with its thermal energy or temperature. The vibrations are unassociated with each other and are randomly oriented. However, it is possible *in theory* that for one instant, enough of 10^{27} atoms in the brick would have their vibrations aligned for the brick to spontaneously jump up slightly above the table. Using thermodynamics and statistical mechanics, it's possible to calculate the odds of this happening. It turns out that even with trillions of bricks (enough to build a house for all the people on Earth) being watched for billions of years, the odds of this happening even once are infinitesimal. But it *could* happen and perhaps if we could watch all the brick-sized objects on all the planets in the universe for trillions of years, it *would* happen.

7. The horizon distance uses a simple geometric argument of the tangent distance from a particular height above the curved surface. Refraction increases the distance by about 10 percent. For sailors observing each other from the top of a 100-foot mast of a tall ship, the horizon distance is about 50 miles, still just a small fraction of the size of the largest bodies of water. Logically, the observation of a horizon shows only that the Earth is curved, not that it is spherical. Greek astronomers used the observation of the shape of the shadow of the Earth cast on the Moon during a lunar eclipse to argue that the Earth was a sphere.

8. The classic text is *Gravitation* by K. S. Thorne, C. Misner, and C. Wheeler (1973), New York: W. H. Freeman, and a gentler introduction can be found in *Black Holes: A Traveller's Guide* by C. Pickover (1998), New York: John Wiley.

9. Changing cosmic expansion plays interesting games with the visible universe. The initially decelerating expansion (caused by dark matter) means that galaxies that were beyond our horizon when they formed could enter our horizon subsequently and so become visible. However, the recent acceleration of the expansion rate (caused by dark energy) may subsequently remove some of

them from view again. The best technical explanation of these complex effects is "Expanding Confusion: Common Misconceptions of Cosmological Horizons and the Superluminal Expansion of the Universe" by T. M. Davis and C. H. Lineweaver (2004), *Publications of the Astronomical Society of Australia* 21: 97–109.

10. *Stargazer: The Life and Times of the Telescope* by F. Watson (2005), Cambridge, MA: Perseus Books Group.

11. "Introduction: Mars Science Laboratory: The Next Generation of Mars Landers," by M. K. Lockwood (2006), and thirteen subsequent articles in the special issue of *Journal of Spacecraft and Rockets* 43: 257.

12. The winner of the essay competition was announced in a NASA press release on May 25, 2009. It can be found at http://www.nasa.gov/mission_pages/msl/msl-20090527.html.

13. Large-scale evidence for water is usually a channel carved by liquid or a place where sediment has been deposited, like an alluvial fan. Evidence of past water is most persuasive if there are sedimentary minerals or formations. On Mars, phyllosilicates are of particular interest; these minerals are made of interlocking tetrahedrals of silicon and oxygen atoms that form into sheets. Common examples are mica, talc, and various forms of clay. Phyllosilicates and sulfates on Mars show all the signs of having been formed in the presence of water. See "Phyllosilicates on Mars and Implications for Early Martian Climate," by F. Poulet et al. (2005), *Nature* 438: 623–27.

14. Curiosity will cover more ground than Spirit or Opportunity, roaming up to 10 miles or more from its landing spot. Its big wheels can roll over obstacles nearly 3 feet high. About 125 Watts of power will come from a Plutonium generator, allowing it to work through the night and through the Martian winter, unlike the solar-powered Spirit and Opportunity.

15. Details abstracted from "Mars Science Laboratory Entry, Descent and Landing System Overview" by R. Prakash et al. (2008), *Aerospace Conference*, IEEE 2008, pp. 1–18.

16. A summary updated in March 2011 is "Mars Science Laboratory Fact Sheet," NASA Jet Propulsion Laboratory document JPL 400-1416.

17. "Did Life Exist on Mars? Search for Organic and Inorganic Signatures, One of the Goals for SAM (Sample Analysis at Mars)" by M. Cabane et al. (2004), *Advances in Space Research* 33: 2240–45.

18. The 2009 movie *Avatar* has grossed over $2.3 billion worldwide to date, so it's an interesting thought that Cameron and Fox Studios could have funded the entire Mars Science Laboratory project. See the NASA press release at http://marsprogram.jpl.nasa.gov/msl/news/index.cfm?FuseAction=ShowNews&NewsID=1116.

19. "Comparative Study of Different Methodologies for Quantitative Rock Analysis by Laser-Induced Breakdown Spectroscopy in a Simulated Martian Atmosphere," by B. Salle et al. (2006), *Spectrochimica Acta Part B-Atomic Spectroscopy* 61: 301–13.

20. "Preservation of Martian Organic and Environmental Records: Final Report of the Mars Biosignature Working Group," by R. E. Summons et al. (2011), *Astrobiology* 11: 157–81.

21. See the NASA Jet Propulsion Laboratory press release "NASA Mars Rover will Check for Ingredients of Life," from January 18, 2011, found online at http://www.jpl.nasa.gov/news/news.cfm?release=2011-018.

22. NASA is the major source of funding for planetary science, and most of its ambitious missions are collaborative with the European Space Agency. NASA has three tiers of interplanetary spacecraft: Discovery missions that cost $100–200 million, New Frontier missions at around $500 million each, and "flagship" missions costing over a billion dollars. All these missions must fit into a budget envelope of about $1.5 billion per year. Priorities are set by the planetary science research community in a process called the Decadal Survey (astronomy and astrophysics has its own Decadal Survey). The most recent survey ran to 400 pages and contained 200 position papers representing the views of 1,700 scientists; see the 2011 publication *Visions and Voyages for Planetary Science in the Decade 2013–2022*, Washington, DC: National Academy of Sciences.

23. *A Passion for Mars: Intrepid Explorers of the Red Planet* by A. Chaikin (2008), New York: Abrams.

24. "Europa Jupiter System Mission: A Joint Endeavor by ESA and NASA," by the Joint Jupiter Science Definition Team (2009), report JPL D-48440 and ESA-SRE(2008)1.

25. There are substantial uncertainties and contingencies involved in missions of this scale. Europa won out over Titan in NASA's deliberations, but the agency reserves the right to change the priorities downstream. NASA and ESA plan synchronized and complementary missions to reach all four Galilean moons, but ESA will compete its Ganymede orbiter against an X-ray telescope and a gravity wave observatory in 2013, and so may withdraw from the partnership. As a result, NASA has a contingency plan of sending its Europa orbiter as a standalone mission.

26. A more direct approach might be taken by the Russian space agency Roscosmos, which hopes to deploy a lander with a drill or an impactor, either of which can sample pristine ice found 10 meters or more below the heavily irradiated surface. Earlier ideas like a hydrobot that could melt through the ice and explore the ocean below are infeasible.

27. "Hydrated Salt Minerals on Ganymede's Surface: Evidence for an Ocean Below," by T. B. McCord et al. (2001), *Science* 292: 1523–25.

28. Newton's law of gravity says that a planet exerts the same force on a star that it orbits as the star exerts on the planet. Rather than the planet orbiting the stationary star, both objects orbit a common center of gravity (which is much closer to the star than the planet). It is this "reflex motion" of the star that provides telltale evidence of the unseen planet. The star will show a sinusoidal Doppler shift with amplitude proportional to the planet mass and a period equal to the orbital period of the planet. For a "Jupiter" orbiting a "Sun" the period is twelve years and the Doppler shift is a small, but detectable, 11 meters per second. An Earth orbiting a Sun seen from afar would have the tiny Doppler shift of 9 centimeters per second, crawling speed.

29. *The Crowded Universe: The Race to Find Life Beyond Earth* by A. Boss (2009), New York: Basic Books. For most of the hundreds of exoplanets

known, the faint planet is not observed directly, so the method does not require a large telescope since photons from the parent star are abundant. A one-meter telescope was used to discover the first exoplanet, orbiting the bright star 51 Pegasi. Rather, the major requirement is spectroscopic precision, since the star's reflex motion is tens of thousands of times slower than the speed of light, setting the fractional variation in wavelength that must be detected in terms of the Doppler effect.

30. Timing is everything in transit detection as well. Not only must thousands of stars be observed simultaneously to find the small percentage of suitably aligned systems, but a Jupiter or an Earth will only pass in front of their parent stars for a tiny percentage of the orbital time. Huge amounts of data have to be gathered to catch these rare events. The paper that laid out the details was "The Photometric Method of Detecting Other Planetary Systems," by W. J. Borucki and A. L. Summers (1984), *Icarus* 58: 121, building on the original idea in "A Two-Color Method for Detection of Extra-Solar Planetary Systems," by F. Rosenblatt (1971), *Icarus* 14: 71–93.

31. "Detection of Planetary Transits Around a Sun-like Star," by D. Charbonneau et al. (2000), *Astrophysical Journal Letters* 529: 45.

32. "Characteristics of Planetary Candidates Observed by Kepler, II: Analysis of the First Four Months of Data," by W. J. Borucki et al. (2011), *Astrophysical Journal* 736: 19–78.

33. As with point-and-shoot digital cameras, the figure of merit for an astronomical camera is not simply the number of pixels, although more pixels always increase the efficiency of any survey. It is also governed by magnification in the camera, which yields a trade-off between the area of sky covered with each picture of the sky and the resolution or the finest details that can be seen in the picture. Better resolution not only means crisper images, it also means a deeper view of the sky. Gaia is looking for tiny motions and angular displacements so its science goals are governed by resolution, which means each pixel in the camera covers as small a region of sky as possible to take advantage of the space environment where light from a distant object is not deviated by the Earth's atmosphere. The result is that the field of view in each picture is modest and Gaia must make many exposures to survey the sky. See "The Three Dimensional Universe with Gaia," edited by C. Turon, K. S. O'Flaherty, and M.A.C. Perryman (2006), ESA Special Publication SP-576.

34. Gaia is in the same distant orbit as WMAP because both satellites were designed to map the entire sky. Lagrange points are named after the eighteenth-century French mathematician who discovered five equilibrium points in the Earth-Sun system, or in any gravitational system of two massive bodies. L2 is a million miles from Earth in the direction away from the Sun, and any satellite at L2 rotates with the Earth in a one-year orbit. With the Sun shielded by the Earth, Gaia can make uninterrupted observations of the entire sky over the course of a year. L2 is unstable, so any spacecraft there must make occasional adjustments to stay in the correct location.

35. "The Ghost of a Dwarf Galaxy: Fossils of the Hierarchical Formation of the Nearby Spiral Galaxy NGC 5907" by D. Martinez-Delgado et al. (2008), *Astrophysical Journal* 689: 184–93.

Sagan, C., and Druyan, Ann. 1997. "Gettysburg and Now." *Billions and Billions: Thoughts on Life and Death at the Brink of the Millennium*. New York: Random House, pp. 192–203.

Sagan, C., and Salpeter, E. E. 1976. "Particles, Environments, and Possible Ecologies in the Jovian Atmosphere." *Astrophysical Journal Supplement* 32, pp. 737–55.

Sarantakes, N. E. 2005. "Cold War Pop Culture and the Image of U.S. Foreign Policy: The Perspective of the Original Star Trek Series." *Journal of Cold War Studies* 7.4, pp. 74–103.

Schenk, P., et al. 2001. "The Mountains of Io: Global and Geological Perspectives from Voyager and Galileo." *Journal of Geophysical Research* 106, pp. 33201–22.

Shatner, W., and Fischer, D. 2008. *Up Till Now: The Autobiography*. New York: Thomas Dunne Books/St. Martin's Griffin.

Smith, B. A., et al. 1979. "The Jupiter System Through the Eyes of Voyager 1." *Science* 204, pp. 951–57.

Swift, D. W. 1997. *Voyager Tales: Personal Views of the Grand Tour*. Reston, VA: AIAA (American Institute of Aeronautics and Astronautics).

Wachhorst, W. 2000. *The Dream of Spaceflight: Essays on the Near Edge of Infinity*. New York: Basic Books.

Wagner, J., and Lundeen, J. 1998. *Deep Space and Sacred Time:* Star Trek *in the American Mythos*. Westport, CT and London: Praeger.

Zubrin, R. 1999. *Entering Space: Creating a Spacefaring Civilization*. New York: Tarcher Putnam.

Chapter 5: Cassini

Brooks, R. A. 2002. *Flesh and Machines: How Robots Will Change Us*. New York: Pantheon Books.

Brown, R., Lebreton, J. P., and Waite, J. H. 2009. *Titan from Cassini-Huygens*. New York: Springer.

Carson, R. 1998. *The Edge of the Sea*, 1955 edition, introduction by Sue Hubbell. Boston and New York: Houghton Mifflin.

Clarke, A. C., and Bonestell, C. 1972. *Beyond Jupiter: The Worlds of Tomorrow*. Boston and Toronto: Little, Brown.

Coustenis, A., and Taylor, F. W. 2008. *Titan: Exploring an Earthlike World*, 2nd ed. London and Singapore: World Scientific Publishing.

Eiseley, L. 1994. *The Star Thrower*. New York: Harvest/Harcourt Brace.

Esposito, L. 2006. *Planetary Rings* (Cambridge Planetary Science Series). Cambridge: Cambridge University Press.

Harland, D. M. 2002. *Mission to Saturn: Cassini and the Huygens Probe*. Berlin and London: Springer–Verlag.

Harland, D. M. 2007. *Cassini at Saturn: Huygens Results*. Chichester, UK: Springer-Praxis.

Helmreich, S. 2009. *Alien Ocean: Anthropological Voyages in Microbial Seas*. Berkeley and Los Angeles: University of California Press.

Henry, H. and Taylor, A. 2009. "Re-Thinking *Apollo*: Envisioning Environmentalism in Space," in *Space Culture and Travel: From Apollo to Space Tourism*, ed. D. Bell and M. Parker. Oxford and Malden, MA: Wiley/Blackwell.

Jones, T., and Stofan, E. 2008. *Planetology: Unlocking the Secrets of the Solar System*. Washington, DC: National Geographic.

Kurzweil, Ray. 2005. *The Singularity Is Near: When Humans Transcend Biology*. New York: Viking.

Lorenz, R., and Mitton, J. 2008. *Titan Unveiled: Saturn's Mysterious Moon Explored*. Princeton, NJ: Princeton University Press.

Lovett, L., Horvath, J., and Cuzzi, J. 2006. *Saturn: A New View*, foreword by K. S. Robinson. New York: Harry N. Abrams.

Matson, D. L., Spilker, L. J., and LeBreton, J. P. 2002. "The Cassini/Huygens Mission to the Saturnian System." *Space Science Reviews* 104, pp. 1–58.

Miller, R., and Durant, F. C. III. 2001. *The Art of Chesley Bonestell*, with M. H. Schuetz. London: Paper Tiger.

Miner, E. D., Wessen, R. R., and Cuzzi, J. N. 2006. *Planetary Ring Systems*. New York: Springer-Praxis.

Sofan, E. R., et al. 2007. "The Lakes of Titan." *Nature* 445, pp. 61–64.

Spilker, L. J. 1997. *Passage to a Ringed World: The Cassini-Huygens Mission to Saturn and Titan*. NASA SP–533. Washington, DC: NASA.

Wachhorst, W. 2000. *The Dream of Spaceflight: Essays on the Near Edge of Infinity*, foreword by Buzz Aldrin. New York: Basic Books.

Chapter 6: Stardust

Burnham, R. 2000. *Great Comets*. Cambridge: Cambridge University Press.

Chown, M. 2001. *The Magic Furnace: The Search for the Origins of Atoms*. Oxford: Oxford University Press.

Crovisier J., and Encrenaz, T. 2000. *Comet Science: The Study of Remnants from the Birth of the Solar System*. Cambridge: Cambridge University Press.

Davis, A. M. (ed.) 2005. *Meteorites, Comets, and Planets*. Oxford and London: Elsevier Ltd.

Gaume, L. A. (2008). "Riddles in Fundamental Physics." *Leonardo* 41, no. 3 , (June(, pp. 245–51.

Gribbin, J., and Gribbin, M. 2000. *Stardust: Supernovae and Life—The Cosmic Connection*. New Haven, CT and London: Yale University Press.

Levy, D. H. 2003. *Impact Jupiter: The Crash of Comet Shoemaker–Levy 9*. New York: Basic Books.

Lewis, J. S. 1997. *Mining the Sky: Untold Riches from the Asteroids, Comets, and Planets*. New York: Perseus Books.

Sagan, C., and Druyan, A. 1997. *Comet*. New York: Ballantine Books.

Thomas, P. J., Chyba, C. F., and McKay, C. P. 1997. *Comets and the Origin and Evolution of Life*. New York: Springer-Verlag.

Wallerstein, G., et al. 1999. "Synthesis of the Elements in Stars: Forty Years of Progress." *Reviews of Modern Physics* 69, pp. 995–1084.

Chapter 7: SOHO

Brennan, M. 1983. *The Stars and the Stones: Ancient Art and Astronomy in Ireland*. London: Thames and Hudson.
Carlowicz, M. J., and Lopez, R. E. 2002. *Storms from the Sun: The Emerging Science of Space Weather*. Washington, DC: National Academies Press.
Clark, S. 2007. *The Sun Kings: The Unexpected Tragedy of Richard Carrington and the Tale of How Modern Astronomy Began*. Princeton, NJ: Princeton University Press.
Eddy, J. A. 2009. *The Sun, the Earth, and Near-Earth Space: A Guide to the Sun–Earth System*. Washington, DC: NASA.
Freeman, J. W. 2001. *Storms in Space*. Cambridge: Cambridge University Press.
Golub, L., and Pasachoff, J. M. 2001. *Nearest Star: The Surprising Science of Our Sun*. Cambridge, MA and London: Harvard University Press.
Krupp, E. C. 1997. *Skywatchers, Shamans and Kings: Astronomy and the Archaeology of Power*. New York: John Wiley & Sons.
Lilensten, J., and Bornarel, J. 2006. *Space Weather, Environment and Societies*. Dordrecht: Springer.
Moldwin, M. 2008. *An Introduction to Space Weather*. Cambridge: Cambridge University Press.
Odenwald, S. 2001. *The 23rd Cycle: Learning to Live with a Stormy Star*. New York: Columbia University Press.
Poppe, B. P., and Jordan, K. P. 2006. *Sentinels of the Sun: Forecasting Space Weather*. Boulder, CO: Johnson Books.
Stout, G., and Stout, M. 2008. *Newgrange*. Cork: Cork University Press.
Varese, L. 1972. *Varese: A Looking Glass Diary*. New York: W. W. Norton.

Chapter 8: Hipparcos

Aveni, Anthony. 2008. *People and the Sky: Our Ancestors and the Cosmos*. New York: Thames and Hudson.
Boyd, B. 2009. *On the Origin of Stories: Evolution, Cognition, and Fiction*. Cambridge, MA, and London: Belknap Press of Harvard University Press.
Evans, J. 1998. *The History and Practice of Ancient Astronomy*. New York and Oxford: Oxford University Press.
Garbedian, H. G. 1939. *Albert Einstein: Maker of Universes*. New York and London: Funk and Wagnalls.
Hirschfield, A. W. 2001. *Parallax: The Race to Measure the Cosmos*. New York: Henry Holt.
Hoskin, Michael. 1999. "Astronomy in Antiquity." *The Cambridge Concise History of Astronomy*, ed. Michael Hoskin. Cambridge: Cambridge University Press, pp. 18–47.
Jin, Wenjing, Imants, Platais, and Perryman, Michael A. C. 2008. *A Giant Step: From Milli- to Micro-Arcsecond Astrometry*. Cambridge: Cambridge University Press.

Kanas, N. 2007. *Star Maps: History, Artistry, and Cartography*. New York: Springer/Praxis.

Meller, Harald. 2004. "Star Search." *National Geographic* 205, no. 1 (January), pp. 76–87.

Nova: Cracking the Maya Code. 2008. DVD, directed by David Lebrun Nova Production with Night Fire Films and ARTE France. WGBH Educational Foundation.

Perryman, M. 2009. *Astronomical Applications of Astrometry: Ten Years of Exploitation of the Hipparcos Satellite Data*. Cambridge: Cambridge University Press.

Perryman, M. 2010. *The Making of History's Greatest Star Map*. New York: Springer.

Ridpath, I. 1988. *Star Tales*. New York: Universe Books.

Schaefer, Bradley. 2005. "The Epoch of the Constellations on the Farnese Atlas and Their Origin in Hipparchus's Lost Catalogue." *JHA (Journal for the History of Astronomy)* 36, pp. 167–96.

Van Leeuwen, F. 2007. *Hipparcos, The New Reduction of the Raw Data*. New York: Springer.

Chapter 9: Spitzer

Bova, B. 2001. *The Story of Light*. Naperville, IL: Sourcebooks.

Gross, M. 2002. *Light and Life*. Oxford: Oxford University Press.

Haswell, C. A. 2010. *Transiting Exoplanets*. Cambridge: Cambridge University Press and the Open University.

Koslow, T. 2007. *The Silent Deep: The Discovery, Ecology and Conservation of the Deep Sea*. Chicago: University of Chicago Press.

Land, M. F., and Nilsson, D.–E. 2002. *Animal Eyes*. Oxford: Oxford University Press.

Nouvian, C. (ed.) 2007. *The Deep: The Extraordinary Creatures of the Abyss*. Chicago and London: University of Chicago Press.

Parker, A. 2003. *In the Blink of an Eye: How Vision Sparked the Big Bang of Evolution*. New York: Basic Books.

Rieke, G. 2009. "History of Infrared Telescopes and Astronomy." *Experimental Astronomy* 25, pp. 125–41.

Seager, S. (ed.) 2011. *Exoplanets*. Tucson: University of Arizona Press.

Sobel, M. I. 1987. *Light*. Chicago and London: University of Chicago Press.

Werner, M., et al. 2004. "The Spitzer Space Telescope Mission." *Astrophysical Journal Supplement* 154, pp. 1–9.

Werner, M. 2009. "How the Spitzer Space Telescope Unveils the Unseen Cosmos." *Astronomy* 37, no. 3 (March), pp. 44–52.

Chapter 10: Chandra

Begelman, M., and Rees, M. 2010. *Gravity's Fatal Attraction: Black Holes in the Universe*. Cambridge: Cambridge University Press.

Fabian, A., Pounds, K., and Blandford, R. 2004. *Frontiers of X-Ray Astronomy*. Cambridge: Cambridge University Press.
Frank, J., King, A., and Raine, D. 2002. *Accretion Power in Astrophysics*. Cambridge: Cambridge University Press.
Gerson, E. S. 2004. "X-ray Mania: The X Ray in Advertising, Circa 1895." *RadioGraphics: The Journal of Continuing Medical Education in Radiology* 24, pp. 544–51.
Henderson, L. D. 2009. "The Image and Imagination of the Fourth Dimension in Twentieth-Century Art and Culture." *Configurations* 17, no. 1 (Winter), pp. 131–60.
Knight, N. 1986. "'The New Light': X Rays and Medical Futurism," in *Imagining Tomorrow: History, Technology, and the American Future*, ed. J. J. Corn. Cambridge, MA, and London: MIT Press, pp. 10–34.
Lienhard, J. H. 2003. *Inventing Modern: Growing Up with X-Rays, Skyscrapers, and Tailfins*. Oxford: Oxford University Press.
Melia, F. 2007. *The Galactic Supermassive Black Hole*. Princeton, NJ: Princeton University Press.
Schlegel, E. M. 2002. *The Restless Universe: Understanding the X-Ray Universe in the Age of Chandra and Newton*. Oxford: Oxford University Press.
Thorne, K. S. 1995. *Black Holes and Time Warps: Einstein's Outrageous Legacy*, foreword by Stephen Hawking. New York: W.W. Norton.
Trümper, J., and Hasinger, G. (ed.) 2008. *The Universe in X-rays*. New York: Springer.
Tucker, W., and Tucker, K. 2001. *Revealing the Universe: The Making of the Chandra X-ray Observatory*. Cambridge, MA: Harvard University Press.
Weaver, K. 2005. *The Violent Universe: Joyrides Through the X-Ray Cosmos*. Baltimore, MD: Johns Hopkins University Press.

Chapter 11: HST

Brown, R. (ed.) 2008. *Hubble 2007: Science Year in Review*. Greenbelt, MD: NASA Goddard Spaceflight Center.
Christensen, L. L., and Fosbury, R. 2006. *Hubble: 15 Years of Discovery*. New York: Springer Science.
Christian, C. A., and Kinney, A. 1999. "The Public Impact of Hubble Space Telescope." Space Telescope Science Institute.
Christianson, G. 1995. *Edwin Hubble: Mariner of the Nebulae*. New York: Farrar, Straus and Giroux.
Dalcanton, J. J. 2009. "18 Years of Science with the Hubble Space Telescope." *Nature* 457, pp. 41–50.
Devorkin, D., and Smith, R.W. 2008. *Hubble: Imaging Space and Time*. Washington, DC: National Geographic.
Gibberman, S. R. 1991. *Star Trek: An Annotated Guide to Resources*. Jefferson, NC, and London: McFarland & Company.
Hoyle, F. 1950. *The Nature of the Universe*. New York: Harper and Brothers.
Hubble, E. 1958. *The Realm of the Nebulae*. New York: Dover Publications.

Kanipe, J. 2006. *Chasing Hubble's Shadows: The Search for Galaxies at the Edge of Time*. New York: Hill and Wang.
Kessler, E. 2006. "Spacescapes: Romantic Aesthetics and the Hubble Space Telescope Images," PhD dissertation, University of Chicago.
Lawrence, John Shelton (2010). "Star Trek as American Monomyth," in *Star Trek as Myth: Essays on Symbol and Archetype at the Final Frontier*, ed. Matthew Wilhelm Kappell, Jefferson, NC: McFarland & Company, pp. 93–111.
Levy, D. H. 1995. *Impact Jupiter: The Crash of Comet Shoemaker–Levy 9*. Cambridge, MA: Basic Books.
Shatner, W., and Fisher, D. 2008. *Up Till Now: The Autobiography*. New York: Thomas Dunne Books/St. Martin's Griffin.
Wagner, J., and Lundeen, J. 1998. *Deep Space and Sacred Time:* Star Trek *in the American Mythos*. West Port, CT: Praeger.
Zimmerman, R. 2008. *The Universe in a Mirror: The Saga of the Hubble Space Telescope and the Visionaries Who Built It*. Princeton, NJ and Oxford: Princeton University Press.

Chapter 12: WMAP

Balbi, A. 2008. *The Music of the Big Bang: The Cosmic Microwave Background and the New Cosmology*. Berlin and Heidelberg: Springer/Verlag.
Bartusiak, M. 2010. *The Day We Found the Universe*. New York: Vintage/Random House.
Chaplin, W. J. 2006. *Music of the Sun: The Story of Helioseismology*. Oxford: Oneworld Publications.
Collins, N., and d'Escrivan, J. 2007. *The Cambridge Companion to Electronic Music*. Cambridge: Cambridge University Press.
Farrell, J. 2005. *The Day Without Yesterday: Lemaitre, Einstein, and the Birth of Modern Cosmology*. New York: Thunder Mouth Press.
Guth, A. 1998. *The Inflationary Universe*. New York: Basic Books.
Heller, M. 1996. *Lemaître, the Big Bang, and the Quantum Universe*. Pachart History of Astronomy Series No. 10. Tucson, AZ: Pachart Publishing House.
James, J. 1995. *Music of the Spheres: Music, Science, and the Natural Order of the Universe*. New York: Springer.
Jeans, J. 1937. *The Mysterious Universe*. Cambridge: Cambridge University Press.
Kragh, H. S. 2007. *Conceptions of Cosmos: From Myths to the Accelerating Universe*. Oxford: Oxford University Press.
Lemaître, G. 1931. "The Beginning of the World from the Point of View of Quantum Theory." *Nature* 127, p. 706.
Lemaître, G. 1950. *The Primeval Atom: An Essay on Cosmogony*. Translated by Betty and Serge Korff. Toronto and New York: D. Van Nostrand.
Lightman, A. P. 2006. *The Discoveries: Great Breakthroughs in 20th Century Science, Including the Original Papers*. New York: Vintage.
Luminet, J.–P. 2008. *The Wraparound Universe*. Translated by Eric Novak. Wellesley, MA: A. K. Peters.
Manning, P. 2004. *Electronic and Computer Music*. Oxford: Oxford University Press.

Partridge, B. 1995. *3K: The Cosmic Microwave Background Radiation*. Cambridge: Cambridge University Press.
Peebles, P., Page Jr., L., and Partridge, B. 2009. *Finding the Big Bang*. Cambridge: Cambridge University Press.
Scaruffi, P. 2003. *A History of Rock Music 1951–2000*. Lincoln, NE: iUniverse.
Vecchierello, H. 1934. *Einstein and Relativity: Lemaître and the Expanding Universe*. Paterson, NJ: St. Anthony Guild Press Franciscan Monastery.

Chapter 13: Conclusion

Boss, A. 2009. *The Crowded Universe: The Race to Find Life Beyond Earth*. New York: Basic Books.
Cartledge, P. 1997. *Democritus (The Great Philosophers)*. London: Routledge.
Chaikin, A. 2008. *A Passion for Mars: Intrepid Explorers of the Red Planet*. New York: Abrams.
Gardner, J. P., et al. 2006. "The James Webb Space Telescope." *Space Science Reviews* 123, pp. 485–606.
Hand, E. 2009. "The Test of Inflation." *Nature* 458, pp. 820–24.
Lockwood, M. K. 2006. "Introduction: Mars Science Laboratory: The Next Generation of Mars Landers," and 13 subsequent articles in the special issue, *Journal of Spacecraft and Rockets* 43, p. 257.
Loeb, A. 2006. "The Dark Ages of the Universe." *Scientific American* (November), pp. 46–53.
Sorabji, R. 1988. *Matter, Space, and Motion*. Ithaca, NY: Cornell University Press.
Watson, F. J. 2005. *Stargazer: The Life and Times of the Telescope*. Cambridge, MA: Perseus Books Group.

Index

References to figures on pages are in *italics*.